Medicinal Plants

Medicinal Plants

Moshrafuddin Ahmed

Vice-Principal

and

Head, Department of Botany

Goalpara College

Goalpara, Assam

MJP PUBLISHERS

MJP PUBLISHERS

© Publishers, 2024   47, Nallathambi Street
All rights reserved   Triplicane
Printed and bound in India   Chennai 600 005

Publisher : J.C. Pillai

Preface

The history of medicine and surgery dates back perhaps to the origin of the human race, but as no mode of recording events existed in prehistoric times, there are no data on the methods of treatment practised in that period. In those days, the subject of human suffering and its alleviation was intimately associated with religion, myth and magic. In addition, there must have been certain rational prescriptions. Whenever the curiosity of the present day man probes into the past and brings to light even fragmentary information on the ingenious methods of our ancestors, it makes a fascinating study.

Herbalists today believe to help people build their good health with the help of natural sources. Herbs are considered to be food rather than medicine because they are complete, all-natural and pure, as nature intended. When herbs are taken, the body starts to get cleansed—it starts purifying itself. Unlike chemically synthesized, highly concentrated drugs that may produce many side effects, herbs can effectively realign the body's defences. Herbs do not produce instant cures, but rather offer a way to put the body in proper tune with nature. For thousands of years, humans have used herbs. Herbs have been used as flavours in foods, as perfumes, as disinfectants to protect us against germs and as medicines to heal sickness.

This book provides information about more than 523 species of medicinal plants and gives their pictures, botanical and local names, distribution, propagation, plant parts, chemical constituents and their uses.

The language used is simple and the subject matter is fully illustrated. I have tried to integrate a chemical factory comprising varied chemicals including alkaloids, glycoside, saponins, resin, oleoresin, sesquiterpene lactones, oils etc.

I thank the researchers, authors of books, publishers and editors whose publications have helped me in drafting this work.

I express my indebtedness to Prof. N. Das (my supervisor, retired Professor in Botany, Gauhati University, Guwahati). I would like to thank Prof. S. Sharma, Head, Department of Botany, Gauhati University, Prof. P.J. Handique, Department of Biotechnology, Gauhati University, Prof. C. Baruah, Prof. S. Sarma, Department of Botany, Gauhati University, and Mr. S. Basumatary, Scientist, Birbal Sahani Institute of Palaeobotany, Lucknow, for their helpful suggestions and encouragement.

I wish to thank my wife Nazma, son Shaquib and daughter Peenaz, for their consistent support and encouragement during the preparation of this book.

I appreciate the all-round co-operation and support of Mr. C. Sajeesh Kumar, Managing Editor, and the editorial team of MJP Publishers, Chennai, for publishing this book with patience, care and interest. Constructive suggestions, if any, are welcome.

Moshrafuddin Ahmed

Contents

Introduction

The history of medicine in India can be traced to the Vedic period. The Rig Veda, perhaps the oldest repository of human knowledge, written about 4500–1600 BC claims about 99 medicinal plants, the Yajur Veda, 82 plants and Sama Veda, 100 plants. Atharvana Veda deals with 288 plants almost all having medicinal ingredients used to cure deadly diseases. Kalpsutra describes about 519 plants (Kaushik and Dhiman 2000). According to Hindu literature, Nagarjuna was a learned person who was the inventor of Kajli (a compound of sulphur and mercury) and art of calcinations (Bhasma). Historical evidence of the use of traditional herbal medicines for women and child health care indicates that there has been a synergistic relation between the herbal medicines and the rural and child health care. Over the last few decades, this has faded and use of the modern medicines has increased phenomenally (Kaushik and Dhiman, 2000). Bhoja Prabandha, a treatise written about 980 AD contains a reference to inhalation of medicaments before surgical operations and an anaesthetic called Sammohini is said to have been used in the time of Buddha.

The people of Garo hills, who were deprived of the fruits of modern medicinal research mainly due to the backwardness of the area and poor communication system, have been largely depending on local herbs and shoots to combat various diseases. The present day plants for drug manufacture like *Holarrhena antidysentrica, Rauvolfia serpentina, Costus specious, Adhatoda zeylanica, Sida* sp., *Phyllanthus* sp., etc. are found in the wilds of Garo Hills (Rao, 1981; Kumar *et al.* 1982).

Medicinal plants are the local heritage with global importance. Humans are endowed with a rich wealth of medicinal plants. Herbs have always been the principal form of medicine in India. People in Europe, North America and Australia are consulting trained herbal professionals and using plant medicines. Medicinal plants also play an important role in the lives of rural people, particularly in remote parts of developing countries with few health care facilities. Plants have been used in the traditional health care system from time immemorial, particularly among tribal communities. The World Health Organisation (WHO) has listed 20,000 medicinal plants globally (Gupta and Chadha, 1995), India's contribution (Singh, 2000) being only about 15–20%. According to the WHO estimates, about 80% of the population in the developing countries depends directly on plants for its medicinal values (Pareek, 1996; Mukhapadhyay, 1998). In India, about 2000 drugs used are of plant origin (Dikshit, 1999). In the last few decades over-exploitation of forest resources has led to species loss. As a result, 20–25% of existing plant species in India have become endangered. Medicinal plants are now under great threat due to excessive collection or exploitation. The degree of threat to natural population of medicinal plants has increased because more than 90% of medicinal plants for herbal industries in India and also for export is drawn from natural habitat (Dhar *et al.* 1999).

As in all ancient cultures, the older systems of medicine saw a period of decline and then neglect with the advent of "modern science". One major impact of this neglect has been that the old wisdom has been relegated to tribal cultures and other forest-based communities who still derive the benefits of this ancient heritage. Unfortunately, in the process, not only has a lot of knowledge been lost, the system of exploitation of this medicinal plant wealth became such that most species have been unscientifically harvested, leading to near extinction of some and rendering others critically endangered. Realization of the importance of scientific cultivation techniques and exploitation of our medicinal wealth dawned lately. It has gained renewed impetus in the recent past especially after the Convention on Biological Diversity, which emphasized the importance of resources in different countries and the need for its conservation. This paved the way to the development of organizations like the National Medicinal Plants Board, set up in the year 2000. This board is undertaking the vital task of formation of policies for economic betterment by generating income and employment, particularly in remote areas and among forest-based communities. This can be realized while creating systems for sustainable cultivation and harvesting of vital medicinal plant species.

In earlier days when people lived in forests, they destroyed them either to use the timber or to make space for agriculture. Wood has been the dominant heating fuel and construction material for housing and ships and in this century vast quantities are also being used for paper production. Paper products use 25% of the world's harvest. In the last 5,000 years, humans have reduced forests from roughly 50% of the earth's land surface to less than 20%. If deforestation continues at the present rates, Thailand will have no forest left in 25 years, Philippines in less than 20 years and Nepal in 15 years. Many of the large areas of grassland in the world, such as the savannas of Africa, the steppes of Eastern Europe and Russia, the pampas of Argentina, and at least some of prairies of North America, were forested thickly before human disturbance. In the drier areas of the world such as North Africa, Greece, Italy, and Australia, the deforested areas have subsequently been overgrazed, and have lost soil so rapidly that they have turned to desert (desertification). The UN in 2000 reported that half of all land in South Asia has lost its agricultural potential because of desertification. Native woods account for 2% of Scotland's land mass and the original biodiverse woodlands of Scotland are believed to have covered 80% of the land, 2000 years ago. By the end of World War I, this figure was down to 5% (Scotsman Native Woods Report, August 26, 2000, The Scotsman Newspaper, Edinburgh). The clearing of tropical forests across the earth has been occurring on a large-scale for many centuries leading to deforestation, which involves the cutting down, burning, and damaging of forests. The loss of tropical rainforests is more profound than other types of forests (John Roper, 1999).

Deforestation in the tropical areas of the world is following a course similar to the earlier clearing of the forests in Europe and North America. Since 1950, the world's population has more than doubled with the fastest population growth in the tropics. In order to meet demands of the increasing population, the forest resources are destroyed, especially the tropical rainforests of the world. Even with tropical deforestation at an all-time high, tropical hardwood process continues to climb as world demand for tropical hardwoods continues to grow. A single teak log for example can bring as much as $ 20,000. Annual world consumption of tropical hardwoods is now more 250 million cubic metres or over 100 billion board feet, per year. All the primary forests in India, Sri Lanka and Bangladesh have gone, Ivory Coast's forests are essentially non-existent and Nigeria's forests have been decimated as well. Amazon rainforests are being cleared on a vast scale for settlements, logging, gold mining, petroleum, cattle ranching, sugar cane (for gasohol), large hydro dams, and charcoal for smelting. Scientists estimate that until 10,000 years ago, the world had 6 billion acres of tropical rainforests. By 1950, we had little less than 2.8 billion acres of rainforest. It was then being cut down at the rate of about 10 to 15 million acres per year. Today we have less than 1.5 millon acres left, and we are clearing this remaining rainforest at the rate of 30 to 50 million acres per year, two to three times as rapidly as just a few decades ago. Scientists project that the rate of tropical deforestation will continue to increase for the next 10 to 15 years until there is not enough forest left to sustain the rate of cutting (John Roper, 1999).

The growing population, human greed and lack of knowledge about ecological balance of nature, have destroyed the environment in such a manner that there is already a vast degradation of soil, water, air, biodiversity and even light/temperature (Barthakur, 1998). It is observed that the total forest area has been showing a decline in recent years due to uncontrolled increase in population and the resulted felling of trees, which leads to clearing large tracts of forest for the habitat and other requirements. This has affected the ecological scenario as all types of forests have been cleared due to unauthorized encroachment. In a study undertaken by the Forest Survey of India (FSI) in 1999, it is estimated that during 1987–97, an area of about 0.13 million ha had been affected by shifting cultivation is the state of Assam and 0.18 million ha are currently affected by shifting cultivation in Meghalaya. In this context, it may be noted that in the large areas of developing countries there has been similar trend in the destruction of forests, (Postel and Heise, 1990; Lanly, 1983).

NORTH-EASTERN INDIA

North-eastern India is inherently very different from other parts of India in terms of its complexity and diversity and hence its problems are also different. The biggest cause of forest degradation in the north-east is probably shifting cultivation. Empirical evidences in the form of satellite imageries show that shifting cultivation has emerged as the single largest source of forest degradation in the north-east. This is more problematic in the areas where there is still primary forest left as it directly affects the biodiversity of these areas. Repeated shifting also leads to the growth of bamboo or grasses and reduces the diversity of the region. It is also believed to increase soil erosion and desiccation of water source. The paradigm of forest management in India had a primary timber orientation, which is itself, a legacy of the British rule. The Indian Forest Department was formed with the notion that the state required an assured supply of timber through the practice of "scientific" forestry. This continued in the post-independence phase with the rationale only changing slightly namely previously in the British days timber was extracted for the expansion of railways and now it was to further the goal of national industrialization. The main reason for deforestation is the settlement of more number of refugees migrated from the neighbouring countries thereby resulting in population explosion (Bhakta, 1992; Taher and Ahmed, 2001).

The ethnic elements, historically, comprise the Mon-Khamer and Tibeto-Burman sub families of the " Indo-Chinese Linguistic family". British occupied Garo Hills in 1872. Prior to this, the East India Company administered it, since 1963. It became part of Assam in 1874, then Khasi Hills came under British rule in 1883. Meghalaya was curved out from Assam as an autonomous Hill State in April 1970, comprising the Garo Hills, Khasi and Jaintia Hills (*Env. Hand Book of N.E. Meghalaya*, 1988).

Climate

Though located in the tropical belt having a broadly tropical monsoon climate because of its higher elevation in the central upland, it enjoys an equable climate. It has four characteristic seasons (i) the hot season from April to July, (ii) the rainy season from June to September, (iii) cold season from late November to March and (iv) season of the retreating South-West monsoon from October to early November. The Garo Hills, due to its lower elevations face weather conditions almost similar to the Brahmaputra valley with certain variations in both rainfall pattern and the temperature.

United Khasi and Jaintia Hills rise in the south abruptly from the plains of Sylhet, now in Bangladesh, and more gradually in the north from the plains of Kamrup, in Assam. The latitude of the Hills is approximately 250 to 260° N, and the longitude between 91° and 93° E, the plateau being bounded on the north by the district of Assam known as Kamrup and Nowgong, on the east by the Kapili river, on the south by Sylhet, and on the west by the Garo Hills. The average altitude of the tableland is about 1,200 m above sea level, but it contains 23 prominent peaks, the highest of which is the Shillong peak, which rises to some 1,960 m.

Shillong, the capital of Meghalaya, is about 1,500 m above sea level. The higher ranges have been denuded of trees and are now covered with grass. For centuries the trees were cut down for household fuel and for feeding the iron-smelters' furnaces. But there are still numerous sacred groves throughout the land. One such grove is in the areas surrounding the Shillong Peak, which is believed to have been the home of the God of Shillong, after whom the town was renamed by the British. These groves still have fine timber trees, rhododendrons, rare orchids, wild cinnamon, and various flowering shrubs. Lavender and ferns grow in profusion, together with medicinal and edible herbs.

Rainfall

Rainfall is heavy throughout the area; it is heaviest during the monsoons, and in winter. Cherrapunjee and its neighbourhood receive the heaviest rainfall in the world, with an average of 1,250 cm per annum. Winters are severe in the uplands with heavy frosts and occasional snow and sleet. But the cold season is generally bright and sunny. Summers are mild in the higher regions, with an average temperature of about 18°C, and sometimes there are fairly cold days even in summer. Very heavy mists and fogs are frequent in the mountains, particularly in the Cherra area. In general the lower slopes enjoy a much milder climate, suitable for the growth of various subtropical vegetation.

Meghalaya holds a world record in receiving the highest rainfall. Meghalaya has an average annual rainfall of about 2600 mm, 2500–3000 mm at Shillong, 3000 mm at Jowai and around 3300 mm at Mawphlang.

The total forest area in the state is 8,510 sq. kms, with only 993 sq. kms, under the control of the State Government, the rest under the District Councils and private managements. The principal timber species are sal, teak, titachap, gomari, bola, pine, birch and makri sal. Garo Hills, in general, has subtropical wet hill-type vegetation with evergreen trees but in the border, lower areas have moist deciduous type of vegetation with Sal (*Shorea robusta*) as the principal tree.

The flora of Meghalaya is rich both in extent and diversity of species. Numerous medicinal plants and other economically important plants live in this region. Recently, it has been noted that increasing biotic influences, including socio-economic development and unrestrained commercial exploitation of forest wealth have threatened the survival of genetic resources, amounting to a great loss of national heritage. The forests have been over-exploited due to road and other infrastructural development and as a result natural forest stands are becoming less day by day. It has also been observed that some important roadside medicinal plants (*Rauvolfia serpentina, Holarrhena antidysentrica, Costus spinosa, Carcuma aromatica, Emblica officinalis, Dioscorea alata,* etc.) used by the tribal community are highly damaged due to soil erosion at the site of road construction and coal mining (Bera *et al.*, 2006).

STUDY OF MEDICINAL PLANTS

The inhabitants are well-aware of the medicinal properties of plants growing around their surroundings. The information has been passed on from generation to generation as a guarded secret. Each tribe has its own medicinal-man and has its own set of medicines. Efforts were made to collect information from the region and an inventory has been prepared. For identification of plants, different relevant floras, monographs and other works were consulted at the Botanical Survey of India, Shillong, Meghalaya. An alphabetical list of species with botanical name, local name, plant parts, chemical constituents and uses has been prepared. Medicinal values of plants have been ascertained in consultation with Das *et al,* (2003), Maheshwari (2000), and Prajapati *et al.* (2003). The general medicinal uses were known from literature, and indigenous people engaged in traditional health care practices were interviewed for recording their unique knowledge about medicinal values of different plant species with their local names.

Medicinal herbs are a significant source of synthetic and herbal drugs. In the commercial market, medicinal herbs are used as raw drugs, extracts or tinctures. Isolated active constituents are used for applied research. For the last few decades, phytochemistry has been making rapid progress and herbal products are becoming popular. Classical texts of Ayurveda, *Charaka Samhita* and *Sushita,* were written around 1000 BC. The Ayurvedic *Materia Medica* includes 600 medicinal plants along with their therapeutics. Herbs like turmeric, fenugreek, ginger, garlic and holy basil are integral part of Ayurvedic formulations.

The modern pharmaceutical industry also requires a large quantity of authentic plants for the manufacture of drugs. Extraction of active principles and manufacture of drug formulations need sophisticated technology and capital investment. However, systematic cultivation of these plants and their drying/preservation at farm level constitute a promising rural industry with attractive remunerations. These semi-processed plants find a ready market with herb dealers/large pharmaceutical industries. The formulations incorporate single herb or more than two herbs (poly-herbal formulations). Medicinal herbs are considered to be a chemical factory, comprising varied chemical compounds like alkaloids, glycosides, saponins, resins, oleoresins, sesquiterpene lactones, oils, etc. Today there is growing interest in chemical composition of plant-based medicines. Several bioactive constituents have been isolated and studied for pharmacological activity.

The medicinal plants spreading over a total 22,429 sq. km area were frequently surveyed. Several attempts were made for study in different seasons in tribal-dominated areas of Khasi, Jaintia and Garo Hills district of Meghalaya.

The total number of medicinal plants are 593 (Table 1), out of which the highest number of medicinal plants are under the family Euphorbiaceae, Rubiaceae, followed by Fabaceae, Asteraceae, Lauraceae, Rutaceae, Verbenaceae, Rosaceae, Apocynaceae, Solanaceae, Zingiberaceae and Polygonaceae.

Table 1 List of medicinal plants

No.	Binomial name	Family
1.	*Abelmoschus moschatus* Madic.	Malvaceae
2.	*Abroma augusta* (L.) f.	Sterculiaceae
3.	*Abrus precatorius* L.	Fabaceae
4.	*Abutilon indicum* (L.) Sweet.	Malvaceae
5.	*Acacia farnesiana* Willd. Fl. Br.	Mimosaceae
6.	*Acacia indica* L.	Euphorbiaceae
7.	*Acacia pinnata* L. Rare	Mimosaceae
8.	Acalypha indica L.	Euphorbiaceae
9.	*Achyranthes aspera* L.	Amaranthaceae
10.	*Aconitum species*	Ranunculaceae
11.	*Acyranthes porphyristachya* Wall.	Amaranthaceae
12.	*Acorus calamus* L.	Araceae
13.	*Acronychia pedunculata* (L) Mig.	Rutaceae

14.	*Adhatoda vasica* Nees.	Acanthaceae
15.	*Adhatoda zeylanica* Medic.	Acanthaceae
16.	Aegle marmelos Linn.	Rutaceae
17.	*Aesculus assamica* Griff.	Sapindaceae
18.	Ageratum conyzoides L.	Asteraceae
19.	*Agrimonia pilosa* Ledeb.	Rosaceae
20.	*Alangium chinense* (Lour) Harm.	Connaceae
21.	*Albizia lebbeck*, Benth.	Mimosaceae
22.	*Albizia odoratissima* Benth.	Mimosaceae
23.	*Alstonia scholaris* R. Br.	Apocynaceae
24.	*Allium chinense* D. Don.	Liliaceae
25.	*Allium hookeri* Thw.	Liliaceae
26.	*Allium tuberosum* Rttl. Ex. Spreng.	Liliaceae
27.	*Alnus nepalensis* D. Don.	Betulaceae
28.	*Alocasia cucullata* (Lour) Schott.	Araceae
29.	*Alocasia macrorrhiza* (L.) Schott.	Araceae
30.	*Aloe barbadensis* Mill.	Liliaceae
31.	*Alpinia galanga* (L.) Willd.	Zingiberaceae
32.	*Alstonia scholaris* R. Br.	Apocynaceae
33.	*Alternanthera sessilis* L.	Amaranthaceae
34.	*Altingia excelsa* Noronha.	Zingiberaceae
35.	*Amaranthus spinosus* L.	Amaranthaceae
36.	*Ampelocissus latifolia* (Roxb) Plench.	Vitaceae
37.	*Anacardium occidentale* L.	Anacardiaceae
38.	*Anaphalis adnata* DC.	Asteraceae
39.	*Andrographis paniculata* Nees.	Acanthaceae
40.	*Annona muricata* L.	Annonaceae
41.	*Anthocephalus chinensis* (Lank.) Rich. and Walp.	Rubiaceae
42.	*Antidesma acidum* Retz.	Euphorbiaceae
43.	*Antidesma bunius* (L.) Spreng.	Euphorbiaceae
44.	*Antidesma diandrum* (Roxb.) Roth.	Euphorbiaceae
45.	*Aphamixis polystachya* (Wall.) Parker. **Vulnerable**	Meliaceae
46.	*Aquilaria agallocha* Roxb.	Thymelaeaceae

47.	*Ardisia colorata* Roxb. **Rare**	Myrsinaceae
48.	*Ardisia paniculata* Roxb.	Myrsinaceae
49.	*Ardisia solanacea* Roxb.	Myrsinaceae
50.	*Argemone mexicana* L.	Papaveraceae
51.	*Argyreia nervosa* (Burn.f.) Boj.	Convolvulaceae
52.	*Aristolochia indica* L.	Aristolochiaceae
53.	*Aristolochia saccata* Wall. **Rare.**	Aristolochaceae
54.	*Aristolochia tagala* Cham.	Aristolochiaceae
55.	*Artemisia maritima* L.	Asteraceae
56.	*Artemisia nilagirica* (Cl.) Pamp.	Asteraceae
57.	*Artocarpus chaplasha* Roxb.	Moraceae
58.	*Artocarpus heterophyllus* Lam.	Moraceae
59.	*Asclepias curassavica* L.	Asclepiadaceae
60.	*Atropa acuminata* Roye ex. Lindley	Solanaceae
61.	*Averrhoa carambola* L.	Averrhoaceae
62.	*Azadirachta indica* A. Juss.	Meliaceae
63.	*Baccaurea ramiflora* Lour.	Euphorbiaceae
64.	*Bacopa monnieri* L.	Scrophulariaceae
65.	*Baliospermum montanum* (Willd.) Muell. Arg.	Euphorbiaceae
66.	*Berberis aristata* DC.	Berberidaceae
67.	*Barringtonia acutangula* (L.)	Lecythidaceae
68.	*Basella alba* L.	Basellaceae
69.	*Bauhinia purpurea* L.	Caesalpiniaceae
70.	*Bauhinia vahlii* Wt. and Arn. Benth.	Caesalpiniaceae
71.	*Bauhinia variegata* L.	Caesalpiniaceae
72.	*Begonia josephii* A.DC.	Begoniaceae
73.	*Begonia palmata* D. Don.	Begoniaceae
74.	*Begonia roxburghii* A. DC. Prodv.	Begoniaceae
75.	*Berberis wallichiana* DC.	Berberidaceae
76.	*Bergenia ciliata* (Haw) Sternb.	Saxifragaceae
77.	*Bidens pilosa* L.	Asteraceae
78.	*Biophytum sensitivum* (L.) DC.	Oxalidaceae
79.	*Bischofia javanica* Bl.	Euphorbiaceae

80.	*Bixa arellana* Linn.	Bixaceae
81.	*Boehmeria mecrophylla* Don.	Urticaceae
82.	*Boehmeria platyphylla* D. Don.	Urticaceae
83.	*Boenninghousenia albiflora* Reichl ex. Meissn.	Rutaceae
84.	*Boerhavia diffusa* L.	Nyctaginaceae
85.	*Breynia retusa* (Dennst) Alston	Euphorbiaceae
86.	*Bridelia retusa* (L.) Spreng.	Euphorbiaceae
87.	*Bridelia stipularis* (L.) Bl.	Euphorbiaceae
88.	*Brucea mollis* Wall. [Rare]	Simaroubaceae
89.	*Buddleia macrostachya* Benth.	Loganiaceae
90.	*Butea monosperma* (Lam.) Kuntze. **near endangered**	Fabaceae
91.	*Caladium bicolor* (Ait.) Vent.	Araceae
92.	*Callicarpa arborea* Roxb.	Verbenaceae
93.	*Calophyllum inophyllum* L.	Clusiaceae
94.	*Calotropis gigantea* (L.) R. Br.	Asclepiadaceae
95.	*Camellia caduca* CL. Ex. Brandis **Endemic**	Theaceae
96.	*Camellia sinensis* (L.) Kuntz.	Theaceae
97.	*Cannabis sativa* L.	Cannabinaceae
98.	*Canarium strictum* Roxb. **Vulnerable**	Burseraceae
99.	*Canthium dicoccum* (Gaertn.) Merrill.	Rubiaceae
100.	*Canthium parviflorum* Lam.	Rubiaceae
101.	*Careya arborea* Roxb.	Lecythidaceae
102.	*Carica papaya* L.	Caricaceae
103.	*Carpesium nephalense* Less.	Asteraceae
104.	*Casearia vareca* Roxb.	Flacourtiaceae
105.	*Carthamus tinctorius* L.	Asteraceae
106.	*Cassia alata* L.	Caesalpiniaceae
107.	*Cassia fistula* L.	Caesalpiniaceae
108.	*Cassia occidentalis* L.	Caesalpinaceae
109.	*Cassia tora* L.	Caesalpinaceae
110.	Castanea sativa	Fagaceae
111.	*Carallia brachiata* (Lour.)	Rhizosphoraceae
112.	*Cayranthia pedata* (Lam) Juss, ex. Gagnep	Vitaceae

113.	*Celastrus paniculatus* Wild. **Vulnerable**	Celastraceae
114.	*Celtis tetrandra* Roxb.	Ulmaceae
115.	*Centella asiatica* (L.) Urban.	Apiaceae
116.	*Cephaelis ipecacuanha* Rich.	Rubiaceae
117.	*Cheilanthes farinosa* (Kaulf.)	Cheilanthaceae
118.	*Choenmorpha fragrans*	Apocynaceae
119.	*Chenopodium ambrosioides*	Chenopodiaceae
120.	*Cinchona* species	Rubiaceae
121.	*Cinnamomum camphora* Nees.	Lauraceae
122.	*Cinnamomum bejolghota*	Lauraceae
123.	*Cinnamomum granduliferum* Wall.	Lauraceae
124.	*Cinnamomum pauciflorum* Nees. **Endemic/ Rare**	Lauraceae
125.	*Cinnamomum tamala* Nees and Eberm.	Lauraceae
126.	*Cinnamomum zeylanicum* Blume.	Lauraceae
127.	*Cissampelos pareira* L.	Menispermaceae
128.	*Cissus quadrangula* L.	Vitaceae
129.	*Cissus javanica* DC.	Vitaceae
130.	*Cissus repens* Lamk.	Vitaceae
131.	*Clausena heptaphylla* Wt. and Arn.	Rutaceae
132.	*Clausena excavata* Burm.F.	Rutaceae
133.	*Clinopodium umbrossum* (Bieb.) Koch.	Lamiaceae
134.	*Citrus limentta* Risso.	Rutaceae
135.	*Citrus medica* L.	Rutaceae
136.	*Clematis buchaniana* DC. Syst.	Rununculaceae
137.	*Clematis gouriana* DC.	Rununcluceae
138.	*Cleome viscosa* Linn.	Capparidaceae
139.	*Clerodendrum colebrokoianum* Walp.	Verbenaceae
140.	*Clerodendrum fragrans* (Vent.) Willd.	Verbenaceae
141.	*Clerodendrum infortunatum* L.	Verbenaceae
142.	*Clerodendrum serratum* (L.) Spreng.	Verbanaceae
143.	*Clerodendrum viscosum* Wall.	Verbenaceae
144.	*Clerodendrum wallichii* Merrill.	Verbenaceae
145.	Codonopsis javanica (BL) H.K. **Rare**	Campanulaceae

146.	*Coelogyne stricta* (D.Don) Schlechter	Orchidaceae
147.	*Coffea arabica* L.	Rubiaceae
148.	*Coffea bengalansis* Roxb, ex.	Rubiaceae
149.	Coix lacryma Jabi L.	Poaceae
150.	*Colchicum luteum* Baker.	Liliaceae
151.	*Colocasia esculenta* (L.) Achott.	Araceae
152.	*Combretum roxburghii* Roxb.	Combretaceae
153.	*Combretum pilosum* Roxb.	Combretaceae
154.	*Commelina paludosa* Blume.	Commelinaceae
155.	*Corchorus capsularis* L.	Tiliaceae
156.	*Corchorus olitorius* L.	Tiliaceae
157.	*Cordia caudatus* Geisel.	Euphorbiaceae
158.	*Cordia dichotoma* Forst. **Very rare**	Boraginaceae
159.	*Cordia fragrantissima* Kurz. **Very rare**	Boraginaceae
160.	*Corydalis longipes* DC.	Fumariaceae
161.	*Coscinium fenestratum* Colebr.	Menispermaceae
162.	*Costus specious* (Koening) Sm. **Vulnerable**	Costaceae
163.	*Cotoneaster congestus* Baker.	Rosaceae
164.	*Crassocephalum crepidiodes* (Benth.) Moore.	Asteraceae
165.	*Cretaeva nurvala* Buch-Ham. **Very rare**	Capparidaceae
166.	*Crotalaria occulta* Grab.	Fabaceae
167.	*Croton caudatus* Gies.	Euphorbiaceae
168.	*Croton tiglium* L. **Rare**	Euphorbiaceae
169.	*Cryptocarya amygdalina* Nees.	Lauraceae
170.	*Cucurbita moschata* Duch. Ex. Poir.	Cucurbitaceae
171.	*Curcuma amada* Roxb.	Zingiberaceae
172.	*Curcuma angustifolia* Roxb.	Zingiberaceae
173.	*Curcuma aromatica* Salisb.	Zingiberaceae
174.	*Curcuma caesia* Roxb.	Zingiberaceae
175.	*Curcuma longa* Salish. Parod.	Zingiberaceae
176.	*Cymbopogon* species	Gramineae
177.	*Cynodon dactylon* (L.) Pers.	Poaceae
178.	*Dalbergia pinnata* (Lour.) Prain.	Fabaceae

179.	*Dalbergia sisso* Roxb-ex. DC.	Fabaceae
180.	*Dalbergia volubilis* Roxb.	Fabaceae
181.	*Dalhousiea bracteata* Grah.	*Fabaceae*
182.	*Daphne papyracea* Wall.	Thymelaeceae
183.	*Datura metel* L.	Solanaceae
184.	*Datura stramonium* L.	Solanaceae
185.	*Dillenia indica* L.	Dilleniaceae
186.	*Dillenia pentagyna* Roxb.	Dilleniaceae
187.	*Dendrobium nobile* Lindl.	Orchidaceae
188.	*Dendrocnide sinuata* (Bl.) Chew.	Urticaceae
189.	*Dendrophthoe falcata* (L.f.) Etting.	Loranthaceae
190.	*Derris ferruginea* (Roxb.) Benth.	Fabaceae
191.	*Derris trifoliate* Lour.	Fabaceae
192.	*Desmodium gangeticum* (L.) DC.	Fabaceae
193.	*Desmodium heterocarpon* (L.) DC. **Critically rare**	Fabaceae
194.	*Desmodium microphyllum* (Thunb.) DC.	Fabaceae
195.	*Desmodium triangulare* (Retz.) Merr.	Fabaceae
196.	*Desmodium triflorum* (L.) DC.	Fabaceae
197.	*Desmodium triquetrum* (L.) DC.	Fabaceae
198.	*Digitalis purpurea* L.	Scrophulariaceae
199.	*Dioscorea bulbifera* L.	Dioscoreaceae
200.	Dioscorea pentaphylla L. **Endemic**	Dioscoriaceae
201.	*Diospyros lancaefolia* Roxb.	Ebenaceae
202.	*Diospyros malabarica* (Desr.) Kost.	Ebenaceae
203.	*Diospyros montana* Roxb. Hiern.	Ebenaceae
204.	*Diospyros pilosula* (DC.) Hiem. **Extremely Rare**	Ebenaceae
205.	*Disporum cantoniense* (lour) Merr.	Liliaceae
206.	*Drymaria diandra* Bl.	Caryophyllaceae
207.	*Dysoxylum binectariferum* Hk.f.et. Bedd.	Meliaceae
208.	*Eclipta alba* L.	Asteraceae
209.	*Eclipta prostrata* L.	Asteraceae
210.	*Elaeocarpus sphaericus* (Gaertn.) K. Schum. **Vulnerable**	Elaeocarpaceae
211.	*Elatostema integrifolium* (D.Don) Wedd.	Urticaceae

212.	*Elephantopus scaber* L.	Asteraceae
213.	*Elettaria cardamomum* (L.) Maton.	Zigiberaceae
214.	*Elsholtzia blanda* Benth.	Lamiaceae
215.	*Elsholtzia fruticosa* (D. Don.) Rehder.	Lamiaceae
216.	*Embelia ribes* Burn F. **Critically endemic**	Myrsinaceae
217.	*Embelica officinalis* Gaertn.	Euphorbiaceae
218.	*Embelia tsjeriam-cottam* A. DC.	Myrsinaceae
219.	*Engelhardtia spicata* Leschn.ex. Bl.	Jublandaceae
220.	*Ephedra gerardiana* Wallich.	Gnetaceae
221.	*Entada pursaetha* DC.	Mimosaceae
222.	*Erigeron bellidioides* Benth.	Asteraceae
223.	*Erythrina stricta* Roxb.	Fabaceae
224.	*Eryngium foetidum* L.	Apiaceae
225.	*Erythroxylum kunthianum* Wall. ex kung, **Endemic**	Erythroxylaceae
226.	*Euonymus lawsonii* Cl and Pr. **Endemic, rare**	Celastraceae
227.	*Euphorbia antiquorum* L.	Euphorbiaceae
228.	*Euphorbia barnhartii* Croizat.	Euphorbiaceae
229.	*Euphorbia neriifolia* L.	Euphorbiaceae
230.	*Euphorbia nivulia* Buch-Ham.	Euphorbiaceae
231.	*Eupatorium adenophorum* Spreng.	Asteraceae
232.	*Eupatorium odoratum* L.	Asteraceae
233.	*Eupatorium cannabium* L.	Asteraceae
234.	*Euphorbia antiquorum* L.	Euphorbiaceae
235.	*Euphorbia hirta* L.	Euphorbiaceae
236.	*Euphorbia ligularia* Roxb.	Euphorbiaceae
237.	*Euphorbia uniflora* Roxb.	Euphorbiaceae
238.	*Eurya acuminata* DC.	Theaceae
239.	*Eurya japonica* Thumb.	Theaceae
240.	*Eusteralis stellata* (Lour) Panig.	Lamiaceae
241.	*Fagopyrum dibotrys* (D.Don) Hara.	Polygonaceae
242.	*Fagopyrum esculentum* Moench.	Polygonaceae
243.	*Flacourtia jangomas* (Lour.) Raeusch.	Flacourtiaceae
244.	*Ferula narthex* Boiss.	Umbelliferae

245.	*Ficus benghalensis* L.	Moraceae
246.	*Ficus fistulosa* Reinw. Ex Bl.	Moraceae
247.	*Ficus heterophylla* L.f.	Moraceae
248.	*Ficus hispida* Vahl.	Moraceae
249.	*Ficus religiosa* L.	Moraceae
250.	*Ficus rumphii* Bl.	Moraceae
251.	*Flemingia macrophylla* (Willd.) Prain ex. Merr.	Fabaceae
252.	*Fragaria indica* Andr.	Rosaceae
253.	*Galinsoga quadriradiata* Ruiz and Pavon.	Asteraceae
254.	*Garcinia cowa* Roxb. Ex DC.	Clusiaceae
255.	*Garcinia morella* (Gaertn.) Desr.	Clusiaceae
256.	*Gaultheria fragrantissima* Wallich.	Ericaceae
257.	*Gonatanthus pumilus* (Don.)	Ericaceae
258.	*Gentiana kurroo* Royle.	Gentianaceae
259.	*Geranium nepalense* Sweet.	Geraniaceae
260.	*Gleichenia linearis* Clarke.	Glicheniaceae
261.	*Glycyrrhiza glabra* L.	Papilionaceae
262.	*Glochidion lanceolarium* (Roxb) Voight.	Euphorbiaceae
263.	*Gloriosa superba* L. **Vulnerable**	Liliaceae
264.	*Gmelina arborea* Roxb.	Verbenaceae
265.	*Gomphostemma parviflora* Wall.	Lamiaceae
266.	*Gonatanthus pumillus* (Don.)	Araceae
267.	*Gouania tiliaefolia* Lamk.	Rhamnaceae
268.	*Grewia multiflora* Juss.	Tiliaceae
269.	*Gynocardia odorata* R. Br.	Flacourtiaceae
270.	*Hedychium acuminatum* Roscoe	Zingiberaceae
271.	*Hedyotis scandens* Roxb.	Rubiaceae
272.	*Hedyotis verticillata* (L) Lamk.	Rubiaceae
273.	*Helicia excelsa* Bl. **Rare**	Proteaceae
274.	*Hemidesmus indicus* L. Br. Schult.	Asclepiadaceae
275.	*Hemiphriagma heterophyllum* Wall.	Scrophulariaceae
276.	*Hibiscus rosa-sinensis* L.	Malvaceae
277.	*Hibiscus surattensis* L.	Malvaceae

278.	*Hiptage benghalensis* (L.) Kurz.	Malpighiaceae
279.	*Holarrhena antidysenterica* (L.) Wall.	Apocynaceae
280.	*Holmskioldia sanguinea* Retz.	Verbenaceae
281.	*Holboellia latifolia* Wall.	Berberidaceae
282.	*Holoptelea integrifolia* Roxb. Planch **Rare**	Ulmaceae
283.	*Homonoia riparia* Lour.	Euphorbiaceae
284.	*Houttuynia cordata* Thumb.	Sauraceae
285.	*Hoya globulosa* Hk.f	Asclepiadaceae
286.	*Hydnocarpus kurzii* (King) Warb.	Flacourtiaceae
287.	*Hygrophila auriculata* (Schum.) Heine.	Acanthaceae
288.	*Hymennodictyon excelsum* (Roxb.) Wall.	Rubiaceae
289.	*Hyoscyamus niger* L.	Solanaceae
290.	*Hyptianthers stricta* (Roxb.) Wight and Arn.	Rubiaceae
291.	*Hyptis suaveolens* Poit.	Lamiaceae
292.	*Ichnocarpus frutescens* (L.) R. Br. **Endangered**	Apocynaceae
293.	*Ilex khasiana* Purk. **Endemic/Rare**	Aquifoliaceae
294.	*Ilex embeliodes* Hk.f. **Endemic/Rare**	Aquifoliaceae
295.	*Imperata cylindrica* Beauv.	Poaceae
296.	*Indigofera tinctoria* L.	Fabaceae
297.	*Ipomoea alba* L.	Convolvulaceae
298.	*Ipomea nil* (linn) Roth.	Convolvulaceae
299.	*Ipomoea pes-caprae* (l) Sweet.	Convolvulaceae
300.	*Ipomoea pes-tigridis* L.	Convolvulaceae
301.	*Ipomea purpurea* (L.) Roth.	Convolvulaceae
302.	*Ipomoea purga* Heyne.	Convolvulaceae
303.	*Ipomoea quamoclit* L.	Rubiaceae
304.	*Ixora acuminata Roxb.*	Rubiaceae
305.	*Ixora nigricans* R. Br. Ex. J.E.Sm.	Rubiaceae
306.	*Jasminum amplexicaule* Buch. Hamm. ex. Don.	Oleaceae
307.	*Jasminum subtriplinerve* Blume.	Oleaceae
308.	*Justicia gendarussa* Burm. f.	Acanthaceae
309.	*Kaemferia galanga* L.	Zingiberaceae
310.	*Kaemferia rotunda*	Zingiberaceae

311.	*Kalanchoe laciniata* (Lamk.) Pers.	Crassulaceae
312.	*Kalanchoe pinnata* (Lamk.) Pers.	Crassulaceae
313.	*Knemna angustifolia* (Roxb.) Warb.	Myristicaceae
314.	*Kydia calycina* Roxb.	Malvaceae
315.	*Lagerstroemia indica* L. **Rare**	Lythraceae
316.	*Lannea coromandelica* (Houtt.) Merr.	Anacardiaceae
317.	*Lantana camera* L.	Verbenaceae
318.	*Lawsonia inermis* L.	Lythraceae
319.	*Leea indica* (Burm. F.) Merr.	Leeaceae
320.	*Leucas aspera* Spreng.	Lamiaceae
321.	*Leucas cephalotes* Spreng.	Lamiaceae
322.	*Leucas plukereti* (Roth.) Spreng.	Lamiaceae
323.	*Ligustrum lucidum* Ait.	Oleaceae
324.	*Lindera latifolia* H.k.f. **Endemic/Rare**	Lauraceae
325.	*Lindera pulcherrima* (Nees) Benth.	Lauraceae
326.	*Lindernia anagallis* (Burm.f.) permell.	Scrophulariaceae
327.	*Lasia spinosa* (L) Thw.	Araceae
328.	*Litchi chinensis* Sonner.	Sapindaceae
329.	*Litsea citrata* BL.	Lauraceae
330.	*Litsea cubeba* (Lour.) Pers.	Lauraceae
331.	*Litsea lancifolia* (Roxb.) Wall.ex. Hook. F.	Lauraceae
332.	*Litsea monopetala* (Roxb.) Pers.	Lauraceae
333.	*Litsea saliciflora* Roxb. Ex. Nees	Lauraceae
334.	*Lobelia inflata*	Lauraceae
335.	*Lobelia nicotianaefolia* Heyne ex. Roth.	Lobeliaceae
336.	*Lonicera japonica* Thumb.	Caprifoliaceae
337.	*Luvunga scandens* (Roxb) Buch-Ham. **Extremely Rare**	Rutaceae
338.	*Lycopodium clavatum* L. **Vulnerable**	Lycopodiaceae
339.	*Lyonia ovalifolia* (Wall) Druce.	Ericaceae
340.	*Macropanax undulatus* Wall ex. D. Don. Saem.	Araceae
341.	*Macrosolen cochinchinensis* (Lour.) van. Tiegh.	Lorantheceae
342.	*Madhuca indica* Gmel.	Sapotaceae
343.	*Maesa chisia* D. Don.	Myrsinaceae

344.	*Maesa ramentacea* Wall.	Myrsinaceae
345.	*Mahonia acanthifolia* G. Don.	Berberidaceae
346.	*Mahonia nepaulensis* DC.	Berberidaceae
347.	*Mallotus philippensis* (Lam.) Muell. Arg.	Euphorbiaceae
348.	*Melastoma malabathricum* L.	Melastomaceae
349.	*Mangifera indica* L.	Anacardiaceae
350.	*Manihot esculenta* Crantz.	Euphorbiaceae
351.	*Melia azedarach* L.	Meliaceae
352.	*Melodinus monogynus* Roxb.	Apocynaceae
353.	*Mentha arvensis* Linn.	Lamiaceae
354.	*Mentha longifolia* (L.) Huds.	Lamiaceae
355.	*Meriandra benghalensis* Benth.	Lamiaceae
356.	*Merremia umbellata* (L.) Hall. f.	Convolvulaceae
357.	*Merremia vitifolia* (Burm. F) Hall. f.	Convolvulaceae
358.	*Mesua ferrea* L. **near Endemic**	Clusiaceae
359.	*Meyna laxiflora* Robyna.	Rubiaceae
360.	*Michelia champaca* L. **Endangered**	Magnoliaceae
361.	*Micromelum minutum* (Forest,f.) Wt. and Arn.	Rutaceae
362.	*Micromelum pubescens* non BL.	Rutaceae
363.	*Mikania micrantha* Kunth ex. HBK.	Asteraceae
364.	*Miliusa roxburghiana* Hk.f. and Th.	Annonaceae
365.	*Millettia caudata* Baker.	Fabaceae
366.	*Mimosa pudica* L.	Mimosaceae
367.	*Mimusops elengi* L. Baker.	Sapotaceae
368.	*Mitragyna rotundifolia* (Roxb.) O. Kuntze.	Rubiaceae
369.	*Momordica charantia* L.	Cucurbitaceae
370.	*Morinda angustifolia* Roxb.	Rubiaceae
371.	*Moringa oleifera* Lam.	Moringaceae
372.	*Mucuna bracteata* DC.	Fabaceae
373.	*Mucuna pruriens* (L.) DC.	Fabaceae
374.	*Murraya koenigii* (L.) Spreng.	Rutaceae
375.	*Murraya paniculata* (L.) Jack.	Rutaceae
376.	*Mussaenda belilla* Buch Ham.	Rubiaceae

377.	*Mussaenda roxburghii* Hook. F.	Rubiaceae
378.	*Myrica esculenta* Buch. Ham.	Myricaceae
379.	*Myrioneuron nutans* Wall.	Rubiaceae
380.	*Naravelia zeylanica* (L.) DC.	Rununculaceae
381.	*Nardostachys jatamansi* DC.	Valerianceae
382.	*Nepenthes khasiana* Hook. f. **Endemic**	Nepenthaceae
383.	*Nephrolepis cordifolia* (L.) Prest.	Nephrolepidaceae
384.	*Nerium indicum* Mill.	Apocynaceae
385.	*Nicotiana tabacum* L.	Solanaceae
386.	*Nyctanthes arbortristis* Linn.	Oleaceae
387.	*Ocimum basilicum* L.	Lamiaceae
388.	*Ocimum canum* Sims.	Lamiaceae
389.	*Ocimum sanctum* L.	Lamiaceae
390.	*Olax acuminata* Wall ex. Benth.	Olacaceae
391.	*Olea dioica* Roxb.	Oleaceae
392.	*Onosma emodi* Wall.	Boraginaceae
393.	*Operculina turpethum* L.	Convolvulaceae
394.	*Ophiopogon intermedius* D. Don.	Liliaceae
395.	*Ophiorrhiza ochroleuca* Hk.f.	Rubiaceae
396.	*Oreocnide integrifolia* (Gaud) Miq.	Urticaceae
397.	*Oroxylum indicum* (L.) Benth. **Endangered**	Bignoniaceae
398.	*Osbeckia nepalensis* Hook. f.	Melastomataceae
399.	*Osbeckia stellata* Buch-Ham.ex.D.Do.	Melastomataceae
400.	*Oxalis corniculata* L.	Oxalidaceae
401.	*Paederia scandens* (Lour.) Merr.	Rubiaceae
402.	*Pandanus fascicularis* Lamk.	Pandanaceae
403.	*Parabatium micranthum* DC. Pierre.	Apocynaceae
404.	*Paris polyphylla* Sm.	Liliaceae
405.	*Parkia roxburghii* G. Don. **Rare**	Mimosaceae
406.	*Pavetta indica* L.	Rubiaceae
407.	*Pedilanthus tithymaloides* L. Poit.	Euphorbiaceae
408.	*Peganum harmala* L.	Rutaceae
409.	*Pegia nitida* Colebr.	Anacardiaceae

410.	*Peliosanthes teta* Andr.	Liliaceae
411.	*Pergularia daemia* (Forsk.) Chiov.	Asclepiadaceae
412.	*Persea bombycina* (King. Ex. Hk.f.)	Lauraceae
413.	*Persea gamblei* (King. Ex. HK. f.)	Lauraceae
414.	*Phaseolus lunatus* L.	Fabaceae
415.	*Phoebe attennuata* (Nees) Syst.	Lauraceae
416.	*Phoebe lanceolata* (Nees.) Nees.	Lauraceae
417.	*Pholidota imbricata* (Roxb.) Lindl.	Orchidaceae
418.	*Pholidota pallida* Lindl.	Orchidaceae
419.	*Picrorhiza kurroa* Royle ex. Benth.	Scrophulariaceae
420.	*Pinus roxburghi* Sargent.	Coniferae
421.	*Phyllanthus emblica* L.	Euphorbiaceae
422.	*Phyllanthus niruri* L.	Euphorbiaceae
423.	*Phyllunthus parvifolius* Ham.	Euphorbiaceae
424.	*Picrorrhiza kurrooa* Royle ex. Benth.	Scrophulariaceae
425.	*Picrasma javanica* Bl. **Facing extinction**	Simaraubaceae
426.	*Piper betle* L.	Piperaceae
427.	*Piper cubeba* L. f.	Piparaceae
428.	*Piper griffithii* Cas DC.	Piperaceae
429.	*Piper longum* Linn.	Piperaceae
430.	*Piper mullesus* Buch. Ham.	Piperaceae
431.	*Piper nigrum* L.	Piperaceae
432.	*Pittosporum nepaulense* (DC.) Rehder and Wilson	Pittosporaceae
433.	*Plantago major* L.	Plantaginaceae
434.	*Plantago ovata* Forsk.	Plantaginaceae
435.	*Plumbago zeylanica* L.	Plumbaginaceae
436.	*Podophyllum hexandrum* Royle.	Berberidaceae
437.	*Pogostemon purpurascens* Dalz.	Lamiaceae
438.	*Polyalthia longifolia* Benth. and Hk.f.	Annonaceae
439.	*Polygala persicaraefolia* DC.	Polygalaceae
440.	*Polygonum capitatum* Buch. Ham.	Polygonaceae
441.	*Polygonum chinense* L. **Endangered**	Polygonaceae
442.	*Polygonum hydropiper* L.	Polygonaceae

443.	*Polygonum orientale* L.	Polygonaceae
444.	*Polygonum perfoliatum* L.	Polygonaceae
445.	*Polygonum posumbi* Bch. Ham.	Polygonaceae
446.	*Pongamia glabra* Vent.	Fabaceae
447.	*Portulaca oleracea* L.	Portulacaceae
448.	*Pouzolzia frondosa* Don.	Urticaceae
449.	*Potentilla integrifolia* (DC.) Lapeyr.	Rosaceae.
450.	*Pouzolzia sanguinea* (BL) Marr.	Urticaceae
451.	*Pratia begonifolia* Lingl.	Lobeliaceae
452.	*Premna letifolia* Roxb.	Verbenaceae
453.	*Prismatomeris tetrandra* K. Schum.	Rubiaceae
454.	*Prunus cerasoides* D. Don.	Rosaceae
455.	*Prunus ceylanica* (Wight) Miq.	Rosaceae
456.	*Prunus nepaulensis* (Ser) Steud.	Rosaceae
457.	*Prunus persica* (L.) Stokes.	Rosaceae
458.	*Pseuderanthemum palatiferum* (Wall.) Radlk. Ex.	Acantheceae
459.	*Psidium guajava* L.	Psilotaceae
460.	*Psoralea corylifolia* L.	Papilionaceae
461.	*Psychotria montana* Bl.	Rubiaceae
462.	*Pterocarpus marsupium* Roxb.	Papilionaceae
463.	*Pterocarpus santalinus* L.f.	Fabaceae
464.	*Pueraria lobata* (Willd.) Ohwl.	Fabaceae
465.	*Punica granatum* L.	Punicaceae
466.	*Pyrus pashia* D.Don.	Rosaceae
467.	*Quercus acutissima* Carruth.	Fabaceae
468.	*Randia longiflora* Lamk.	Rubiaceae
469.	*Rauvolfia densiflora* (Wall.) Benth ex. Hook. f. **Critically endangered**	Apocynaceae
470.	*Rauvolfia serpentina* (L.) Benth. Ex. Kurz. **Endangered**	Apocynaceae
471.	*Rhododendron arboreum* Sm.	Ericaceae
472.	*Rheum emodi* Wallich.	Polygonaceae
473.	*Rhus javanica* L.	Anacardiaceae
474.	*Rhus succedanea* (non L.) Gamble.	Anacardiaceae

475.	*Ricinus communis* L.	Euphorbiaceae
476.	*Rosa macrophylla* Lindl.	Rosaceae
477.	*Rourea minor* (Gaertn.) Alston.	Cannaraceae
478.	*Rubus alceifolius* Poir.	Rosaceae
479.	*Rubus ellipticus* Sm.	Rosaceae
480.	*Rubus niveus* Thunb.	Rosaceae
481.	*Rubus ulmifolicus* Schott.	Rosaceae
482.	*Rumex marimus*	Polygonaceae
483.	*Rananculus diffusus* DC.	Rananculaceae
484.	*Rhynchoglossum obliquum* Bl.	Gesneriaceae
485.	*Sabia lanceolata* Colebr.	Sabiaceae
486.	*Saccharum officinarum* L.	Poaceae
487.	*Saccharum spontaneum* L.	Poaceae
488.	*Salix tetrasperma* Roxb.	Salicaceae
489.	*Santalum album* L.	Santalaceae
490.	*Sapindus rarak* DC. **Rare**	Sapindaceae
491.	*Saprosma ternatum* Hook. f.	Rubiaceae
492.	*Sarcandra glabra* (Thunb) Nakai.	Chloranthaceae
493.	*Saraca asoka* (Roxb.) de Wild. **Endangered**	Caesalpiniaceae
494.	*Saraca indica* L.	Caesalpiniaceae
495.	*Saussurea lappa* Clarke.	Asteraceae
496.	*Schefflera venulosa* (Wright and Arn.) Harms.	Araliaceae
497.	*Schima khasiana*, **Endemic/Rare**	Theaceae
498.	*Schima wallichii* L.	Theaceae
499.	*Scoparia dulcis* L.	Scrophulariaceae
500.	*Scutellaria discolors* Wall.ex. Benth.	Lamiaceae
501.	*Securinega virosa* (Roxb.) Baillon.	Euphorbiaceae
502.	*Sedium multicaule* Wall.	Crassulaceae
503.	*Siegesbeckia orientalis* L.	Asteraceae
504.	*Semecarpus anacardium* L. f.	Anacardiaceae
505.	*Shorea robusta* Gaertn. F. **Near threatened**	Dipterocarpaceae
506.	*Sida acuta* Burm. f.	Malvaceae
507.	*Sida cordifolia* Linn.	Malvaceae

508.	*Sida rhombifolia* Linn.	Malvaceae
509.	*Skimmia laureola* (DC) Sieb. and ex. Zuoc. Ex. Walp.	Rutaceae
510.	*Smilax ovalifolia* Roxb.	Labaceae
511.	*Solanum erianthum* D. Don.	Solanaceae
512.	*Solanum indicum* L.	Solanaceae
513.	*Solanum khasianum* C.B. Clarke.	Solanaceae
514.	*Solanum melongena* L.	Solanaceae
515.	*Solanum nigrum* L.	Solanaceae
516.	*Solanum surattense* Burm. f.	Solanaceae
517.	*Solanum torvum* Sw.	Solanaceae
518.	*Solanum tuberosum* L.	Solanaceae
519.	*Sonerila maculata* Roxb	Melastomataceae
520.	*Spondias pinnata* (L.f.) kurz.	Anacardiceae
521.	*Stachytarpheta jamaicensis* (L.) Vabl.	Verbenaceae
522.	*Stephania japonica* (Thunb.) Miers.	Menispermaceae
523.	*Sterculia coccinea* Roxb.	Sterculiaceae
524.	*Sterculia urens* Roxb.	Sterculiaceae
525.	*Sterculia villosa* Roxb. Cor.	Sterculiaceae
526.	*Stereopermum chelonoides* DC.	Bignoniaceae
527.	*Strobilanthes divaricatus* T. Anders.	Acantheceae
528.	*Strobilanthes flaccidifolius* Nees.	Acantheceae
529.	*Strychnos wallichiana* Steud.	Loganiceae
530.	*Swertia chirata* Kar.	Gentianaceae
531.	*Swertia dilatata* Wall.	Gentianaceae
532.	*Swertia nervosa*	Gentianaceae
533.	*Swertia purpurascens* Wall.	Gentianaceae
534.	*Symplocos lucida* (Thunb.)	Symplocaceae
535.	*Symplocos paniculata* Wall.	Symplocaceae
536.	*Symplocos racemosa* Roxb.	Symplocaceae
537.	*Symplocos theaefolia* D.Don.	Symplocaceae
538.	*Syzygium aromaticum* (L.) Merrill and Perry	Myrtaceae
539.	*Syzygium cumini* (L.) Skeels.	Myrtaceae
540.	*Syzygium jambos* (L.) Alston.	Myrtaceae

541.	*Tabernaemantana divaricata* (L.) R. Br.	Apocynaceae
542.	*Tamarindus indica* L.	Caesalpiniaceae
543.	*Terminalia arjuna* Roxb. Near threatened	Combretaceae
544.	*Terminalia bellirica* (Gaertn.) Roxb.	Combretaceae
545.	*Terminalia chebula* Retz.	Combretaceae
546.	*Terminalia citrina* Gaertn. Flem.	Combretaceae
547.	*Thalictrum javanicum* Bl.	Rununculaceae
548.	*Thunbergia grandiflora* (Rottb.) Roxb.	Acanthaceae
549.	*Tinospora cordifolia* (Willd).	Menispermaceae
550.	*Tithonia diversifolia* A. Gray.	Asteraceae
551.	*Toddalia asiatica* (L.) Lam.	Rutaceae
552.	*Toona ciliata* Roemm.	Meliaceae
553.	*Trachelospermum lucidum* (D. Don.) K. Schum.	Apocynaceae
554.	*Trema orientalis* Bl.	Ulmaceae
555.	*Trevesia palmata* Roxb.	Araliaceae
556.	*Trianthema portulacastrum* L.	Aizoaceae
557.	*Tribulus terrestris* L.	Zygophyllaceae
558.	*Trichosanthes dioica* Roxb.	Cucurbitaceae
559.	*Trichosanthes tricuspidata* Lour.	Cucurbitaceae
560.	*Tylophora indica* (Burm.f.) Merr.	Asclepiadaceae
561.	*Urena lobata* L.	Malvaceae
562.	*Urginea indica* Kunth.	Liliaceae
563.	*Vallaris solanacea* (Roth.) O. Kuntze.	Apocynaceae
564.	*Valeriana hardwickii* Wall ex. Roxb.	Valerianaceae
565.	*Valeriana officinalis* L.	Valerianaceae
566.	*Vanda roxburghii* R. Br. **Endangered**	Orchidaceae
567.	*Ventilago madraspatana* Gaerth.	Rhamnaceae
568.	*Vernonia anthelmintica* Willd.	Asteraceae
569.	*Vernonia cineria* Less.	Asteraceae
570.	*Viburnum colebrookianum* Wall. ex Cl.	Caprifoliaceae
571.	*Viburnum foetidum* Wall.	Caprifoliaceae
572.	*Viscum monoicum* Roxb. ex.	Loranthaceae
573.	*Vitex negundo* L.	Verbenaceae

574.	*Vitex peduncularis* Wall ex. Schauer.	Verbenaceae
575.	*Vitex trifolia* Linn.	Verbenaceae
576.	*Vitis barbata* Wall.	Vitaceae
577.	*Viola serpens* Wall.	Violaceae
578.	*Wendlandia tinctoria* (Roxb.) DC.	Rubiaceae
579.	*Withania somnifera* Dunal.	Solanaceae
580.	*Woodfordia fructicosa* Kurz.Syn.	Lythraceae
581.	*Wrightia arborea* (dennst.) Mad.	Apocynaceae
582.	*Xanthium strumarium* L.	Asterraceae
583.	*Xeromphis spinosa* (Thunb) Keay.	Rubiaceae
584.	*Xylosma longifolium* Clos.	Flacourtiaceae
585.	*Zantedeschia aethiopica* (L.) Spreng.	Araceae
586.	*Zanthoxylum armatum* DC.	Rutaceae
587.	*Zanthoxylum khasianum* Hk.f	Rutaceae
588.	*Zea mays* L.	Poaceae
589.	*Zingiber officinale* Roscoe.	Zingiberaceae
590.	*Zizyphus jujuba* (Lamk.)	Rhamnaceae
591.	*Zizyphus mauritiana* Lamk.(K)	Rhamnaceae
592.	*Zizyphus mauritiana* Lamk.	Rhamnaceae
593.	*Zizyphus oenoplia* (L.) Mill.	Rhamnaceae

ENDANGERED MEDICINAL PLANTS

Out of 593 medicinal plants (Table 1), *Acorus calamus, Aphamixis polystachya, Canarium strictum, Celastrus paniculatus, Costus speciosus, Eleocarpus sphaericus, Gloriosa superba, Holarrhena antidysenterica* and *Lycopodium clavatum* are **vulnerable**; *Ichnocarpus frutescens, Michelia champaca, Oroxylon indicum, Polygonum chinense, Rauvolfia serpentina, Saraca asoca,* and *Vanda roxburghii* are **endangered**; *Rauvofia dessiflora* is **critically endangered**; *Camellia caduca, Cinamomum pauciflorum, Dioscorea pentaphylla, Erythroxylum kunthianum, Euonymus lawsonni, Lindera latifolia, Nepenthes khasiana,* and *Schima khasiana* are **endemic**; Embetiea ribes and Rouvolfia densiflora are **critically endemic**; *Butea monospora* and *Mesua ferrea* are **near endemic**; *Acacia pinnata, Ardisia cordata, Aristolochia saccate, Croton tiglium, Codonopsis javanica, Hedyotis verticillata, Helicia excelsa, Holoptelea integrifolia, Ilex embelioides, Lagerstroemia indica* and *Parkia roxburghii* are **rare**; *Cordia dichotoma, Cordia fragrantissima* and *Cretaeva nurvala* are **very rare**, *Desmodium heterocarpon* is **critically rare**, *Shorea robusta* (mother tree) and *Terminalia arjuna* are **near threatened** and *Picasma javanica* is **facing extinction**.

Vulnerable species	09
Endangered species	07
Critically endangered species	01
Endemic species	08
Critically endemic species	01
Near endemic species	02
Rare species	12
Very rare species	03
Extremely rare Species	01
Near threatened species	02
Facing extinction species	01

Vulnerable Species

A taxon is vulnerable when it is not critically endangered or endangered but is facing a high risk of extinction of the wild the wild in the medium-term future. Some examples are:

Acorus calamus

Aphamixis polystachya

Canarium strictum

Celastrus paniculatus

Costus speciosus

Eleocarpus sphaericus

Gloriosa superba

Holarrhena antidysenterica

Lycopodium clavatum

Endangered Species

A taxon is endangered when it is not critically but facing a very high risk of extinction of the wild in the near future. Some examples are:

Ichnocarpus frutescens

Michelia champaca

Oroxylon indicum

Polygonum chinense

Rauvolfia serpentina

Saraca asoca

Vanda roxburghii

Critically Endangered Species

A taxon is critically endangered when it is facing an extremely high risk of extinction of the wild in the immediate future. Some examples are:

Rauvofia densiflora

Embelia ribes

Endemic Species

Endemic species by virtue of their narrow distribution also qualify for conservation. Some examples are:

Camellia caduca

Cinamomum pauciflorum

Dioscorea pentaphylla

Erythroxylum kunthianum

Euonymus lawsonnii

Lindera latifolia

Nepenthes khasiana.

Schima khasiana

Critically Endemic Species

Embelica ribes

Rauvolfia densiflora

Near Endemic Species

Butea monospora

Mesua ferrea

Rare Species

Acacia pinnata

Ardisia cordata

Aristolochia saccata

Brucea mollis

Codonopsis javanica

Croton tiglium

Hedyotis verticillata

Helicia excelsa

Holoptelea integrifolia

Ilex embelioides

Lagerstroemia indica

Parkia roxburghii

Very Rare Species

Cordia dichotoma

Cordia fragrantissima

Cretaeva nurvala

Critically Rare Species

Desmodium heterocarpon

Extremely Rare

Diospyros pilosula

Luvunga scandens

Near Threatened Species

Taxa which do not qualify for conservation but which are close to qualifying for vulnerable. Some examples are:

Shorea robusta (mother trees).

Terminalia arjuna

Facing Extinction Species

A taxon is facing extinction in the wild.

Picasma javanica

Abelmoschus moschatus Madic.

Family   Malvaceae

Indian names   Musk mallow (English); *Latakasturika* (Sanskrit); *Latakasturi, Maskdana* (Hindi); *Kadukasturi* (Kannada); *Latakasturi* (Malayalam), *Kattukasturi*; *Vettrilaikkasturi* (Tamil); *Kasturibendavittu* (Telugu); *Dhokrakanda* (Assamese).

Description   An erect hispid annual herb or undershrub up to 60–180 cm in height; leaves simple, of varying shapes, usually palmately 3–7 lobed, lobes narrow-acute or obovate, crenate, serrate, hairy on both surfaces; flowers large, solitary, axillary; epicalyx segment, linear, calyx up to 3 cm long, 5-toothed, stellate, tomentose outside. Petals 6–8 cm long, obovate, yellow with purple centre, with spathaceous calyx; fruits hairy, capsular; seeds many, black or greyish brown, accented.

Flowering and fruiting   July–February

Distribution   Common and cultivated throughout India, forests of Meghalaya.

Propagation   By seeds

Parts used   Roots, leaves and seeds

Chemical constituents   Leaves, flowers and seeds of the plant contain β-sitosterol and its glucoside, campesterol, ergosterol, higher fatty acids, esters of acetic acid and cyandin 3-glucoside.

Uses   Seeds aphrodisiac, antispasmodic, carminative, diuretic, stomachic and tonic, used as an inhalant in hoarseness and dryness of the throat, it is useful in nervous disorders and in hysteria; leaves and roots are used for gonorrhoea and veneral diseases. Seeds rubbed to paste to cure itch and bites of serpents.

Abroama augusta (L.) F.

Family   Sterculiaceae

Indian names   Devil's cotton (English); *Pivari, Pisacakarpasa, Yosini* (Sanskrit); *Ulatkambal* (Hindi); *Ulatkambal* (Bengali); *Divvahatti, Melpundigida, Sivapputtutti,* (Kannada); *Sivapputtutti* (Tamil), *Kasturibendavittu* (Telugu); *Gorukhia-Karai, Bon-Kopahi* (Assamese); *Dieng-tyrkhum* (Khasi).

Description   Shrubs; branches usually horizontal, young parts tomentose; leaves ovate, acuminate, cordate or truncate, at base dentate or sub-entire, sparsely hairy or glabrescent above and tomentose along nerves beneath; pedicels jointed, bracts deciduous, lanceolate; sepals lanceolate. Persistent; petals caducous; staminal tube short; capsule truncate at apex, septicidal.

Flowering and fruiting   June–November

Distribution   Indo-Malaya and throughout India, usually cultivated in Meghalaya, noticed as an undergrowth in open places in evergreen forests.

Propagation   By seeds

Parts used   Root, bark, stem and leaves

Chemical constituents   Stem bark contains sitosterol and friedelin. Seeds contain fixed oil composed of linoleic, oleic, hexadecenoic, palmitic and stearic acids. Roots contain a fixed oil, resins, an alkaloid in minute quantity and water-soluble bases.

Uses   The root and bark are uterine tonic; they are useful in congestive and nervous dysmenorrhoea, amenorrhoea, sterility and other menstrual disorders. Root powder is an abortifacient and antifertility agent. Leaves are useful in treating uterine disorders, diabetes, rheumatic pain of joints, and headache with sinusitis. Infusion of leaves and stem has been found to be efficacious in gonorrhoea.

Abrus precatorius L.

Family   Fabaceae

Indian names   Jequirity, Indian liquorice (English); *Gunja* (Sanskrit); *Ghugachi, Chirami, Ratti, Guncaci* (Hindi); *Gurugunii* (Kannada); *Kunni* (Malayalam); *Kuntumani* (Tamil), *Guruginja* (Telugu); *Ralurmani* (Assamese); *Kunch* (Manipuri).

Description   Woody climber, profusely branched, glabrous or shiny, silky, leaflets paripinnate in 10–20 opposite pairs, linear or linear oblong, minutely apiculate, glabrous above, shiny, silky beneath. Flowers reddish or dull white, clustered in dense pedunculate racemes. Fruits pod.

Flowering and fruiting   September–January

Distribution   Nearly throughout India, frequent in Meghalaya in shady localities; climbing on small bushes and shrubs.

Propagation   By seeds

Parts used   Seeds, leaves and roots

Chemical constituents   Roots and leaves contain glycyrrhizin; precol, abrol, abrasine and precasine from roots. Gallic acid, abrine, hypaphorine, alanine, serine, valine, choline, trigonelline, precatorine, and methyl ester are obtained from seeds.

Uses   Seeds purgative, emetic, tonic, antiphlogistic, and used for nervous disorders, paste as local application in stiffness of shoulder joints. Leaf decoction is used for cough, cold, colic pain; fresh juice applied in painful swelling in leucoderma. Root tonic is diuretic, emetic, alexeteric, used in treatment of gonorrhoea, jaundice.

Abutilon indicum (L.) Sweet

Family   Malvaceae

Indian names   Country mallow (English); *Atibala* (Sanskrit); *Kanghi* (Hindi); *Tutti* (Kannada); *Velluram* (Malayalam); *Perumtutti* (Tamil); *Patari* (Assamese); *Petari* (Bengali); *Kunch* (Manipuri).

Description   Shrubs 3–5 m high, branchlets tomentose, leaves, ovate, angular, orbicular-angular, acute or acuminate, cordate at base, irregularly dentate, 5–7 nerved from base, tomentose on both surfaces, greyish beneath, petiole 1–5 cm long; flowers yellow, pedicels joined near the tip, calyx tomentose, united at base, lobes ovate, petals yellow, obovate, ovary densely hairy, schizocarp depressed, discoid, globose, mucronate, stellately hairy.

Flowering and fruiting   August–February

Distribution   Commonly seen throughout the warmer parts of India; in lower altitudes, in comparatively dry open habitats, in Jhum areas.

Propagation   By seeds

Parts used   Seeds, leaves, bark and root

Chemical constituents   Leaves contain tannin, organic acids and traces of asparagine and ash containing alkaline sulphate, chlorides, magnesium phosphate and calcium carbonate. Roots also contain asparagine.

Uses   Seeds in piles, laxative, expectorant, in chronic cystitis, gleet and gonorrhoea; leaves demulcent, locally applied for boils and ulcers and as a fomentation to painful parts of the body; bark: astringent and diuretic.

Acacia farnesiana Willd. Fl. Br.

Family   Mimosaceae

Indian names   Cassie flower (English); *Vutkhadira* (Sanskrit); *Gandh babul, Vilayati kikar* (Hindi); *Guye babul* (Bengali); *Ladivel, Kadivel* (Tamil); *Kasturi tuma* (Telugu); *Babul* (Assamese); *Brangdaru* (Garo).

Description   A thorny bush or small tree, 4.5 m high, branches spread out from the main trunk. Bark with brown streaks. Leaves bipinnate with stipular spines; flowers bright yellow, exceedingly sweet scented in globose heads, calyx and corolla 5-toothed.

Flowering and fruiting   September–February

Distribution   Burma, Bangladesh, North-East India; common at lower elevations along forest margins and secondary forests, particularly along river banks associated with *Croton caudatus, Uncaria* sp. etc.

Propagation   By seeds

Parts used   Leaves, bark and pods

Chemical constituents It contains eicosane, benzyl alcohol, farnesol, geranyl acetate, linalool, nerolide, and pinitol. The ripe pods contain gallic acid, its derivatives and tannin, aromadendrin, kaempferol, naringenin and their glucosides while apigenin and rutin glucosides have been found in the leaves.

Uses Decoction of the bark together with ginger is an astringent wash for the teeth, and so it is useful in the bleeding of gums, etc. Tender leaves are crushed with little water and swallowed to treat gonorrhoea.

Acacia indica L.

Family Mimosaceae

Indian names *Karuvelai* (Tamil); *Karuvelam* (Malayalam); Remsu (Garo).

Description A small tree, with branchlets and leaf-rachises densely covered with fine grey hairs, the latter with several glands. Bark grey-black, cracked into quandrangular pieces, 2 mm thick. Flowers sessile, white, in spikes, fruit pods, 7–12 cm long with a triangular beak at the apex, seeds many.

Distribution Throughout India

Propagation By seeds

Parts used Pods, leaves and bark

Chemical constituents Same as *Acacia farnesiana*

Uses Leaves and bark are astringent to bowels. Leaves are used to cure bronchitis, piles and for healing fractures. Bark is used to cure cough, bronchitis, diarrhoea, dysentery, leucoderma and urinary infections. The gum is an astringent to bowels, cures leprosy, urinary and vaginal discharges (Ayurveda).

Acacia pinnata L. [Rare]

Family   Mimosaceae

Indian name   *Sirengki* (Garo).

Description   Large, prickly climbers, prickles more or less recurved; bark dark brown, young pubescent; common rachis up to 25 cm long with a large cuplike gland near the base and 2–4 smaller ones near the tip, usually prickly; pinnae 4–10 cm long, pinnules 16–50 pairs, 7–10 × 1–2 mm, often overlapping, linear oblong, unequally broad, sharply acute, oblique at base; heads dull white, 0.8–1.2 cm across, 1–4 together in terminal leafy panicles; pods 10–15 × 2–4 cm, strap-shaped, reddish to dark brown, 8–14 seeded.

Flowering and fruiting   July–December

Distribution   Indo-Malaya, tropical Africa and throughout India.

Propagation   By seeds

Parts used   Bark

Chemical constituents   The bark contains saponin. The sugar identified is glucose, arabinose and rhamnose.

Uses   Bark is used as fish poison.

Acalypha indica L.

Family   Euphorbiaceae

Indian names   *Harittamanjiari* (Sanskrit); *Kuppi* (Hindi); *Muktabarsi* (Bengali); *Vanchi Kanto* (Gujrati); *Kuppamani* (Malayalam); *Khokhali* (Marathi); *Indramaris* (Oriya); *Kuppichettu* (Tamil); *Adasaramu* (Telugu).

Description   An annual herb, up to 75 cm high. Leaves 3–8 cm long, ovate, thin usually 3-nerved; margins of the leaves toothed; leaf-stalks longer than leaves. Flowers in axillary erect spikes;

female flowers supported by conspicuous wedge-shaped bracts, male flowers, minute, borne towards the top of the spike. Fruits small, hairy, concealed in the bracts.

Distribution   Throughout India

Propagation   By seeds

Parts used   Roots and leaves

Chemical constituents   Constituents of drug Acalypha.

Uses   The plant is useful in curing bronchitis, asthma, pneumonia and rheumatism. Roots and leaves have laxative properties. Juice of the leaves is considered as an efficient emetic. Fresh leaves are useful for ulcers.

Acorus calamus L. [Vulnerable]

Family   Araceae

Indian names   Flagroot, Sweet cane, Sweet grass (English); *Bacha, Bodhaniya, Galani* (Sanskrit); *Bach, Ghorbach* (Hindi); *Vasa, Baje gida* (Kannada); *Vaembu, Vayambhu* (Malayalam), *Vashambu, Vasambu* (Tamil); *Vadaja, Vasa* (Telugu); *Bach* (Assamese).

Description   A semi-aquatic rhizomatous perennial herb, rhizome creeping, much branched, as thick as the middle finger, cylindrical or slightly compressed, light brown or pinkish brown externally, white and spongy within, leaves bright green, distichous, ensiform, base equitant, thickened in the middle, margin wavy; flowers light brown densely packed in sessile cylindrical spadix; fruits oblong turbinate berries with a pyramidal top; seeds free, pendent from the apex of the cells.

Distribution   Throughout India, in areas elevated upto 1,800 m in marshes.

Propagation   By rhizome

Part used   Rhizome

Chemical constituents   The important constituents are asarone and its α-isomer. Other constituents are calamenol; calamene, calamenone, methyl eugenol, eugenol and α–pinene and camphene, presence of small quantities of palmitic, heptulic and butyric acids, asaronaldehyde, calamol, calamone and azulene has also been reported. Sesquiterpenic ketones like acorone, calarene, calacone, calacorene, acorenone, acolamone, isocolamone, epishyobunone, shyobunone, isoshyobunone and acorager macrone, and alcohols like isocalamendiol and preisocalamendiol are also present.

Uses   The rhizome is acrid, bitter, thermogenic, aromatic, intellect promoting emetic, laxative, carminative, stomachic, anthelmintic, emmenagogue, diuretic, alexeteric, expectorant, adodyne, antispasmodic, aphrodisiac, anticonvulsant, resuscitative, anti-inflammatory, sudoric, antipyretic, sialagogue, insecticidal, tranquilizing, nervine tonic, sedative and tonic. It is useful in witiated conditions of vata and kapha, hoarseness, colic flatulence, dyspepsia, helminthiasis, amenorrhoea, dysmenorrhoea, nephropathy, calculi, strangury, cough, bronchitis, odontalgia, pectoralgia, hepatodynia, otalgia, inflammations, gout, epilepsy, delirium, convulsions, depression and other mental disorders, tumours, dysentery, hyperdipsia, haemorrhoids, intermittent fevers, skin diseases, numbness and general debility.

Achyranthes aspera L.

Family Amarantheceae

Indian names Prickly chaff flower (English); *Apamarga* (Sanskrit); *Chirchira, Chirchitta, Lat jira* (Hindi); *Apang* (Bengali); *Utranigida, Uttaraanne* (Kannada); *Kadaladi* (Malayalam); *Chirukadaladi, Naaurivi* (Tamil); *Apamarganu, Uttareeni* (Telugu); *Monongganchi* (Garo); *Aghadam aghara* (Marathi).

Description Herbs, about 1 m to 2 m high. Stems erect, pubescent, swollen at the nodes. Leaves opposite, short-petioled, margins undulate; flowers numerous, stiffly deflected against the pubescent rachis in elongate terminal spike, 20–30 cm long, urticle oblong-cylindrical, enclosed in the hardened perianth, brown. Seeds oblong ovoid.

Distribution Throughout India, common in Meghalaya.

Propagation By seeds

Parts used Entire plant, especially the seeds and roots

Chemical constituents The root contains triterpenoid saponins, betaine, achyranthine, hentriacontane, and ecdysterone. The dried dehusked seeds contain amino acids.

Uses The whole plant especially the roots are characterized by their anti-inflammatory and uterine-stimulant activity, are prescribed in the rheumatism lumbago, osteodynia, dysuria, post-partum haematometra and dysmenorrhoea. A decoction of plant is useful in pneumonia and renal dropsy, ophthalmia, dysentery. The benzene extract of stem bark showed significant abortifacient activity. The whole plant is reported to dilate the blood vessels, lower the blood pressure, depress the heart, and increase the rate of amplitude of respiration. It is also useful in dropsy, asthma and cough. The leaves are used to cure gonorrhoea and excessive perspiration. Its alcoholic and aqueous extract showed antibiotic action against *Micrococcus pyogens* var. aureus and *Escherichia coli*. The roots are useful in treatment of cancer, stomach troubles, and stones in the bladder.

Achyranthes porphyristachya Wall.

Family   Amaranthaceae

Indian names   *Lat Jiara, Chirchita* (Assamese); *Olti-Koro* (Naga).

Description   An erect herb with pubescent branches, attaining a length up to 1 m. Leaves opposite variable, elliptic obovate or sub-orbicular, softly adpressed, hairy; flowers greenish, stiffy, deflexed, in simple or panicled pubescent spikes which elongate into fruits.

Flowering and fruiting   During rainy season

Distribution   Throughout India

Propagation   By seeds

Parts used   Roots and seeds

Chemical constituents   The roots contain triterpenoid saponins, achyranthine, hentriacontane and two glycosides of oleanolic acid have been reported. The dried seeds contain amino acids.

Uses   Astringent, diuretic, alterative and antiperiodic, purgative, emetic and anthelmintic. A decoction of the plant is useful in treating pneumonia, renal dropsy, ophthalmia and dysentery. The benzene extract of stem bark showed significant abortifacient activity. The whole plant is reported to dilate the blood vessels, lower the blood pressure, depress the heart, and increase the rate and amplitude of respiration. It is also useful for dropsy, asthma and cough. The roots are useful in treatment of cancer, stomach troubles, and stones in the bladder.

Acronychia pedunculata (L) Mig.

Family   Rutaceae

Indian names   *Bhoothaali* (Kannada); *Vidukanali, Orilatheeppettimaram* (Malayalam); *Vidukanalei, Muttanari* (Tamil); *Dieng sohphlang* (Khasi); *Bolgrak, Bol thimatchi* (Garo); *Loajan* (Assamese).

Description   Small unarmed trees; bark greyish-brown, minutely warty, leaves, oblanceolate, oblong-lanceolate or obovate-lanceolate, obtuse or subacute, base narrowed, cuneate, entire, margins narrowly recurved; panicles, flowers greenish-white, sepals minute, lanceolate; petals long white, oblong, reflexed, silky inside; stamens 8, erect, disk orange red, ovary silky; fruit shortly beaked, 0–8 cm.

Flowering and fruiting   June–November

Distribution   Indo-Malaya, throughout India; frequent in evergreen forests particularly in the slopes, associated with *Itea macrophylla, Castanopsis indica*, etc. as a sub-storey tree.

Propagation   By seeds

Parts used   Roots, twigs, stem bark and leaves

Chemical constituents   The stem bark contains acrovestone, acronylin, bauereno, potassium oxalates and traces of alkaloids. The root bark yields 6-dimethylacronulin and acrovestone.

Uses   The roots are utilized in the therapy of rheumatism, lumbago, pain in the limbs, post-partum blood stasis, furunculosis impetigo and snakebite. The torrified roots or leaves are effective for stomache, for dyspepsis in parturients. A poultice made of heated leaves and a wash with a decoction of the trunk bark are useful for treating furunculosis and impetigo.

Adhatoda vasica Nees.

Family   Acanthaceae

Indian names   *Amalaka, Arusak* (Sanskrit); *Adosa* (Hindi); *Bahaka* (Assamese); *Khumsal* (Garo); *Vasaka* (Bengali); *Alduso* (Gujrati); *Adusoge* (Kannada); *Adalodakam* (Malayalam); *Adulsa* (Marathi); *Adhatodai* (Tamil); *Adamkabu, Adasaram* (Telegu).

Description   Small evergreen, subherbaceous bush. The leaves are 10 to 16 cms in length, minutely pubescent and broadly lanceolate. The inflorescence is dense, short pedunculate, bracteate and spike terminal. The corolla is large and white with lower lip streaked purple. The fruit is a 4-seeded small capsule.

Flowering and fruiting   January–June

Distribution   Grows all over the plains of India and in the lower Himalayan ranges.

Propagation   By stem cutting

Chemical constituents   The leaves of the plant contain as essential oil and alkaloids vasicine and vasicinone. The roots are known to contain vasicinolone, vasicol, peganine and 2-glucosyl-oxychalcone. The flowers contain β-sitosterol D-glucoside, kaempferol, its glycosides and quercetin.

Uses   Bronchitis and bronchial asthma, local bleeding and thrombocytopenia and pyorrhoea. The leaf extract has been used for treatment of bronchitis and asthma for many centuries. It relieves cough and breathlessness. It is also prescribed commonly in Ayurveda for bleeding due to idiopathic thrombocytopenic purpura, local bleeding due to peptic ulcer, piles menorrhagia, etc. large doses of fresh juice of leaves have been used in tuberculosis. Its local use gives relief in pyorrhoea and in bleeding gums.

Adhatoda zeylanica Medic.

Family   Acanthaceae

Indian names   *Vajidantakahaatarusha, Vasaka* (Sanskrit); *Vasak, Adusa* (Hindi); *Aadusouge, Aadu muttada gida* (Kannada); *Adel-odagam, Atalotakam* (Malayalam); *Atatotai, Atatotai ilai* (Tamil); *Addasaramu, Addasarapaku* (Telugu).

Description   Shrubs up to 3 m high; bark grey or greyish brown; leaves 7–20 × 2.5–8 cm, lanceolate, elliptic-lanceolate, acuminate or acute, base cuneate, attenuate, narrowed to the petiole; spikes 5–10 cm long; flowers 2.5–4 cm long, white or greenish white; capsules 2 cm long.

Flowering and fruiting   January–June

Distribution   Southeast Asia; throughout India, commonly seen in Meghalaya, usually cultivated at lower elevations.

Propagation   By vegetative method

Parts used   Leaves, roots, flowers, bark and fruits

Chemical constituents   Vasicine is the active principle obtained from root/bark and leaves.

Uses   Expectorant, diuretic, antispasmodic and alternative. It is an important constituent of cough mixtures. The plant is considered efficacious in preliminary diseases. It has antiseptic properties and it is reported that a good insecticide can be obtained from its bark. The leaf extract is used in tonsillitis, influenza, fever, cough, headache, giddiness, bodyache and as a diuretic.

Aegle marmelos Linn.

Family   Rutaceae

Indian names   Bael fruit, Bengal quince (English); *Adhararuha, Asholam* (Sanskrit); *Bael sripal, Bello* (Hindi); *Bilva, Bilvapatre* (Kannada); *Koovalam, Vilvam* (Malayalam); *Vilvam, Kuvilam* (Tamil); *Bilvamu, Bilvapandu* (Telugu); *Bel, Bael* (Bengali).

Description   Large deciduous trees with a lax, oblong, oval crown; bark reddish brown or brownish grey, often with thorns at the base; leaves up to 15 cm long; leaflets elliptic, elliptic-lanceolate, oblanceolate, acuminate, base narrowed, glabrous or glabrescent; flowers 2–2.5 cm across, yellowish; calyx obscurely lobed, rounded; petals oblong, deflexed; 1–5 cm long; stamens basifixed, anther nearly equalling the length of the filaments, stigma club-shaped; berries 5–10 cm in diameter, pulpy.

Flowering and fruiting   March–May

Distribution   Indo-Malaya, not truly wild in Meghalaya.

Propagation   By seeds and air layering

Parts used   Roots, leaves and fruits

Chemical constituents   Several biochemical constituents namely alkaloids, coumarin and steroid have been isolated from different parts of the trees. They are skimmianine, aegelin, γ-sitosterol, aegelenine, myrlene, ethyl cinnamide, lupeol, alloimperatorin, imperatorin, marmesin, β-sitosterol, dietamine, marmin, umbelliferone, marmelosin, rutacine. Roots of the tree have been found to contain psoralin, xanthotoxin, scopoletin and tembamide.

Uses   It is widely used as antipyretic, antiscorbutic, astringent, digestive, febrifuge, laxative, odoriferous, restorative, stimulant and tonic. It is also useful in the treatment of fever, bronchitis, constipation, hypochondriasis-melancholia, palpitation of the heart, diabetes, debility, diarrhoea, dysentery and piles. The oil extracted from seeds shows antibacterial activity against bacteria, fungi, virus, pests and protozoans. The polyphenol extracts of tree showed antitumorous and antimutagenic activity. Oral administration of pyranocoumarin isolated from the seeds offers significant protection against carcinogens in mice. Oral administration of pyranocoumarin isolated from the seeds offers protection against aspirin-induced gastric ulcers in rats and stress-induced gastric ulcers in rats and guinea pigs. Aqueous extraction of roots and root bark promoted healing and also helped to overcome the depression caused by dexamethasone.

Aesculus assamica Griff.

Family   Sapindaceae

Indian names   Horse chestnut (English); *Singuri* (Assamese); *Bol-rimmu* (Garo); *Dieng dula* (Khasi).

Description   Middle sized tree with a very attractive, sub-globose crown; bark greyish and warty; leaves 30–60 cm across; petiole 13–30 cm long; leaflets 14–30 cm, oblanceolate, obovate, elliptic, long acuminate, base narrowed to the short petiole panicles 30–75 cm, oblong, conical, erect, flowers white with yellow throat, 2–2.5 cm across; calyx campanulate, 0.6–0.8 cm long, downy, petals downy below; stamens 2–3 cm long; fruits brown, rugose, 5–9 cm long.

Flowering and fruiting   January–August

Distribution   Indo-Malaya; confined to North-East India; very common in deciduous forests of Meghalaya, mostly along watercourses, associated with *Trewia nudiflora, Eleaocarpus rugosus, Vatica lanceofolia*.

Propagation   By seeds

Parts used   Seeds, leaves and bark

Chemical constituents   Horse chestnut contains triterpenoids, saponins, coumarins and flavonoids. Aescin, the main active constituent has anti-inflammatory properties.

Uses   Horse chesnut is an astringent, anti-inflammatory and helps to tone the vein walls, which when distended may become varicose, haemorrhoidal otherwise problematic. An oil extracted from the seeds has been used as an external treatment for rheumatism. A decoction of the leaves has been given in cases of whooping cough.

Ageratum conyzoides L.

Family   Asteraceae

Indian names   Goat weed, White weed, Appa grass (English); *Visamustih* (Sanskrit); *Jangli pudina, Visadodi* (Hindi); *Uralgidda* (Kannada); *Muryampacha* (Malayalam); *Poompillu, Sinnapoompillu* (Tamil); *Pokabanthi* (Telugu); *Gendelabon* (Assamese); *Vaihlenhlo* (Mizoram); *Khongjai napi* (Manipuri).

Description   Plant erect herbs or shrubs, hairy up to 1 m high. Leaves 1–10 × 0.75 cm, ovate, rhomboid, obtuse or acute, base subtracted or acute, 3- nerved, cernate. Heads 0.4–0.7 cm across, white, bluish or violet. Heads corymbose or panicled, homogenous, involucres campanulate, bracts 2–3 seriate, involucre lanceolate. Corolla slightly longer, brown, tubular, equal, regular, 5-cleft. Anthers appendaged, base obtuse. Style arms elongate, obtuse. Achenes 5 angled; pappus 5 short, free or connate, scales 10–20 narrowed, unequal.

Flowering and fruiting   Nearly throughout the year

Distribution   Species about 16, native of tropical America, common throughout India; abundant in Meghalaya at all elevations forming gregarious patches in road sides, forest cleared areas, new plantations, and forest roads.

Propagation   By seeds

Parts used   Leaves

Uses   Crushed leaf juice used for cuts and wounds as homeostatic.

Agrimonia pilosa Ledeb.

Family   Rosaceae

Indian names   Hairy agrimony (English); *Lesukuria, Velu* (Hindi); *Lynniong-tynning, Roishing* (Arunachali).

Description   Hairy herb with perennial and woody roots, clothed with spreading soft hairs. Leaves interrupted, imparipinnate with very small leaflets between largest ones. Leaflets sessile, obovate to elliptic, flowers, yellow in terminal spike like racemes. Petals 5, sepals 5, stamens about 15 adnated to the mouth of calyx tubes. Carpels included in the calyx tube.

Distribution   Roadsides, grassy areas in lowlands and mountains found in Korea, Japan, China, Siberia, Eastern Europe and North-East India.

Propagation   By seeds

Parts used   Roots and leaves

Chemical constituents   Agrimonolides, tannins, phenylpropanoids, coumarins, flavonoids and triterpenes.

Uses   The root is chewed to ameliorate toothache, and is used as an astringent, anthelmintic, diuretic and tonic and as a remedy for cough. It has antitumour, antidysentry action, used in taenia, boils and eczema.

Alangium chinense (Lour) Harm.

Family   Connaceae

Indian name   *Phagrang* (Garo).

Description   Small trees, 10 m high, bark dark grey or brown, branches horizontally spreading; leaves ovate, ovate-orbicular, acuminate, base truncate, often angled or shortly lobed, dark green above, pale beneath, 3–7 nerved from base; hairy along nerves; cymes up to 5 cm long; flowers 1.5–2.5 cm long, white, narrow, oblong in bud; calyx minutely toothed; petals strap-shaped; drupe 0.7–1.3 cm long; dark purple when ripe, 2-seeded.

Flowering and fruiting   April–October

Distribution   Indo-Malaya; throughout Northern India; common in Meghalaya at lower elevations, particularly along forest edges, forest roadsides and secondary forests. Usually associated with *Grewia microcos*.

Propagation   By seeds

Parts used   Roots and fruits

Chemical constituents   The root bark contains the following alkaloids: alangine A and B, alangicine, dimethylpsychotrine, marckine, marckindine, lamarckine, psychotrine, tubulosine, cephaeline emetine, protoemetinol and alangiosterol. The root bark also contains fatty acid composition—myristic, palmitic, oleic, linoleic, and resin acids. The seeds contain the alkaloids cephaeline, *N*-methylcephaeline, and alangiside. The seeds contain the alkaloids emetine, cephaeline, psychotrine, *N*-methylcephaeline and alangiside.

Uses   The roots are acrid, astringent, emollient, anthelmintic, thermogenic, diuretic and purgative. Root bark is an antidote for several poisons. The roots are useful for external application in acute case of rheumatism, leprosy and inflammation and for external and internal application in the case of bites of rabies dogs. Fruits are sweet, cooling and purgative, and are useful in treating burning sensation and haemorrhages.

Albizia lebbeck Benth.

Family   Mimosaceae

Indian names   Siris tree, East indian walnut (English); *Sirisah, Bhandi* (Sanskrit); *Siris, Garso* (Hindi); *Hombage, Begemara* (Kannada); *Nenmenivaka, Vaka* (Malayalam); *Vakai, Siridam* (Tamil); *Dirisana* (Telugu); *Gisim Khilchi* (Garo).

Description   A large spreading tree up to 25 m high, bark grey, usually cracked, young parts white, pubescent, pinnae 6–10 pairs, leaflets 5–15 pairs, oblique-oblong, arbicular, usually with midrib eccentric towards upper half, a large gland present near the base of rachis and an elevated medium sized gland near the base of last pair of pinnae, umbel few flowered, peduncle long, slender staminal tube slightly shorter than corolla tube. Pod linear oblong, obtuse, thickened with sutures, bluntly pointed, pale yellow, smooth, shiny, reticulately veined above the seed; seeds 4–12, pale brown, ellipsid, compressed.

Flowering and fruiting   April–December

Disrtibution   North-East India; frequent in Meghalaya in lower elevation in deciduous and open forests associated with *Vitex penduncularis, Terminalia bellarica, Lagerstroemia parviflora*, etc.

Propagation   By seeds

Parts used   Bark, flowers and seeds

Chemical constituents   Bark yields tannins of condensed type, viz. D-catechin, isomers of leucocyanidin and melacacidin and a new leucoantho-cyanidin and lebbecacidin. Seeds give crude protein, calcium, phosphorus, iron, niacin, and ascorbic acid. The amino acid composition of the protein arginine, histidine, leucine and isoleucine, lysine, methionine, phenylalanine, threonine, tyrosine, and valine. The flowers contain lupeol, α- and β-amyrin and a pigment similar to crocetin.

Uses   Bark and flowers in decoction showed antiasthmatic and antianaphylactic activities (due to inhibition of sensitisation). It is used to cure bronchitis, piles and vata (Ayurveda). Seeds are used to cure tuberculosis and snake bite (Unani).

Albizia odoratissima Benth.

Family   Mimosaceae

Indian names   Black siris (English); *Bhusirisah* (Sanskrit); *Bassein bersa* (Hindi); *Kala Siris* (Hindi); *Bilbhara* (Kannada); *Pulivaka, Nellivaka, Karivaka, Kunnivaka* (Malayalam); *Karuvakai, Sittilavakai* (Tamil); *Cinduga, Sirisi, Telsu* (Telugu); *Hiharu* (Garo).

Description   Medium sized unarmed trees about 20 m in height, with dark coloured young shoots, grey, rough, irregularly cracked bark with dark patches, leaves abruptly pinnate, alternate, main rachis with a gland on the upper side near its basal part and often with a similar or a semicordate one at the base, obtuse or rounded at the apex and dark green. Flowers in heads or in corymbiform spreading panicles; fruits shortly stalked pods, brown, slightly reticulately veined; seeds flat yellow.

Flowering and fruiting   April–November

Distribution   Throughout India; common in Meghalaya in lower elevations.

Propagation   By seeds

Parts used   Bark

Chemical constituents   The bark yields tannins of condensed type, viz. D-catechin, isomers of leucocyanidin and melacacidin and a new leucoantho-cyanidin called lebbecacidin. It also gives friedelin and β-sitosterol.

Uses   The bark is an astrigent, acrid, cooling, depurative and expectorant, and is useful in ulcers, leprosy, skin diseases, cough, bronchitis, diabetes and burning sensation.

Alstonia scholaris R. Br.

Family   Apocynaceae

Indian names   Dita bark (English); *Saptaparna* (Sanskrit); *Chhotin* (Hindi); *Elelehale* (Kannada); *Pala, Mangalappala* (Malayalam); *Satvin* (Marathi); *Palaigh* (Telugu); *Palai* (Tamil); *Chatiana* (Oriya); *Thalasan* (Garo); *Chhaiten* (Assamese); *Chhattim* (Bengali).

Description   Lofty trees, 20–40 m high with pyramidal, ovoid crown and symmetrical branches, buttressed or fluted at base; bark grey, vertically fissured, lenticellate, leaves obovate-oblanceolate, oblanceolate-elliptic, rounded or obtuse, emarginate, base narrowed, attenuate, dark green above, glaucous beneath, lateral nerves slender, parallel; panicles pyramidal, 0.8–1.2 cm across; corolla constricted at middle; follicles linear, 1–6 cm long, drooping.

Flowering and fruiting   January–July

Distribution   Tropics from Africa to Australia; throughout India; frequent in Meghalaya at lower elevations in deciduous forests and forest margins and the secondary forests of unclassified nature.

Propagation   By seeds

Parts used   Bark, leaves, milky exudates

Chemical constituents   The bark contains alkaloids: ditaine, echitenine, echitamine and echitamidine together with triterpenes: α-amyrin and lupeol.

Uses   The bark is bitter, astringent, acrid, thermogenic, digestive, laxative, anthelmintic, febrifuge, antipyretic, depurative, stomachic and cardiotonic. It is useful in fevers, malarial fevers, abdominal disorders, diarrhoea, dysentery, dyspepsis, skin diseases, tumours, chronic and foul ulcers, asthma, bronchitis, cardiopathy, and debility. The tender leaves are used in the form of poultice for ulcers with foul discharges.

Allium chinense D. Don.

Family   Liliaceae

Indian names   Rakkyo (English); *Ja-ut* (Khasi); *Tej anglasing, Sangtem lashing* (Naga).

Description   An annual herb, leaves basal, narrowly linear, scape tall, slender, head lax flowered, and pedicel much longer, campanulate red-purple flowers, head, few or many flowered.

Distribution   Throughout India and common in entire North-East India.

Propagation   By bulb

Parts used   Bulb

Chemical constituents   Essential oil obtained from the bulb contains allicin, diallyl disulphide, allyl propyl disulphide and other sulphur compounds; volatile oil, aldehydes, vinyl sulphide and vitamin C.

Uses   Bulb is grounded and boiled in mustard oil and rubbed on body to reduce fever. It is also used to cure stomach ache.

Allium hookeri Thw.

Family Liliaceae

Indian name *Yenua-napakpi* (Mizoram); *Maroinapapkpi* (Manipuri).

Description Bulbous herb; slender, leaves basal linear, membranous, shorter than the tall subtrigonous scape, 1-nerved, head globose laxly many-flowered, pedicels much longer than the stellate white flowers, sepals linear, acuminate, about equalling the filiform filaments. Bulb clothed with long narrow membranous sheaths, spathe with a long tail.

Distribution In North-East Region, common in Meghalaya.

Propagation By seeds and bulbs

Parts used Leaves and bulbs

Chemical constituents The leaves and bulbs contain suphur compounds, saponins and bitter substances. The seeds yield saponins and alkaloids.

Uses Lukewarm juice is massaged on body to relieve body ache. The leaves and bulbs possess antibacterial properties. They are useful for the treatment of haemoptysis, epistasis, cough, sore throat, asthma and haematoma.

Allium tuberosum Rttl. Ex. Spreng.

Family Liliaceae

Indian names Chinese chives, Garlic chives (English); *Piat* (Khasi); *Jyllang* (Jaintia).

Description Herbs, leaves 4–5 basal, erect, narrow, linear, flat, tall compressed or trigonous above, pedicels much longer than the small white or pink stellate flowers, sepals oblong–

lanceolate, filaments simple, linear, included, connate below and ovary perigynous, style short.

Distribution   China, Japan, Western Himalaya, Khasi hills.

Propagation   By bulbs

Parts used   Leaves

Chemical constituents   Same as *Allium hookeri.*

Uses   Decoction is given for urinary trouble and is reported to be a good diuretic.

Alnus nepalensis D. Don.

Family   Betulaceae

Indian names   *Utis* (Hindi); *Udis* (Assamese); *Rupuo pareng* (Mizo), *Udis* (Arunachali); *Intounglong* (Naga).

Description   Middle sized trees; bark grey; greyish brown; warty, horizontally lenticellate; leaves broadly elliptic, obovate-elliptic, obtuse or acute, base narrowed or rounded; glaucous beneath; male spikes up to 25 cm long; slender; panicled, drooping, yellow, female spikes short, cylindric; fruiting spikes, oblong, blackish brown when ripe.

Flowering and fruiting   September–April

Distribution   Temperate Himalayas to Burma, confined to higher elevations in Meghalaya along Shillong–Jowai belt.

Propagation   By seeds

Parts used   Bark

Chemical constituents   Alder contains lignins, tannin, emodin and glycosides.

Uses   The drying action of the decoction of the bark helps to contract the mucous membrane and reduce inflammation, Bark paste is taken to cure stomach ache and dysentery.

Alocasia cucullata (Lour) Schott.

Family   Araceae

Indian names   *Manaka* (Sanskrit); *Mankanda* (Hindi); *Maanaka* (Kannada); *Marambu* (Malayalam); *Merukankilangu* (Tamil); *Charakanda* (Telugu); *Alu* (Marathi); *Tharang* (Garo), *Kashu* (Assamese).

Description   An annual herb, leaves sub-petiolate, broadly ovate-cordate, petiole very long, peduncles shorter, sub-solitary; root stock 30–50 cm with many suckers, spathe 15–30 cm in fleshy tube. Spadix shorter than the stalks; bearing male flowers above and female below. Berry ovoid, red when ripe.

Distribution   Grows wild in forest and mountains

Propagation   By rhizome

Parts used   Leaves

Chemical constituents   Oxalic acid, calcium oxalate, flavonoids, cyanogenic, glycosides, alocasin, β-sitosterol, campesterol, fucosterol, amino acid, citric acid, malic acid, ascorbic acid, succinic acid.

Uses   Leaves are eaten to cure bodyache.

Alocasia macrorrhiza (L.) Schott.

Family   Araceae

Indian name   Big rooted taro, Kopeh root, Gaint alocasia, Giant taro (English); *Hastikarni manaka* (Sanskrit); *Mankanda* (Hindi); *Maanaka, Marasani marasakage* (Kannada); *Marambu, Cheema chembu* (Malayalam); *Merukankilangu* (Tamil), *Chara kanda* (Telugu); *Mankachu* (Bengali); *Thahasom* (Garo); *Boro mankachu* (Assamese).

Description   A robust herb 2 m in length with about 1 m long leaves; cultivated for its sub-erect thick root stocks which are used as vegetable; leaf blades 60 cm –1 m long; erect or spreading, ovate, sagittate with 1.35 m long sheathing petioles; flowers fragrant, spadix and spathe or 25–30 cm long peduncle.

A

Distribution   Found wild and cultivated all over India

Propagation   By rhizomes

Parts used   Leaves

Chemical constituents   Same as *Alocasia cucullata*.

Uses   The rootstock is pungent, fragrant, cooling, useful in inflammation and diseases of abdomen. After boiling it is given as a mild laxative and is useful as diuretic and also in piles. The leaves are used as astringent. The juice of the petiole is used for ear pain, toothache and coughs. The chopped roots and leaves are used as rubefacient. The leaf is used against tumours.

Aloe barbadensis Mill.

Family   Liliaceae

Indian names   Curacao, Jaffarabad aloe (English); *Indian aloe; Kumari, Ghartkumari* (Sanskrit); *Ghikumar, Ghikumari* (Hindi); *Kathaligia* (Kannada); *Kattarvazha* (Malayalam); *Kattazhai, Sottrukattazhai* (Tamil) *Kalabanda* (Telugu); *Gwapatta* (Assamese).

Description   Perennial with short stem and shallow root system; leaves fleshy in rosettes, sessile, often crowded with thorny prickles on the margins, convex below, 45–60 cm long tapering to a blunt point, surface pale green with irregular white blotches, flowers fall off as the racemes elongate, perianth reddish yellow, cylindrical stamens equalling the perianth; fruits loculicidal capsule.

Distribution   Found throughout India, especially in North-East Region.

Propagation   By suckers

Parts used   Leaves

Chemical constituents   The leaves contain barbaloin, chrysopanol, glycosides and the aglycone, and emodin. The mucilage of the leaves contains glucose, galactose, mannose and galacturonic acid in addition to an unidentified aldopentose and a protein with 18 amino acids.

Uses   Stomachic tonic in small doses but purgative in large doses. Emmenagogue and anthelmintic. Juice of the leaf mixed with water of rice wash is applied in headache and giddiness. Fresh juice of the leaves is cathartic and cooling; it is used in eye troubles and spleen and liver ailments. It is useful for X-ray burns, dermatitis, cutaneous leishmaniosis and other disorders of the skin. It is an important constituent of a large number of Ayurvedic preparations; it is used in veterinary medicine.

Alpinia galanga (L.) Willd.

Family   Zingiberaceae

Indian names   Siamese ginger (English); *Elaparni, Kulanjan* (Sanskrit); *Kulanjan* (Hindi); *Dhumarasmi, Dumbarasme* (Kannada); *Aratta, Cittaratta* (Malayalam); *Arattai, Ardubam* (Tamil); *Kachoramu* (Telugu); *Koshta kulinjan* (Marathi).

Description   Perennial herbs, 1–2 m high; rhizome cylindrical, stout, aromatic, covered with scales. Leaves alternate, lanceolate, lower part surrounding the stem, the upper surface glabrous and shining. Inflorescence in terminal dense raceme 20–30 cm long, flowers white, tip veined and red. Fruit globose or ovoid.

Distribution   Throughout Eastern Himalayas and North-East Region.

Propagation   By rhizomes

Parts used   Rhizome

Chemical constituents   The rhizome produces an essential oil consisting of cineol, methyl cinnamate and flavones, galangin, alpirin, 4-methoxy flavone.

Uses   The rhizome is an antibacterial agent and a digestive. It is indicated in the treatment of dyspepsis, flatulence, vomiting, gastralgia, colic, diarrhoea and malaria fever. It is aromatic, stimulant and bitter, used as a stomachic and carminative in rheumatic and catarrhal infections as well as respiratory troubles in children. Paste is applied on body to reduce temperature; Used as cure for leucoderma and piles. It has an important action on bronchioles.

Alternanthera sessilis L.

Family   Amaranthaceae

Indian names   *Matsyaksi, Patturah* (Sanskrit); *Gudrisag* (Hindi); *Honugonesoppu* (Kannada); *Minannani, Ponnannani, Ponnankannikkira* (Malayalam); *Ponnannkannikkirai* (Tamil), *Ponnagantikura* (Telugu); *Senchi* (Garo).

Description   Plants herbs, usually prostrate; leaves opposite, oblong, lanceolate or elliptic, obtuse or subacute. Flowers small, white, capitate; heads axillary often clustered; sepals unequal, anterior and 2 posterior flattened; 2 lateral innermost concave. Stamens 2–5, filaments short, connate into a short cup with or without interposed staminodes, another 1-celled. Ovary orbicular or ovoid; stigma subsessile, capitellate, ovule 1, pendulous, long obcordate, margins often winged or thickened. Seed inverse, lenticular, testa coriaceous; embryos annular, cotyledons narrow.

Distribution   About 16 species, distributed in tropical and subtropical areas, throughout hotter regions of India and Ceylon in damp places, ascending the Himalayas to 4000 feet, and in all warm countries.

Propagation   By seeds

Parts used   Whole plant

Chemical constituents α-sitosterol, stigmasterol, campesterol, 2, 4-methylene cycloartenol, lupeol.

Uses   The plant is bitter, astringent, acrid, cooling, constipating and febrifuge and is useful in vitiated conditions of kapha and pitta, burning sensation, diarrhoea, skin disease and fever.

Altingia excelsa Noronha.

Family   Zingiberaceae

Indian names   Rasamala (English); *Ashmapushpa, Chala* (Sanskrit); *Silaras* (Hindi); *Silaras* (Kannada); *Rasamala* (Malayalam); *Neriyurishippal, Erikacu* (Tamil); *Shilarasamu* (Telugu); *Jutili* (Assamese).

Description   A magnificient, large decidious tree common in North-East Region. The wood is reddish-brown to brown with a smooth feel.

Distribution   Meghalaya and adjoining states.

Uses   Stimulant, expectorant, anodyne, autophlogistic.

Amaranthus spinosus L.

Family   Amaranthaceae

Indian names   Prickly amaranth (English); *Tenduliyah* (Sanskrit); *Cauleyi, Kateli* (Hindi); *Mulluhalavesoppu* (Kannada); *Mullanicita, Cerucia* (Malayalam); *Mullukkirai* (Tamil); *Mullutotakura* (Telugu); *Chenduri* (Garo), *Khudna kata* (Assamese).

Description   Plant erect, glabrous, plant varies in colour from green to red and purple; stem erect, 2 feet, hard, spinous leaves 1.5–4 on base cuneate; petiole slender, equalling the blade; leaf axis with 5 spines, leaves long petioled, ovate or oblong, obtuse, flower in axillary clusters and long, dense or laxed, spikes, bracts setaceous equalling or exceeding the sepals, sepals of male acuminate and of female obtuse and apiculate; stigma 2. Utricle with a thickened top; seed black, shining, border obtuse, not thickened. 5 Utricle rugose nearly equalling the sepals.

Distribution   Various tropical countries, throughout India and Ceylon, in waste places, field and gardens.

Propagation   By seeds

Parts used   Whole plant

Chemical constituent   Higher alkanes and their methyl derivatives, higher aliphatic alcohols, acids and esters, amino acids, cholesterol, α-spinasterol.

Uses   The plant is sweet, cooling, laxative, diuretic, stomachic, and febrifuge. It is useful in vitiated conditions of pitta, bronchitis, haemorrhoids, boils, burns, fever, anaemia and general debility. The roots are thermogenic and haemostatic.

Ampelocissus latifolia (Roxb.) Plench.

Family   Vitaceae

Indian names   *Amlavetasah* (Sanskrit); *Bhiranya, Panibel* (Hindi); *Bilee hambu, Kaadu drakshi* (Kannada); *Vllia-pira-pitica* (Malayalam); *Kattukodimindri* (Tamil); *Adavi draksha* (Telugu); *Kumarlata* (Bengali).

Description   Glabrous extensive climbers; young shoots glaucous and often purplish; leaves ovate-acute or shortly acuminate, base cordate or truncate; panicles up to 10 cm long; flowers reddish-brown, berries ellipsoid, black.

Flowering and fruiting   May–September

Distribution   Nearly throughout India, usually at lower elevation of Jaintia hills, Meghalaya.

Propagation   By seeds

Parts used   Roots

Uses   Root juice is diuretic, aspirant and a blood purifier. It is used for eye trouble and ulcers.

Anacardium occidentale L.

Family   Anacardiaceae

Indian names   Cashewnut tree (English); *Vikkabijah, Vikkaphalah,Venamtah* (Sanskrit); *Kaju* (Hindi); *Gerumata, Tutakagetu* (Kannada); *Kasumavu* (Malayalam); *Muntiri* (Tamil); *Gidlmavrldi* (Telugu); *Garo-Checunut, Kaju* (Assamese).

Description   Plants small or middle sized trees, up to 10 m high; crown oval, spreading, usually branching from base, bark brown, nearly smooth with vertical fissures; panicles corymbose; flowers 1–2 cm across, white or yellow with red streaks; nuts reniform, 2–3 cm long, greyish, hard, on a fleshy swollen, obconic, yellow to red pseudocarp.

Flowering and fruiting   February–June

Distribution   Native of America, cultivated nearly throughout India in hotter parts, often run wild; in Meghalaya much cultivated in Garo Hills.

Propagation   By seeds

Parts used   Roots, bark, leaves and fruits

Chemical constituents   The essential amino acids are as follows: arginine, histidine, lysine, tryptophan, and valine. Other amino acids present are alanine, aspartic acid, cystine, glycine, serine and tyrosine.

Uses   The roots are considered to be purgative. The bark has alternative properties and is used along with the inflorescence for the treatment of snake bite. The bark and leaves are useful in odontalgia. Fruits are acrid, sweet, thermogenic, aphrodisiac, trichogenous and anthelmintic, and are useful in skin diseases, dysentery, and anorexia.

Anaphalis adnata DC.

Family   Asteraceae

Indian name   *Bugla* (Hindi).

Description   Stout woody undershrubs up to 1.5 m high; leaves oblong, elliptic, oblong-lanceolate, acute or obtuse, base attenuate or semi- amplexicaul; heads 0.4–0.6 cm across, white achenes; minute, heads in globose clusters, numerous in terminal bracteate compound corymb, Flowers white; pappus white.

Flowering and fruiting   August–January

Distribution   Temperate Himalayas and Burma; common in pine forests as undergrowth and in wastelands.

Propagation   By seeds

Parts used   Flowers and leaves

Uses   The flower heads and hairs are used to check bleeding. The dried heads and leaves are made into wicks.

Annona muricata L.

Family   Annonaceae

Indian names   Guanabana, Graviola, Soursop (English); *Mullu raamana phala, Mullu ramaphala,* (Kannada); *Mullancakka, Mullanchukka* (Malayalam); *Mulluccitta, Pulippala* (Tamil); *Lakshmana phalamu* (Telugu); *Soursop, Pengsung* (Garo).

Description   Small tree 4–7 m tall. Leaves alternate, petiolate, the blades leathery and oblong-lanceolate. Flowers trimerous, sepals and petals fleshy, and greenish in colour. Fruit fleshy, syncarp with a green exocarp covered with conspicuous long pseudo-spines, mescocarp somewhat fibrous juicy, with sweet-sour flesh surrounding several large smooth black seeds.

Distribution   Cultivated at lower elevation

Propagation   By seeds and vegetative method

Parts used   Leaves and fruits

Chemical constituents   Lipids, reticuline, monotetrahydrofuran, acetogenins, β-sitosterol, reticuline, muricine and tannins.

Uses   Leaves are used in treating stomach ailments. Antimalarial, smooth muscle relaxant, uterine stimulant, cytotoxin, cardiac depressant, antifungal, hypertensive.

Anthocephalus chinensis (Lank.) Rich. and Walp.

Family   Rubiaceae

Indian names   Wild chinchona (English); *Kadambah* (Sanskrit); *Kadamb* (Hindi); *Kaduavalatige* (Kannada); *Katamba* (Malayalam); *Vellaikkatampu* (Tamil); *Rudrakskamba* (Telugu); *Mi-Bol* (Garo).

Description   Large tree up to 35 m high with straight, cylindric and spreading, horizontal branches forming a conical crown; bark smooth, blackish-brown or dark brown; young parts lenticellate, leaves 15–25 × 6–12 cm, ovate, oblong-elliptic or broadly ovate-lanceolate, shortly acuminate or acute, base rounded, cuneate or obtuse, margin entire, usually undulate, glabrous; heads 4–6 cm in diameter; flowers greenish-yellow or orange-red; corolla lobes oblong, style exerted; syncarp fleshy, orange, 4–7 cm in diameter.

Flowering and fruiting   December–October

Distribution   Indo-Malaya; nearly throughout India; common in Meghalaya at lower elevation in tropical forests; sometimes planted.

Propagation   By seeds

Parts used   Bark, leaves and fruits

Chemical constituents   Cadambine, isocadambine, the leaves also produce hentriacontanol and α-sitosterol. The dried stem bark gave cadambagenic acid and quinovic acid, and four saponins.

Uses   The bark is bitter, astringent, acrid, cooling, anodyne, anti-inflammatory, digestive, carminative alexeteric, febrifuge, diuretic and tonic. It is useful in wounds, ulcers and debility.

Antidesma acidum Retz.

Family   Euphorbiaceae

Indian name   Arobakh (Garo).

Indian name   *Amli, Sharschanti* (Hindi); *Bili komme, Hulimajjige kolu* (Kannada); *Abutenga, Nekhon tenga* (Assamese); *Chouding, Dieng japew* (Khasi); *Gummodi pella, Gurrapujalli* (Telugu).

Description   Small trees 6–8 m high with an ovoid, spreading crown; bark rough, greyish; leaves variable in size and shape, 1.5–10 × 0.7–3 cm, oblanceolate, obovate-elliptic, oblong-elliptic, acute–obtuse, base narrowed, cuneate, rusty, pubescent along nerves beneath; recemes 1.5–6 cm long, slender; flowers greenish-yellow, minute; fruits 0.4–0.5 cm long, ellipsoid, purplish-red.

Flowering and fruiting   June–December

Distribution   Throughout India, common in Meghalaya particularly at lower elevations often as a sal forest undergrowth. Leaves are eaten in Garo Hills.

Propagation   By seeds

Parts used   Leaves

Uses   Leaves are useful for joint pains

Antidesma bunius (L.) Spreng.

Family   Euphorbiaceae

Indian names   Bignay, Chinese laurel (English); *Himalcheri* (Hindi); *Kareekomme, Naayikomme* (Kannada); *Mail-kombi, Karivetti* (Malayalam); *Nolaiali, Nolaidali* (Tamil); *Janupullari* (Telugu); *Bol aborak* (Garo); *Bor heloch* (Assamese).

Description   A small evergreen tree up to 10 m high, young parts hairy. Bark blackish brown, crown dense, oval, young parts tomentose. Leaves oblong-lanceolate, elliptic, shortly acuminate, obtuse, often mucronate, entire, coriaceous, glossy, green and glabrescent. Flowers in simple or racemes, pubescent spikes, male flowers, calyx copular, sparsely pubescent, disc glabrous, fleshy, stamens-3, pistilode. Female flowers; calyx truncate, persistent ovary, glabrous, stigmas 3–4 fruits ellipsoid compressed, red, turning black.

Flowering and fruiting   April–October

Distribution   Indo-Malaya; Entire North-Eastern India, commonest in Antidesma in Meghalaya.

Propagation   By seeds

Parts used   Leaves

Uses   Boiled leaf decoction is used to bathe the patients to cure joint pains and also used in syphilitic ulcers. It is diaphoretic and also given for indigestion. The stem and root bark are reported to possess cyanophoric compounds which are considered to be poisonous in nature.

Antidesma diandrum (Roxb.) Roth.

Family   Euphorbiaceae

Indian names   Amli, Amblu (Hindi); *Meginakadusoppu, Sanakulipa* (Kannada); *Aburok* (Garo); *Abutenga* (Assamese); *Gurrapujalli, Manchipulleri* (Telugu).

Description   Small tree or shrubs; bark pale brown, young parts tomentose; leaves elliptic-lanceolate, elliptic-oblanceolate, caudate-acuminate, base rounded or narrowed, glabrous except nerves; spikes often panicled, flowers greenish-purple; fruits ovoid or obovoid, red.

Flowering and fruiting   April–October

Distribution   Bangladesh and North-East Region, commonly seen in Meghalaya in bamboo forests at lower elevations.

Propagation   By seeds

Parts used   Leaves

Chemical constituents   Same as *Antidesma bunius.*

Uses   Leaves are useful in joint pains and syphilitic ulcers.

Aphanamixis polystachya (Wall.) Parker [Vulnerable]

Family   Meliaceae

Indian names   Rohituka tree (English); *Rohitakah* (Sanskrit); *Harin-bara* (Hindi); *Mullumuntala* (Kannada); *Cemmaram* (Malayalam); *Malampuluvam* (Tamil); *Sevamanu* (Telugu); *Akawatong* (Naga); *Heriank khoi* (Mizo); *Sahata* (Kuki).

Description   Middle sized trees 10–15 m high; crown dense, umbrella shaped hemispheric; bark blackish-brown, rough, scaly; exfoliating in circular flakes, wood reddish brown, leaves large, imparipinnately compound, 30–75 cm long; leaflets opposite, 7–19, 8–25 × 3.5–8 cm, elliptic-oblong or oblong-lanceolate, ovate-obliquely lanceolate, acuminate or acute, base oblique, rounded, cuneate, entire, glabrous, male panicles/female racemes up to 30 cm long; flowers yellow or dull white 3–5 mm across; male flowers numerous in axillary panicles, female or bisexual flowers larger than male in axillary or supra-axillary solitary spikes, calyx 5-partite, petals 3; stamens 6, included; capsules yellow when ripe, ovoid or obovoid, 3–4 cm long, 3-valved; seeds red, arillate. Fruits globular, smooth, yellow when ripe, opening by 3-valved; seeds with scarlet aril.

Flowering and fruiting   October–March

Distribution   Tropics of chiefly Indo-Malaya; nearly throughout India.

Parts used   Bark and Seeds

Chemical constituents   A Limonoid, rahitukine has been isolated from the seeds. Petroleum extract of the dried bark gave a new triterpenoid, aphanomixin.

Uses   The bark is acrid, astringent, bitter, vulnerary, digestive, anthelmintic, depurative, urinary astringent, ophthalmic. It is useful in splenomegaly, liver disorders, tumour, ulcer, dyspepsia, intestinal worms, skin diseases, leprosy, diabetes, ophthalmopathy, jaundice, haemorrhoids, burning sensation, rheumatoid arthritis and leucorrhoea; seeds are acrid, refrigerant, laxative, anodyne and anthelmintic. They are useful in ulcers, ophthalmopathy, otopathy, myalgin, skin diseases, intestinal worms, burning sensation and vitiated conditions of *vata*.

Aquilaria agallocha Roxb.

Family   Thymelaeaceae

Indian names   Aloe wood, Eagle wood (English); *Aguruh, Kranaguruh* (Sanskrit); *Agar* (Hindi); *Hrisnagaru* (Kannada); *Akil, Karakil* (Malayalam); *Agar, Agalicandanam* (Tamil); *Krsnagaru* (Telugu); *Bol-Agar* (Garo); *Sasi pat* (Assamese).

Description   A large evergreen tree, about 21 m in height, with straight and elongated crown; bark white or greyish white, warty; leaves linear-lanceolate, oblong-elliptic or oblanceolate, caudate, acuminate, base acute, 5–9 cm long, silky, glossy and faintly parallel nerved; flowers small, greenish on very slender shortly peduncled umbels on younger branches; perianth about 5 mm long, slightly hairy outside, stamens alternate the perianth, filaments red at the apex, ovary tomentose, subsessile, villous, 2 celled, stigma larger, subsessile; fruit slightly compressed, yellowish tomentose capsules, 4–5 cm, obovoid.

Flowering and fruiting   June–October

Distribution   Indo-Burma, confined to North-East India; frequent in Meghalaya in dense tropical evergreen forests below 800 m.

Propagation   By seeds

Parts used Fragrant resinous wood and oil

Uses Juice of fresh plant diuretic, alternative, anodyne and hypnotic. Seeds are laxative, nauseant, expectorant, and sedative combining the action of castor oil and cannabis. Fresh root has an anodyne effect. Ayurveda uses it for eye disease also. Unani system uses for purifying blood. Milky juice is used to relieve blisters, and heal indolent ulcers. Population of this plant is getting fast depletion due to the illegal felling for the much-priced agar, commercially known as Aloe or eagle wood.

Ardisia colorata Roxb. [Rare]

Family Myrsinaceae

Indian name *Bol-simbal* (Garo).

Description Small tree or shrubs, up to 10 m height; bark grey, warty, minutely lenticellate; leaves 8–30 × 2.8 cm, oblong-lanceolate, oblanceolate, acuminate or acute, base cuneate, glabrous, entire; panicles up to 15 cm long, peduncles flattened; flowers pink or reddish white, 0.6–0.7 cm across; fruits 0.5–0.6 cm in diameter, globose, pink.

Flowering and fruiting December–May

Distribution Indo-Malaya; confined to North-East India rather, in Meghalaya, in dense tropical evergreen forests at lower elevations (below 800 m).

Parts used Roots and bark

Uses Bark paste is used as antidote in snakebite. The plant shows anti-acetylcholine activity. The roots are used for dropsy and for internal injuries. They are also used in stomach ache especially after child-birth, as a febrifuge in diarrhoea.

Ardisia paniculata Roxb.

Family   Myrsinaceae

Indian name   *Dieng-soh-botul* (Khasi).

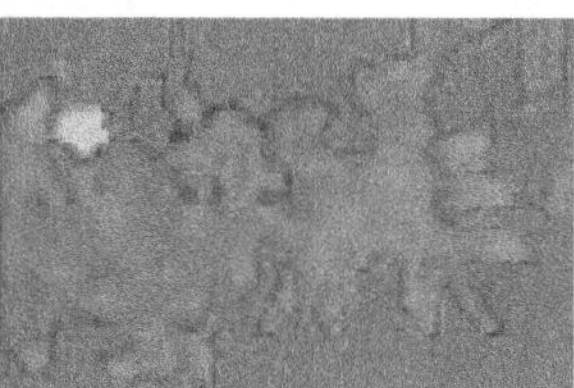

Description   Shrubs up to 5 m high; leaves obovate oblong or elliptic, acute or acuminate, glabrous; panicles terminal, up to 30 cm long; flowers reddish-pink, 1–1.5 cm across; fruits longitudinally ribbed, 0.5–0.6 cm in diameter.

Flowering and fruiting   March–December

Distribution   Indo-Malaya, throughout India, occasional in Meghalaya in deciduous and mixed evergreen forests.

Propagation   By seeds

Parts used   Roots and bark

Uses   Bark paste is used as antidote in snakebite. The plant shows anti-acetylcholine activity. The roots are used for dropsy and for internal injuries.

Ardisia solanacea Roxb.

Family   Myrsinaceae

Indian names   *Bisi* (Hindi); *Bodhina gida, Shuli* (Kannada); *Kaka-njara* (Malayalam); *Manipudbam, Kozhikkottai* (Tamil); *Kondamayur, Dudumara* (Telugu); *Kadna, Katapenga* (Marathi).

Description　Shrubs up to 5 m high; leaves obovate, oblong or elliptic, acute or acuminate, entire, coriaceous. Flowers 1–1.5 cm across, pinkish; calyx segments ovate, ciliate, black dotted; corolla lobes black dotted; filaments short; ovary subglobose; fruits subglobose, 0.7–0.9 cm in diameter, blackish.

Flowering and fruiting　June–January

Distribution　Indo-Malaya; throughout India; frequent in Meghalaya in deciduous forests at lower elevations.

Propagation　By seeds

Parts used　Roots and bark

Uses　The plant shows anti-acetylcholine activity. The roots are used for dropsy and for internal injuries. They are also used for stomach ache especially after childbirth, as a febrifuge for diarrhoea and for rheumatism.

Argemone mexicana L.

Family　Papaveraceae

Indian names　Mexican poppy, Yellow Mexican poppy, Prickly poppy (English); *Svamaksiri, Brahmadani* (Sanskrit); *Satyanasi, Bhatbhamt* (Hindi); *Arasina, Datturigida* (Kannada); *Ponnummattu* (Malayalam); *Kutiyotti* (Tamil); *Brahmadandicettu* (Telugu).

Description　A strong branched, prickly annual; 60–90 cm high with yellow latex; leaves simple, sessile, spiny, sinuate-pinnatifid, variegated white. Spinous, veins white, flowers large, bright yellow, terminal on short leafy branches; fruits prickly capsules, oblong-ovoid, opening by 4–6 valves; seeds numerous.

Distribution   Throughout India in hotter parts.

Propagation   By seeds

Parts used   Whole plant

Chemical constituents   Allocryptopine, berberine, chelerythrine, coptisine, cryptopine, protopine, the flowers contain isorhamentin, isorhamnetin-3-glucoside and isorhamnetin-3, 7-diglucoside.

Uses   The plant is bitter, acrid, cooling, diuretic, depurative, anodyne, anthelmint, stomachic and sedative. The roots are useful for skin diseases, leprosy, pruritus, inflammations, all types of poisoning, constipation, flatulence, colic and malarial fever and vesicular calculus. The leaves are useful in cough, wounds, ulcers and skin diseases. The seeds are useful in cough, asthma, rheumatalgia, colic fever and flatulence. The latex is useful in dropsy, jaundice, skin diseases, leprosy, blisters, indolent ulcers, conjunctivitis, inflammations, burning sensation and malarial fever.

Argyreia nervosa (Burn. f.) Boj.

Family   Convolvulaceae

Indian names   Elephant creeper (English); *Vrddhadarukah, Bastantri* (Sanskrit); *Samandar-ka-pat* (Hindi); *Candrapada* (Kannada); *Marututari* (Malayalam); *Samuttirappachai* (Tamil); *Candrapada* (Telugu); *Jatap-masi* (Khasi); *Uritsisomli* (Mizo).

Description   A scandent shrub with woody white, tomentous stems; leaves simple, 9–22 × 7–17 cm, long petioled, large, ovate, cordate, glabrous above, persistently white tomentose beneath; peduncles long; flowers, large, purple, silky-pubescent in long peduncled cymes, sub-capitate;

bracts large, ovate, lanceolate, acute; flowers in axillary capitate cymes; calyx segments elliptic–oblong; corolla white tinged with light rose; fruits dry, 2–3 cm in diameter, yellow or orange yellow, globose, apiculate.

Flowering and fruiting   August–December

Distribution   Throughout India, in areas up to 800 m elevation. All over North-Eastern India common in Meghalaya at lower elevations in deciduous forests, secondary forests and grasslands.

Propagation   By seeds

Parts used   Roots and leaves

Chemical constituents   Tannin, and amber-coloured acid-resin soluble in ether, benzole and partly soluble in alkaline solutions.

Uses   Root juice is used for alleviating rheumatism. Leaf paste is applied on boils and other skin diseases. Powdered root is given in syphilis. It is also an alternative and nerve tonic; powdered root is soaked in juice of *Asparagas racemosus* and dried and given as a tonic which improves intellect, strengthens body and prevents effect of age. In the indigenous system of medicine, the leaves are emollient and varicant. They are used externally in the treatment of ringworm, asthma, itch and other skin diseases and internally to cure boils, and swellings. The leaves are also used as a local stimulant and rubefacient.

Aristolochia indica L.

Family   Aristolochiaceae

Indian names   *Ishwarrimul* (Sanskrit); *Isharmul, Ishwari* (Hindi); *Utamanlelaye* (Tamil); *Guntaganjeera* (Telugu); *Sapasani* (Marathi); *Isurmul* (Bengali); *Arkmul* (Gujrati).

Description   A twining shrub, stems woody below, slender and soft above. Main rootstock woody; leaves variable in size and shape from linear, to obovate; flowers small, greenish-white, in small bunches arising from the axils of the leaves; Flowers purple, characteristically trumpet-shaped; fruit about 5 cm long, longer than broad, opening from below upwards into six valves.

Distribution   The plant occurs in northern India from Kashmir to Kumaon.

Propagation   By seeds and vegetative method

Parts used   Leaves and flowers

Chemical constituents   The leaves and flowers contains ceryl alcohol, aristolochic acid and aristo red. Santonin is contained in the young leaves and flower heads.

Uses   The drug is useful in fever and dropsy, stimulant and of expulsion of worms.

Aristolochia saccata Wall. [Rare]

Family   Aristilochiaceae

Indian names   *Rekangbatelaung* (Garo); *Krah-Lahil* (Manipuri); *Rikangbateloung* (Arunachali).

Description   Lianes, stems with deflexed hairs; leaves 15–30 × 5–8 cm, oblong-lanceolate, linear-oblong, gradually tapering, base cordate, hairy along nerves above, brown silky beneath; cymes branched, up to 5 cm long; flowers 4–5 cm long; whitish or yellowish white with reddish streaks.

Flowering and fruiting   August–March

Distribution   Central and eastern Himalayas; rare in Meghalaya usually in dense evergreen forests in moist, shady localities.

Propagation   By seeds and vegetative method

Parts used   Root, rhizome and leaves

Uses   Root juice is useful for rheumatism; leaf paste is used for boils and other skin diseases. Powdered root is useful for syphilis and used as a nerve tonic. The leaves are emollient and varicant; they are used externally in the treatment of ringworm, eczema, itching and other skin diseases and internally to cure boils, and swellings.

Aristolochia tagala Cham.

Family   Aristilochiaceae

Indian names   Dodda eeshvari balli, Gattada eeshvari (Kannada); *Aadutheendapalai* (Tamil); *Nallaeeshvara* (Telugu); *Panpipuli* (Assamese); *Krah-kshung* (Manipuri), *Belikol* (Garo).

Description   Twiners; stem glabrous, sulcate; roots stout somewhat aromatic, leaves 8–18 × 4–10 cm, ovate, chordate, ovate-oblong or ovate-lanceolate, acute or acuminate, base widely cordate, 5–7 nerved, membranous; flowered cymes, Perianth lip villous, 4–5 cm long; capsules 4–6 cm long, obtusely triangular, ribbed.

Flowering and fruiting   April–December

Distribution   Indo-Malaya; Himalayas and South India, frequent in Meghalaya, climbing over other bushes.

Propagation   By seeds

Parts used   Root and leaves

Uses   Paste of leaves and roots is used in rheumatic pains, tooth ache, stomach ache and snake bite. The roots are tonic, carminative and emmenagogue in the indigenous system of medicine.

Artemisia maritima L.

Family   Asteraceae

Indian names   *Gadadhar* (Sanskrit); *Kirmala, Chhuhri Arjun* (Hindi); *Jatara kriminaashini* (Kannada); *Kirmani ova* (Marathi); *Phillar* (Punjabi); *Murni* (Kashmiri).

Description   A stout, much-branched, perennial aromatic shrub, about 1 m high. Leaves 2–5 cm long, whitish in colour, divided into numerous fine linear segments; upper leaves simple undivided, linear. Flower heads small, in short spikes.

Distribution   In Northern India from Kashmir to Kumaon about 2000 to 3000 m altitude.

Propagation   By seeds

Parts used   Leaves and flower heads

Chemical constituents   The chief constituent is Santonin. The plant contains a crystalline pentacyclic alcohol, fermentol, stigmasterol, β-sitosterol, and α-amyrin and its acetate.

Uses   The chief use of the drug is for expulsion of worms from stomach. The worms are pushed to large bowels, from where they are removed by a purgative. The drug is also useful for fevers and dropsy, and as a stimulant.

Artemisia nilagirica (Cl.) Pamp.

Family   *Asteraceae*

Indian names   Indian wormwood, Fleabana (English); *Damanakah, Damanah* (Sanskrit); *Nagdona, Davana* (Hindi); *Urigattiga* (Kannada); *Makkippuvu, Masipatri* (Malayalam); *Makkippu* (Tamil); *Masipatri* (Telugu); *Sangien* (Garo); *Pina, Ditivati* (Naga); *Nilum* (Arunachali); *Khel-byak, Kay-kaung* (Manipuri).

Description An aromatic perennial under shrub, often gregarious, hairy, pubescent or tomentose, stem paniculately branched; leaves ovate, lobed, lanceolate or 1–2 pinnately partite with stipule-like lobes at the base, pubescent above, white tomentose beneath, lobes acute, irregularly serrate or lobulate; heads ovoid or sub-globose, clustered or in panicled racemes; flowers in subglobose heads, in spicate or suberect or horizontal panicled racemes; outer flowers female, very slender, inner disk flowers fertile, bisexual; bracts ovate or oblong, margins scarious; fruits oblong-ellipsoid, minute achenes.

Flowering and fruiting August–February

Distribution Indo-Malaya; throughout India, very common in Meghalaya along forest margins, forest roads and wastelands, also as a pine forest undergrowth.

Parts used Whole plant

Chemical constituents The plant has p-cymene, β-thujone, azulene, and a crystalline pentacyclic alcohol, fernenol, stigmasterol, β-sitosterol, and α-amyrin and its acetate.

Uses Leaves and flower tops are bitter, astringent, acrid, thermogenic, aromatic, anodyne, anti-inflammatory, diuretic, appetizer, digestive, stomachic and febrifuge. It is useful in asthma and brain diseases. It is used as tonic, anthelmintic and expectorant. Juice is applied on forehead to reduce headache, and giddiness; decoction of the leaf is taken to cure piles and also as a diuretic. Leaves show anti-bacterial and anti-fungal activities. The herb is considered to be an emmenagogue, anthelmintic and stomachic. The plant is also used as a febrifuge and as an inferior substitute for cinchona in fevers. It possesses antilithic and alexipharmic properties and assists in parturition. A weak decoction is given to children suffering from measles. It is used externally for skin dieases and ulcers. An infusion of the leaves and flowering tops is administered in nervous and spasmodic afflictions and asthma. The leaves are applied as haemostatic and to alleviate the burning sensation in conjunctivitis. The roots are used as tonic and antiseptic. The plant is also employed to keep away fleas and other insects.

Artocarpus heterophyllus Lam.

Family Moraceae

Indian names Jackfruit tree (English); *Panasah* (Sanskrit); *Kathal, Cakki* (Hindi); *Halasu* (Kannada); *Plavu, Pilavu* (Malayalam); *Palamaram* (Tamil); *Panasa* (Telugu); *Kathal* (Assamese); *Kathal* (Bengali); *Armu* (Garo).

Description A large monoecious evergreen tree, 18–28 m in height, bark black mottled with green, rough with warty excrescences, scaly with white patches; leaves 5–20 × 4–10 cm, obovate-oblong, obtuse, retuse or shortly acute, cuneate, coriaceous; stipules 2–4 cm long, glabrescent; receptacles greenish, obovate, cylindric, mature ones 30–75 cm long, sharply pointed; male flowers crowded on cylindrical receptacles—female flowers crowded on globose receptacles-both cauliflorous; fruits multiple, large fleshy, globose or oblong, covered with tubercles; seeds oval with a membranous testa; cotyledons unequal.

Flowering and fruiting   February–October

Distribution   Throughtout India, cultivated in major parts of Asia; occasionally wild in southern parts of Garo Hills.

Propagation   By seeds

Parts used   Roots, leaves, fruits, seeds, wood, latex

Chemical constituents   Fruit contains β-carotene. A dipeptide, aurantiamide acetate, has been isolated from the seeds. Latex from fruits contains cycloartenone, cycloartenol, α-sitosterol, leucine, tyrosine and valine. The hardwood contains morin, artocarpin, and artocarpanone. Bark contains betulinic acid, cycloartenol and cycloartenone. The leaves contain cycloartenol, β-sitosterol and tannins. The root contains β-sitosterol, ursolic acid, cycloartenone and artoflavonone.

Uses   The leaves are useful in fever, boils, wounds, skin diseases; the roots are credited with antidiarrhoeal property. The unripe fruits are acrid, astringent, carminative and tonic. The ripe fruits are sweet, cooling, oleaginous, laxative and tonic, and are useful in treating ulcers.

Asclepias curassavica L.

Family   Asclepidaceae

Indian names   Blood flower, Swallow wort (English); *Kakanasika, Kakatundi* (Sanskrit); *Kakatundi* (Hindi); *Hulugilu gida, Kaakatundi* (Kannada); *Ariyaman* (Tamil); *Jilledu mandaara* (Telugu); *Krishnachura* (Manipuri).

Description   An under shrub; leaves lanceolate or elliptic acuminate; flowers bright orange red in umbrellate cymes; calyx glandular; corolla rotate, 5 lobed, reflexed coronal scales adnate to staminal column; ovary of 2 distinct carpels.

Distribution   Some parts of India and throughout North-East Region.

Propagation   By seeds

Parts used   Roots and leaves

Chemical constituents   Asclepiadin, asclepin, calactin, calotropin.

Uses   The root is regarded as purgative and consequently astringent. It is also regarded as a remedy for piles and gonorrhoea. Paste is applied as an antidote for poisonous insects. The leaf juice is antidysenteric, anthelmintic and is also used against haemorrhages and gonorrhoea.

Atropa acuminata Roye. ex. Lindley

Family   Solanaceae

Indian names   *Angurshafa, Sagangur* (Hindi); *Yebruj* (Bengali); *Nati belladonna* (Kannada); *Sagangur* (Kashmiri).

Description   An erect, branched perennial herb, 60–90 cm high. Leaves brownish green, 7–15 cm long, narrowed at both ends. Flowers in axils of leaves, solitary or in pairs, about 2.5 cm long, bell-shaped, yellowish brown. Berry roundish, about 1.5 cm across, purple-black.

Distribution   Kashmir, at altitudes of 2000 to 3500 m. It is also cultivated.

Propagation   By seeds

Part used   Leaves and roots

Chemical constituents   The roots contain starch, tannin, asparagine, choline and succinic acid.

Uses   The drug Belladonna is prepared from the dried leaves and other aerial parts of the plant. Belladonna in the treatment of intestinal biliary and colic disorders and are often prescribed with vegetable laxatives. It is useful for asthma, whooping cough, bladder and ureteral spasms. It is considered as a sedative and narcotic. Ethyl alcohol extract (50%) of leaves is antiprotozoal, antiviral and hypoglycaemic.

Averrhoa carambola L.

Family   Averrhoaceae

Indian names   Carambola apple, Coromandel Gooseberry (English); *Karmarangah, Karukah* (Sanskrit); *Kamarakh, Kamaramga* (Hindi); *Derehuli* (Kannada); *Chaturappuli* (Malayalam); *Tamarattai* (Tamil); *Karamonga, Tamaratama* (Telugu); *Amrenga* (Garo).

Description   Small tree up to 10 m high; crown compact, ovoid, bark blackish or dark grey, horizontally wrinkled; leaves 7–18 cm long; leaflets 1–7 × 0.7–3.5 cm, (terminal largest), ovate, ovate-elliptic, rhomboid or oblong-elliptic, acuminate, base truncate or cuneate, often oblong; panicles up to 10 cm across, flowers 3–5 mm long, pinkish-purple; deeply 5-lobed, oblong in outline.

Flowering and fruiting   September–February

Distribution   Throughout India, cultivated and occasionally run wild in lower elevations particularly in sandy soils.

Propagation   By seeds and air layering

Parts used   Leaves and fruits

Chemical constituents   They also contain oxalic acid and potassium oxalate. The presence of fluorine is also reported. A wide range of vitamin content is also recorded from different places in India.

Uses   The leaves are antipuritic, anthelmintic and they are useful in scabes, various types of poisoning, intermittent fevers and intestinal worms. The fruits are sweet, sour, febrifuge and tonic. They are useful in diarrhoea, vomiting, haemorrhoids and hyperdipsia.

Azadirachta indica A. Juss.

Family   Meliaceae

Indian names   *Nimba* (Sanskrit); *Nim* (Hindi); *Nim* (Assamese); *Nim* (Bengali); *Limbro* (Gujrati); *Bevu* (Kanada); *Vepu* (Malayalam); *Nimba* (Marathi); *Nim* (Oriya); *Vembu, Velam* (Tamil), *Konda vepa* (Telugu).

Description   Shady middle-sized trees, up to 25 m tall, with an ovoid, lax crown, bark grey, crocked and rough on mature branches; leaflets 12–16, crenate–serrate, oblique up to 6 × 2 cm. Flowers 2-merous, 10–8 merous; slaminal column, branched at tip, anthers exerted; ovary globose, 3 locular, style elongate, stigmas 3 lobed with a basal rim; drupe globose, pulpy, with a very large seed; 1-seeded, shiny, yellow at maturity.

Flowering and fruiting   March–July

Distribution   Cultivated throughout India

Propagation   By seeds

Parts used   Every part of the plant—bark, root, fruit, seeds, flowers, leaves, gum—possesses medicinal value.

Chemical constituents   The stem bark contains tannin, non-tannin and red dye. The leaves contain nimbin, nimbinene, nimbandiol and quercetin. The presence of β-sitosterol, n-hexacosanol and nonacosane is also reported. The fruit contain gedunine, 7-deacetoxy-7α-hydroxy gedunin, azadirachtin, 7 α-hydroxy-azadirachtin and nimbol.

Uses   Antibacterial, antifungal, astringent, tonic, antiperiodic, vermifuge, demulcent, antiseptic and is used in most of the ailments.

Baccaurea ramiflora Lour.

Family   Euphorbiaceae

Indian names   Lutqua, Rambai (English); *Kataphal, Lutco* (Hindi); *Leteku* (Assamese); *Moktok* (Manipuri); *Pangkai* (Tamil); *Dieng sohmyndong* (Khasi); *Dojuka* (Garo).

Description   Middle sized tree up to 12 m high with a compact, oval crown; bark grey, warty; leaves 8–12 × 4–8 cm, elliptic, oblanceolate-elliptic, acuminate, base narrowed, cuneate, glabrous; racemes up to 12 cm long; flower yellow, 0.2 cm across; bracts lanceolate, fruits ovoid-globose, yellow, pulp rose coloured.

Flowering and fruiting   April–August

Distribution   Indo-Malaya; confined to North-East India, both wild and cultivated in Meghalaya at lower elevations.

Propagation   By seeds

Parts used   Leaves and fruits

Uses   Characteristic on account of their profuse cauliflorous bloom. The fruits are edible and are marketed.

Bacopa monnieri L.

Family   Scrophulariaceae

Indian names   Thyme-leaved gratiola (English); *Saumyalata* (Sanskrit); *Brahmi* (Hindi); *Brahmi* (Assamese); *Barna, Nirbrahmi* (Malayalam); *Ghola* (Marathi); *Nirpirani* (Tamil); *Sambrani cettu* (Telegu).

Description   Herb spreads on ground and its stems and small leaves are succulent, i.e., fleshy; roots arise on the nodes of the stem also; flowers arise in the axils of the leaves and are borne on short pedicels. One of the five sepals is larger than others. The petals are bluish-white in colour and about 1 cm across.

Propagation   By vegetative method

Part used   Whole plant

Chemical constituents   Alkaloid, saponin, bacoside A and B; monniem, betulic acid, D-mannitol, saponin possesses cardiotonic, sedative and spasmodic properties.

Uses   The drug Brahmi, is valued as a tonic for nerves and is prescribed in nervous disorders, mental disease, constipation and as a diuretic to promote urination. Leaf decoction is given to infants in bronchitis; the relief is due to the vomiting and purging brought about by the drug. Leaf juice mixed with petroleum is applied for rheumatism.

Basella alba L.

Family   Basellaceae

Indian names   Indian spinach (English); *Upodika* (Sanskrit); *Pai, Lalbaclu* (Hindi); *Bansali* (Kannada); *Vasalacheera, Basalacheera* (Malayalam); *Pasalakkirai, Sivappu* (Tamil); *Bacali* (Telugu); *Purai* (Garo); *Buna* (Mizo); *Pui* (Arunachali).

Description   A moderate sized deciduous tree with vertically cracked grey bark, wood moderately hard, greyish brown with irregular darker patches; fleshy stems often turning red; leaves broadly ovate to orbicular, shining; flowers bisexual, red or white in lax peduncled spikes with a small bract and 2 adnate fleshy bracteoles longer than perianth; fruits red, white or black globose, utricle enclosed in the perianth.

Distribution   Cultivated throughout India, and throughout North-East Region, sometimes growing near hedge.

Propagation   By seeds and stem cutting

Parts used   Stems and leaves

Chemical constituents   Iodine, fluorine, carotenoids, organic acids, vitamin K.

Uses   Diuretic. Leaves are demulcent and cooling. Leaves are used in catarrhal affliction and to hasten suppuration. Decoction of root relieves vomiting. Also used as a blood purifier in Meghalaya. Herb is used successfully to check malnutrition in children and used as cooling medicine in digestive disorders. It is beneficial in tumours. The Santhals use it against syphilitic ulcers in the nose. A poultice of the leaves is applied to sores, warts and pimples, to hasten suppuration. The leaf juice possesses demulcent, laxative and diuretic properties and is reported to be useful in gonorrhoea balanitis and in catarrhal afflictions. It is useful in urticaria caused by dyspepsia, and in burns and scalds. The herb contains antiviral substances. The fruit juice of the red variety is applied to the eyes in conjunctivitis. The roots are rubefacient and are used as poultice to reduce local swellings; the sap is used for treating acne.

Bauhinia purpurea L.

Family   Caesalpiniaceae

Indian names   *Camarikah, Tamrapuspah* (Sanskrit); *Gairal, Kaliar* (Hindi); *Basavanapadu, Kempukanjivala* (Kannada); *Mandaram, Suvannamandaram* (Malayalam); *Kalavilaichi, Mandarai* (Tamil); *Bodanta, Devakanjanamu* (Telugu); *Megong* (Garo); *Dieng long* (Khasi).

Description   Small or middle sized trees up to 10 m high. Crown sub globose, bark dark brown or ashy brown, greenish brown in young parts, lenticellate and rough; leaves 6–15 × 6–15 cm deeply lobed 1/3 – 1/2 way down, cordate or sub-cordate at base, lobes elongated, oval, narrowed obtuse or acute lomentose, corymblike; flowers 6–9 cm across, buds often pinkish, petals all white; pods 20–25 × 1.5 cm, subfalcate, oblanceolate; 10–16 seeded.

Flowering and fruiting   September–April

Distribution   Indo-Malaya, throughout India, very common in Meghalaya in deciduous secondary forests and forest margins; associated with *Callicarpa arborea, Grewia mico-cos. Mimosa himalayana, Emblica officinalis*, etc.

Propagation   By seeds

Parts used   Roots, barks, flowers and buds

Chemical constituents   α-sitosterol, lupeol, 5,7-dihydroxy- and 5,7-dimethoxyflavanone, L-rhamnopyranosyl, β-D-glucopyranosides.

Uses   The roots and bark are astringent, acrid, cooling, constipating, depurative, anti-inflammatory and styptic. They are useful in diarrhoea, dysentery, skin diseases, leprosy, intestinal worms, tumours, wounds, ulcers, etc. Flowers have laxative and anthelmintic properties.

Bauhinia vahlii Wt. Arn. Benth.

Family   Ceasalpiniaceae

Indian names   *Asmantaka, Malanjhana* (Sanskrit); *Jallur, Malghan* (Hindi); *Anipadu, Hepparige* (Kannada); *Mottanvalli* (Malayalam); *Kattumandarai* (Tamil); *Adattige, Madapu* (Telugu); *Chambuli* (Marathi).

Description   A gigantic climber, often with irregularly fluted stem showing on cross section irregular masses of xylem tissue, arranged roughly in a floral pattern; branchlets often ending in a pair of tendrils; leaves variable in size, sometimes smaller, long or broad, lobes obtuse, rounded, thin but tough, dark green and glabrescent above; flowers white or cream coloured in terminal woolly corymbs or corymbose racemes; calyx tube, slender; 5 toothed, irregularly splitting into two broadly ovate, reflexed lobes; petals obovate or oblanceolate, silky pubescent outside along the back; fertile stamens 3; ovary densely woolly, stipe adnate to calyx tube.

Flowering and fruiting April–January

Distribution Throughout India, nearly entire North-East Region.

Propagation By seeds

Parts used Roots, seeds and leaves

Uses Aphrodisiac, demulcent and tonic.

Bauhinia variegata L.

Family Caesalpiniaceae

Indian names Buddhist bauhinia, Orchid tree (English); *Kancnarah*, *Kovidarah* (Sanskrit); *Kachnar* (Hindi); *Kempumandara* (Kannada); *Mandaram*, *Malayakatti* (Malayalam); *Mandarai*, *Vellaippuvatti* (Tamil); *Devakancanamu* (Telugu); *Bol-migong* (Garo).

Description Small or middle sized deciduous trees up to 8 m high, crown irregularly spreading, often branching from base; bark dark grey or greyish-brown; leaves 6–14 × 5–13 cm, often smaller, lobed 1/3 way down, deeply cordate, lobes rounded glaucous and pubescent; flowers 6–8 cm across, white, pink or purple; posterior petal usually darker or variegated; pods 20–30 × 2–3 cm, dark brown, subfalcate, 10–15 seeded.

Flowering and fruiting February–April

Distribution   Indo-Malaya; throughout India, frequent in Meghalaya in deciduous forests and slopes.

Propagation   By seeds

Parts used   Seeds and leaves

Uses   Seeds are tonic and aphrodisiac. Leaves are demulcent and mucilagin.

Begonia josephii A. DC.

Family   Begoniaceae

Indian name   *Chulan-dara* (Assamese).

Description   Very variable in size and habit, stemless or stem leafy, always easily recognized by the peltate leaves; rootstock of one or few tubers; radical leaves (on petioles 4–10 in.) often 6 in. sometimes nearly regularly ovate, acuminate, acutely 3-lobed or orbicular with numerous acute lobes, acute, serrate or doubly serrate or less often almost entire, usually nearly glabrous but often slightly pubescent on the nerves beneath, sometimes weakly pilose above; stipules ovate, deciduous, glabrous or nearly so; Male flowers 2, caducous; petals 2, smaller; stamens shortly monadelphous, sometimes 8–30; anther obovoid; connective not produced. Female flowers, with perianth 4–6; styles 3 separate, 2 fid near the top, stigma tortuous lunate; capsule including wings, styles persistent, upper margin of the wing horizontal, and narrow and the lower faces below, and broad dehiscing first by 4 lines, two on either side of each of the two narrow wings.

Distribution   Throughout India and common in Meghalaya.

Propagation   By seeds

Parts used   Bulb

Uses   The bulb is useful for stomach ache.

Begonia palmata D. Don.

Family   Begoniaceae.

Indian names   *Jaiew* (Khasi), *Alikchaba-tassla* (Naga).

Description   Perennial herb with thick rhizome or tubers; climber; leaves lobes; caudate-acuminate, distantly serrate, softly pubescent, stipulate, persistent; outer perianth segments pinkish.

Distribution   Throughout North-East India, and Meghalaya in particular.

Propagation   By vegetative method

Parts used   Whole plant

Uses   Plant is useful for liver complications, diarrhoea, fever, stomach ache and tooth ache.

Begonia roxburghii A. DC. Probv.

Family   Begoniaceae

Indian names   *Jaiew* (Khasi); *Kanchal* (Manipuri).

Description   Root fibrous, not tuberous; stem usually 1–3 ft. erect, succulent, glabrous or minutely pubescent when young; leaves 6–9 in. acuminate, glabrous or minutely pubescent on the nerve of both surfaces; petiole 2–5 in., stipules lanceolate, linear; bracts, often few-flowered, producing but one or two fruits from each axil; male flowers with sepals 2, large, glabrous, nearly white; petals usually 2, smaller than the sepals, white or nearly so; stamens about 50. Female with ovary cells 4; placentas very thick, succulent, equally 2-partite; style 4, bifid stigma. Fruit walls very thick, succulent, indehiscent or finally dehiscent at the angles. Seeds shortly ellipsoid, some what obovoid.

Distribution   In Meghalaya up to 1500 m

Propagation   By tubers

Parts used   Tubers

Uses   Tuber paste is taken for diarrhoea and dysentery.

Berberis aristata DC.

Family   Berberidaceae

Indian names   *Daru haridra* (Sanskrit); *Rasaut, Kilmora* (Hindi); *Haldi* (Assamese); *Daru haldi* (Bengali); *Mara darisina* (Malayalam); *Daruhald* (Marathi); *Maramanjal* (Tamil); *Rasvat* (Kashmiri); *Kashmal* (Punjabi).

Description   A large thorny shrub; wood yellow; branches whitish or pale grey; leaves fascicled in axils of branched or simple spines, coriaceous, usually sharp-toothed, veins very fine; Flowers yellow, in short bunches; fruit ovoid, bluish purple; seeds few.

Distribution   In the Himalayas between 2000 and 3000 m; it also grown in the Nilgiri hills in Tamil Nadu.

Propagation   By seeds

Part used   Root, bark, roots and lower stems

Chemical constituents   Alkaloid berberine, oxycanthin and resin.

Uses   Root bark, roots and lower stems are boiled in water, strained and evaporated till the semi-solid mass is obtained; this is called "Rasaut." Rasaut is used in curing fevers, and as laxative and tonic. It is useful in stomach disorders and its use in cholera and acute diarrhoea in experimental rabbits has been confirmed.

Berberis wallichiana DC.

Family   Asteraceae

Indian names   Aallich's barbery, Willich's berberis (English); *Dieng-mat-shynrang* (Khasi).

Description   Thorny shrubs, up to 5 m high; bark greyish brown, branchlets angled; leaves 3–8 × 1–2 cm, lanceolate, oblong-lanceolate, acute, cuneate, spinous, serrulate shining on both surfaces; flowers yellow, 8 mm across; sepals 5 mm long; petals nearly equalling the size of sepals; anthers dehiscing by valves; stigmas subsessile; berries 0.8 cm long, oblong, ellipsoid, deep purple when ripe.

Flowering and fruiting   August–February

Distribution   Eastern Himalayas; frequent in high altitudes in open forests and forest margins, associated with *Ruhus rugosus, R. assamensis, Symplocos spicata, Myrica esculenta*, etc.

Propagation   By seeds

Parts used   Root, bark and fruits

Chemical constituents   Berberine, oxycanthine, berbamine, berberubine, chelidonic acid, bervulcine, columbamine, jatrorrhizine, resin.

Uses   Root bark is a tonic and anti-periodic. Diaphoretic and antipyretic. Fruit and berries are given as mild laxative to children. It is also useful for gastric and duodenal ulcers. As a local application it is used for indolent ulcers. The stems are used in rheumatism. The roots are supposed to have anti-cancer properties.

Bergenia ciliata (Haw.) Sternb.

Family   Saxifragaceae

Indian names   *Pasanabheda* (Sanskrit); *Brohala* (Assamese); *Brahala* (Arunachali).

Description   A perennial herb with stout rootstock; leaves 5–35 cm long coarsely hairy, broadly obvate or elliptic, sparsely denticulate; flowers white, pink or purple in long, cymose panicles; capsules round.

Distribution   Entire North-East Region, common in Meghalaya.

Propagation   By seeds

Parts used   Root, rhizome and leaf

Chemical constituents Root extract contains tannic acid (14.2%), glucose (5.6%) mucilage and wax.

Uses The acetone extract of rhizome is cardiotoxic in higher doses and has depressant action on central nervous system. It has anti-inflammatory and anti-diuretic effect in lower doses. Roots are useful in eye diseases, boils, cuts, diarrhoea, cough and burns.

Bidens pilosa L.

Family Asteraceae

Indian names Blackjack, Beggar tick, Cobblers peg (English); *Kumur, Kumra, Kurei* (Hindi); *Vawkpui-thal, Vawkpuithal* (Mizoram).

Description   Undershrubs up to 1.5 m high; branches angular; leaves 4–12 cm long; leaflets 1–10 × 0.7–5 cm, ovate, ovate-lanceolate, orbicular or elliptic, acute, obtuse or acuminate, sharply crenate-serrate, glabrous, heads 0.8–1 cm across with white ray florets and yellow disc florets; involucral bracts spathulate, oblong, sparsely hairy, often glandular; achenes ribbed, up to 1 cm long, narrowly oblong, black.

Flowering and fruiting   August–February

Distribution   Native of tropical America, common nearly throughout India; abundant in forest margins, forest roadsides, secondary forests, wastelands, etc. in Meghalaya.

Propagation   By seeds

Parts used   Aerial parts

Chemical constituents   Aerial parts contain flavonoids, xanthophylls, volatile oil, acetylene, sterols and tannins.

Uses   As an astringent and diuretic it is employed to treat bladder and kidney problem. Astringency is beneficial in counteracting peptic ulcers, diarrhoea and colitis.

Biophytum sensitivum (L.) DC.

Family   Oxalidaceae

Indian names   *Viparitalajjalu* (Sanskrit); *Lajjalu* (Hindi); *Ban-Naranga* (Garo); *Mukkutti* (Malayalam); *Tintanall* (Tamil); *Pulicenta* (Telugu).

Description   A slender erect annual with a rosette of leaves on top of the stem; leaves abruptly pinnate, sensitive, leaflets opposite, 6–12 pairs, the terminal pair being the largest, oblong, apiculate at the apex, glabrous, pale beneath; flowers yellow, dimorphic, peduncles many, slender up to 10 cm long, fruits ellipsoid, capsules; seeds prominently ridge, transversely striate.

Flowering and fruiting   May–August

Distribution   Throughout the hotter parts of India as weeds in moist shady places. Tropical Asia, Africa and America.

Propagation   By seeds

Parts used   Whole plant

Chemical constituents   Insulin

Uses   The plant is bitter, thermogenic, diuretic, lithontriptic, suppurative, expectorant, stimulant and tonic. It is useful in urinary calculi, hyperdipsia in bilious fever, wounds, asthma, stomach ache and snake bite. A decoction of the leaves is given in asthma. An aqueous extract of the fresh leaves partially inhibits the growth of *Mycobacterium tuberculosis*. The mature leaves are used in diabetes. Powdered seeds are used as a vulnerary.

Bischofia javanica Bl.

Family   Euphorbiaceae

Indian names   *Panila* (Hindi); *Bol-Asri* (Garo); *Kainjal* (Bengali); *Nira, Thirippu* (Malayalam); *Thondi* (Tamil); *Nalupumushti* (Telugu); *Gobranerale* (Kannada).

Description   A large tree up to 30 m tall with abundant clear latex when bruised. Leaves alternate, trifoliate, leaflets ovate-elliptic with toothed margins; flowers minute, unisexual, cream to yellowish, borne in many flowered axillary panicles; fruit is a small globose berry with thin flesh surrounding 3–6 seeds; flowers seen apparently during the summer, but fruits can be found throughout the year.

Flowering and fruiting   May–December

Distribution   Moderate, common from sea level to mid-montane in primary and secondary forests, forest edges, grassy slopes.

Propagation　By seeds and vegetative method

Parts used　Leaves, stem and bark

Chemical constituents　β-amyrin, betulinic acid, ellagic acid, fisetin, friedelin, lutein and glucoside, β-sitosterol and ursolic acid.

Uses　Liquid from the stem is used to treat children who have not walted by two years of age. The bark is used to treat stomach ulcers, mouth ulcers and athelete's foot. The leaves are used in treating stomach ache.

Bixa arellana Linn.

Family　Bixaceae

Indian names　Annatto, Arnato (English); *Sinduri* (Sanskrit); *Sinduriya* (Hindi); *Latakam* (Assamese); *Kappumankala* (Kanada); *Kappamannal* (Malayalam); *Sappira virai* (Tamil); *Jaffracettu* (Telugu).

Description　A small handsome evergreen tree; leaves large, cordate acuminate, glabrous on both surface; flowers white or pink in terminal panicles; flowers white or pinkish; sepals 4–5 free; stamens numerous; fruits reddish brown or green capsules, clothed with soft bristles; ovary superior, unilocular with two opposite parietal placenta; seeds trigonous covered with a red pulp.

Distribution　Widely cultivated throughout India

Propagation　By seeds

Parts used　Root, bark and seeds

Chemical constituents   Bixin

Uses   The root, bark and seeds are antiperiodic, antipyretic and astringent. They are useful in intermittent fevers and gonorrhoea. Pulp sorrounding the seeds is astringent, seeds and roots are used as, astringent and febrifuge. Root bark is antiperiodic and antipyretic. A non-toxic dye, obtained from the pulp is used for colouring edible materials.

Boehmeria macrophylla Don.

Family   Urticaceae

Indian names   *Bara-siauru* (Hindi); *Dieng-sohkhara* (Khasi).

Description   Trees or shrubs; bark brown, leaves opposite, lanceolate, caudate-acuminate, base above hispid rugulose and pustular, beneath softly or hispidly pubescent or glabrate; spikes elongate, pendulous leafless, simple or branched below, clusters with lanceolate bracts; fruits obovate-cuneate, compressed, ciliate with 2–4-toothed neck.

Flowering and fruiting   August–January

Distribution   Sub-tropical Himalayas, frequent in Meghalaya, particulary along the river banks.

Propagation   By seeds

Parts used   Roots

Chemical constituents   Cyanhydric acid, the roots contain the flavonoid rutin. The seeds contain some essential fatty oils.

Uses   The root posses antibacterial and diuretic properties. It is used in the therapy of threatened abortion, pregnancy, metritis and leucorrhoea. The paste is applied on wounds as it heals the wound faster. It is also used in eczema.

Boehmeria platyphylla D. Don.

Family   Urticaceae

Indian name   *Labit-long* (Khasi).

Description   Shrubs with spreading branches; branchlets strigose; leaves ovate, ovate-orbicular, abruptly acuminate, base round or cordate, often oblique, scabrous, dentate; spikes erect or spreading, 10–20 cm long; flowers greenish-white; achenes brownish.

Flowering and fruiting   June–January

Distribution   Indo-Malaya; throughout the Meghalaya in wastelands

Propagation   By seeds

Chemical constituents   It contains alkaloids, triocontanol hentriacontane, α-sitosterol, ursolic acid, 4-dimethoxy-6, 8-dimethyl flavone and α-ecdysone.

Uses   The plant is bitter, astringent, cooling, anthelmintic, diuretic, aphrodisiac, cardiac stimulant, emetic, expectorant, anti-inflammatory, febrifuge, laxative and tonic. It is useful in treating, leucorrhoea, ophthalmia, lumbago, scabies, cardiac disorder, jaundice, anaemia and cough.

Boenninghausenia albiflora Reichl ex. Meissn.

Family   Rutaceae

Indian names   *Pissumar* (Hindi); *Nukmoman* (Manipuri).

Description   A perennial herb with branching stems; leaves alternate, compound, 2–3 pinnate, terminal leaflet large; flowers in terminal, leafy panicled cymes; seeds black.

Distribution   North-East Region and particularly in Meghalaya.

Propagation   By seeds

Parts used   Whole plant

Uses   It kills fleas, lice and other insects. The herb is a mild insecticide and has insect repellent properties. The oil is used as an active repellent against dayflies and may be used for preparing flea-repelling ointments. It also shows antifungal properties.

Boerhavia diffusa L.

Family   Nyctaginaceae

Indian names   *Raktakanda, Punarnava* (Sanskrit); *Punarnava, Santhi, Sant, Survari* (Hindi); *Boridol* (Garo); *Punarvaba* (Bengali); *Tambadi vasu* (Marathi); *Mukkaratti* (Tamil); *Ataki, Atika mamidi* (Telugu).

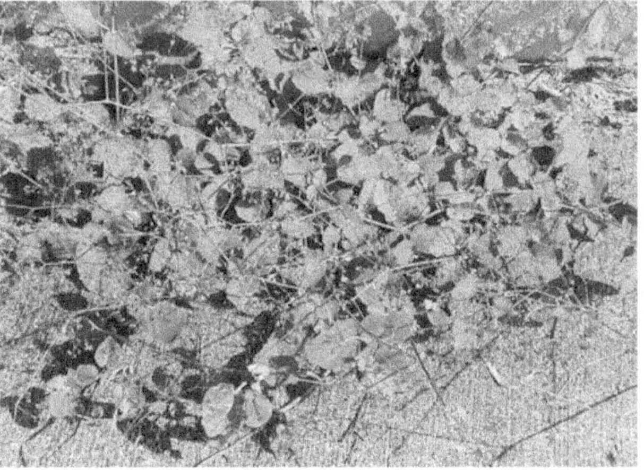

Description   A perennial diffuse herb with stout root stock and many procumbent branches; leaves simple opposite, short-petioled in unequal pairs, ovate-oblong, acute or obtuse, rounded or subcordate at base, glabrous above, and whitish beneath; flowers pale rose coloured, small, short-stalked, in irregular clusters of terminal panicles at the ends of branches; fruits highly viscid, easily detachable, one-seeded, indehiscent with a thin pericarp.

Distribution   Found throughout India; especially in North-Eastern region

Propagation   By seeds

Parts used   Whole plant

Chemical constituents   The plants contains alkaloids, triacontanol, α-sitosterol, ursolic acid, 5,7-dihydroxy-3, 4-dimethoxy-6, 8-dimethyl flavone, and hypoxanthin-9-L-arabinoside, certain hormones and α-ecdysone.

Uses   The plant is bitter, astringent, cooling, anthelmintic, diuretic, aphrodisiac, cardiac stimulant, diaphoretic, emetic, expectorant, anti-inflammatory, febrifuge, laxative and tonic. It is useful in cardiac disorder, jaundice, anaemia, constipation, cough, bronchitis and scabies.

Breynia retusa (Dennst) Alston.

Family   Euphorbiaceae

Indian names   Breyny (English); *Bahupuspa, Kamboji* (Sanskrit); *Lapano* (Hindi); *Kaamboji* (Kannada); *Perin-nirouri* (Malayalam); *Soh-matior-syurang* (Khasi)

Description   Shrubs or small trees with slender branches; leaves 1–3,5 × 0.8–2 cm elliptic, ovate or ovate-elliptic, obtuse, base cuneate or round, glaucous beneath; male flowers usually drooping, yellow, turbinate; female flowers; erect, green; fruit depressed globose, orange yellow or reddish.

Flowering and fruiting   March–September

Distribution   Sri Lanka, Bangladesh and throughout India, common in Meghalaya usually as an undershrub.

Propagation   By seeds

Parts used   Roots

Chemical constituents   White breyny contains curcubitacins, glycosides, volatile oil and tannins.

Uses   It is useful in ulcers, asthma, bronchitis and high blood pressure.

Bridelia retusa (L.) Spreng.

Family   Euphorbiaceae

Indian names   *Ekavira, Mahavira* (Sanskrit); *Ekdaniya, Gadi kuri* (Hindi); *Avugandha, Mullahonne* (Kannada); *Mulluvenga, Mulkaini* (Malayalam); *Adamarudu, Malaivengai* (Tamil); *Bontavegisa, Duriyamaddi* (Telugu); *Khasi* (Garo).

Description   Medium sized tree up to 15 m high with lax crown; leaves 3–4 in., base rarely acute, glabrous above, glabrous and glaucous beneath, or finely pubescent; petiole 1.5 cm, bracts small, obtuse, villous; flowers in clusters of about 0.4 cm diameter, both axillary and in long spikes, pubescent or glabrous, of which the males are very slender; calyx 0.3 in diameter, tube pubescent, lobes usually glabrous; disk of male pulvinate, of female, enclosing the young ovary; fruit of the size of a pea, purple-black, cocci dehiscing.

Flowering and fruiting   May–December

Distribution   Indo-Malaya, common throughout India, occurs in sal and mixed deciduous forests.

Propagation   By seeds

Parts used   Leaves and root

Uses   The leaf and roots are useful for diarrhoea and dysentery.

Bridelia stipularis (L.) Bl.

Family   Euphorbiaceae

Indian names   *Ghonta* (Sanskrit); *Akshatthe balli, Bisila balli* (Kannada); *Donkabuvara, Sirayannemu* (Telugu); *Mouhilika* (Assamese); *Umei tongkrong risan um* (Khasi).

Description   Large climber or scrambling less ever green shrubs; barks brown, branches straight or flexuous, usually fulvous-tomentose; leaves 2–8 × 0.5–5 in., margin sometimes undulate, glabrous rarely acuminate; petiole 0.5–0.9 in., stout, tomentose; clusters of flowers, green, in often very long and panicled spikes; bracts tomentose; the stipular leaves are ovate-lanceolate, sometimes much longer than the flowers; Calyx 0.3 in., in diameter; lobes lanceolate, acuminate. Corolla alike in both sexes; disk of male pulvinate and female, urceolate, with a ring of bristles at the base within; styles with long slender arms; fruit nearly 0.5 in. long, obtuse, bluish black, smooth.

Flowering and fruiting   September–March

Distribution   Tropical Africa to Malaya, throughout India; common in Meghalaya in sal and other mixed deciduous forests.

Propagation   By seeds

Parts used   Bark and flowers

Uses   The bark and flowers are used in wounds, tuberculosis, back pain, tooth ache, rheumatism, stomach ache and general debility.

Brucea mollis Wall. [Rare]

Family   Simaroubaceae

Indian names   Kosam (English); *Dieng lalseienkhlaw* (Jaintia).

Description   Shrubs or small trees; young parts tomentose; bark reddish-brown, leaflets 5–16 × 1–8 cm, ovate- oblong, oblong-lanceolate, acuminate, base rounded or unequal, pubescent on both surface, entire, panicles 10–25 cm across; flowers minute unisexual, dioecious, green, greenish white; calyx minute,3 mm across; petals 1–2 mm long; filaments glabrous; druplets ovoid, orange-red; stone reticulate; seeds compressed rugose, blackish brown.

Flowering and fruiting   March–December

Distribution   South-East Asia, North-East India and Andamans, in Meghalaya. Balphakram and Bagmara in Garo Hills.

Propagation   By seeds

Parts used   Seeds

Chemical constituents   The fruit produces a fatty oil, a kind of glucoside.

Uses   The seeds are used as a parasiticide. They are effective for amoebiasis. They are active in malaria. The oil is extracted from seeds to avoid nausea. A poultice of pounded seeds relieves haemorrhoids.

Buddleia macrostachya Benth.

Family   Loganiaceae.

Indian name   *Ja-long-kren* (Khasi)

Description   Large shrub or a small tree up to 5 m high; leaves lanceolate acuminate or acute, serrate sub-coriaceous upper surface dark-green; flowers rose purple in sub-sessile cymes arranged in dense terminal spikes. Calyx campanulate, persistent; corolla purplish outside, orange within the tube, dilated at base, lobes 4; stamens 4, ovary 3-celled.

Distribution   Throughout North-East particularly in Meghalaya

Propagation   By seeds

Parts used   Leaves

Use   The leaves are useful to cure venereal diseases.

Butea monosperma (Lam.) Kuntze. [near endangered]

Family   Fabaceae

Indian names   Flame of the forest, Bastard teak (English); *Palasha* (Sanskrit); *Palas, Tesu, Dhak* (Hindi); *Bol-uri* (Garo); *Palas* (Bengali); *Palasu, Camata* (Malayalam); *Camata* (Tamil); *Moduga* (Telugu); *Muttagamara* (Kannada); *Palash papra* (Urdu); *Khakro* (Gujrati).

Description   Middle size trees up to 20 m high with a compact, dense crown; bark dark brown or blackish brown, rough; leaves up to 50 cm long; leaflets 12–20 × 10–18 cm, obliquely ovate, rhomboid, terminal ones broadly obovate, acute to subacute, base rounded, glaucous beneath; racemes dense, usually panicled, black or rusty tomentose; flowers bright red, 5–7 cm long, falcate; pods 10–15 × 3–4 cm, oblong, rounded or subacute, silky tomentose, flowers bright orange red, large in rigid racemes; fruits pods; thickened at the sutures; seeds dark brown.

Flowering and fruiting   February–August

Distribution   Indo-Malaya; throughout the hotter parts of India in deciduous forests; rather rare in Meghalaya, occurs in open and secondary forests, usually associated with *Lannea coromandelica*, *Gmelina arborea*, *Emblica officinalis*, etc.

Propagation   By seeds and vegetative method.

Parts used   Bark, leaves, flowers, seeds and gum

Chemical constituents   The composition of the leaves varies with the place of origin. Analysis of the leaves from Jammu contacted crude protein, crude fibre, ether extract. The tree extract showed minerals, calcium and phosphorus.

Uses   The bark is acrid, bitter, astringent, thermogenic, emollient, aphrodisiac, appetiser, digestive, constipating, anthelmintic and tonic. It is useful for diarrhoea, dysentery, haemarrhoids, intestinal worms, bone fractures, renal diseases, ulcer, tumours, flatulence and diabetes. The leaves are aphrodisiac and are useful in curing acne and boils, worm infestations and haemarrhoids. The flowers are astringent, sweet, cooling, constipating, aphrodisiac, diuretic, febrifuge and tonic. They are useful for fever, leprosy, skin diseases, swelling, haemoptysis, arthritis, burning sensation, bone fractures, and are very effecacious in birth control. The seeds are purgative, ophthalmic, anthelmintic, rubefacient, depurative and tonic. They are useful for ringworm, skin diseases, epilepsy, round worm, arthritis and diabetes. The gum is an astringent, haemostatic, depurative and tonic, and is useful in diarrhoea, haemarrhoids, ulcer, general debility, hyperacidity and fever. The ash of the tender branches are useful in abdominal disorders such as flatulence, colic dysentry, etc.

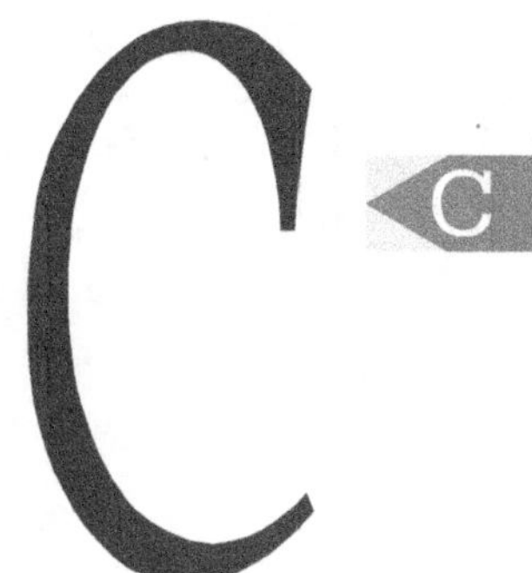

Caladium bicolor (Ait.) Vent.

Family   Araceae

Indian name   *Sam laupi* (Manipuri).

Description   Rhizomatous herb, up to 30 cm in height, with sagittate, ovate, variously coloured, spotted or blotched, thin, long-petioled leaves and light green spathe, violet at the throat.

Distribution   West Indies, South America and throughout the North-East Regions of India.

Propagation   By seeds and vegetative method.

Parts used   Roots

Chemical constituents   Calcium oxalate, L-asparagine, proteolytic enzyme Dumbcain.

Uses   The fresh rhizomes are reported to be used as emetic and purgative.

Callicarpa arborea Roxb.

Family   Verbenaceae

Indian names   *Ghiwala, Kumhar* (Hindi); *Dieng lakhiot* (Khasi); *Mondol* (Manipuri); *Hnahkiah* (Mizoram); *Khimbar, Moskhanchi* (Garo).

Description   Middle sized trees up to 15 m high; branching, usually from base; bark dark grey or brown, usually smooth; leaves 15–30 × 5–15 cm, elliptic-lanceolate, oblong-lanceolate, acuminate, base cuneate, entire or subentire, opposite, ovate, thickly tomentose beneath; peduncle as long as the petiole, cymes panicled, 5–15 cm cross; flowers 0.3–0.4 cm long, pink or purplish; calyx subtruncate, stellate pubescent, corolla long, purplish; fruits 0.2–0.3 cm, across, drupes purple-black, shining when ripe.

Flowering and fruiting   Nearly throughout the year (more during April to September)

Distribution   Indo-Malaya; mostly in Northern India, throughout North-Eastern region up to an altitude of 1800 m abundant in Meghalaya at all elevations in secondary, deciduous and pine forests; associated with *Careya arborea, Glochidion* sp., *Grewia microcos, Rhus* sp. etc. in Garo Hills, Tura.

Propagation   By seeds and stem by vegetative method

Parts used   Leaves and bark

Chemical constituents   Leaves contain C22 – C24 fatty acids, ethyl ester of C23 fatty acid, 7-0-glucuronides of luteolin and apigenin, and ursolic acid, 2 α-hydroxyursolic and crategolic acids.

Uses   Leaf extract is useful in fever, gastric disorders, giddiness, headache, etc. Bark is used for skin diseases and has carminative properties also.

Calophyllum inophyllum L.

Family   Clusiaceae

Indian names   *Nagachampa, Punnaga* (Sanskrit); *Sultannachampa* (Hindi); *Pen-hakhang* (Garo); *Sultanachampa* (Bengali); *Punna* (Malayalam); *Vuma, Hone* (Kannada); *Punnai, pinnay* (Tamil); *Pouna* (Telugu).

Description   Tree up to 25 m tall with robust trunk, which exudes white latex when bruised; leaves opposite, petiolate, thick and shiny with numerous parallel secondary veins; flowers borne in axillary cymes, flowers moderate-sized, white, fragrant, with variable number of perianth parts and yellow anthers; fruit single, purplish-black globoid or ovoid in shape, drupe with a single seed enclosed.

Flowering and fruiting   March–January

Distribution   Indo-Burma; confined to North-East India; occurs sporadically in primary forests, associated with *Castanopsis* sp. and *Quereus* sp.

Propagation   By seeds

Parts used   Leaves, bark and fruits

Chemical constituents   Seeds contain essential and vegetable oils amentoflavone, β-amyrin, apetalodie, sixteen xanthones including buchanaxanthone, butyl citrate, cinnamic acid, costatolides, erucic acid, quercetin, β-sitosterol and stigmasterol.

Uses   The leaves are soaked in water to make an eyewash for removing foreign objects from the eyes; it is also useful for eye pains. The leaves are softened by heating and then applied to sores and cuts. Boiled leaves are used to make a solution used for bathing skin rashes. Oil from fruit is rubbed onto joints to cure rheumatism. Wounds are treated with gum from the bark. Oil is used in massaging rheumatic aches.

Calotropis gigantea (L.) R. Br.

Family   Asclepiadaceae

Indian names   Yercum, Wara (English); *Aditya, Aharbandhava* (Sanskrit); *Lalak, Lalmadar* (Hindi); *Ekka, Yekkadaguda* (Kannada); *Erukku, Vellerikku* (Malayalam); *Aruchunam, Arukkam* (Tamil); *Arkamu, Jilledu* (Telugu).

Description   Large shrubs; bark white, fissured; leaves 3–20 × 3.5–12 cm, oblanceolate, oblong-obovate, acute or obtuse, base cordate, thick, coriaceous, floccose, tomentose when young; corymbs 8–12 cm long; flowers 3–4 cm across, white or bluish purple; corolla ultimately spreading; follicles 6–10 cm long, turgid.

Flowering and fruiting   May–October

Distribution   Indo-Malaya, throughout India, frequent in Meghalaya at lower elevation especially in drier localities below 500 m; Balphakram and Nangalbipra, areas.

Propagation   By seeds

Parts used   Latex, leaves, root bark and flowers

Chemical constituents   Calotropin, uscharin, calotoxin, calactin and gigantin.

Uses   The dried bark of the root is considered an excellent substitute for ipecacuanha for the treatment of dysentery. The fresh leaves are applied as dry fomentation for swelling, pounded leaves are applied on burns. Juice is rubbed on body for body pain and also for skin diseases. The latex is a cardiac stimulant and also a fish poison. The seeds have a bitter toxic substance.

Camellia caduca CL. Ex. Brandis [Endemic]

Family Theaceae

Indian names *Akom* (Garo); *Dieng-tyrnem* (Khasi).

Description Small tree or shrub, bark grey, branchlet reddish-brown; leaves oblong-elliptic, elliptic-lanceolate, acuminate, base cuneate, obtuse or acute, glabrous, except midrib, minutely blistered, crenate, serrate, narrowly revolute; inflorescence generally terminal, sometimes axillary, solitary or 2–3 fascicled, flowers up to 3 cm across, erect or suberect; sepals minutely ciliate, concave; petals obovate, often notched 1–1.5 cm; stamens yellow, outer connate into a tube, tube deeply cleft; ovary silky; style terminal, bi- or trifid; capsule globose, up to 2 cm long.

Flowering and fruiting September–March

Distribution Endemic to Meghalaya; occurs in dense forests associated with *C. caudate*.

Propagation By seeds

Parts used Leaves

Chemical constituents It contains ascorbic, malic and oxalic acids, theophylline, theobromine, xanthine, hypoxanthine, adenine, gums, dextrine and inositol.

Uses The leaves are astringent, bitter, thermogenic, appetiser, digestive, carminative, diuretic, detergent, and nervine tonic. They are useful in hyperdipsia, ophthalmia, inflammations and abdominal disorders.

Camellia sinensis (L.) Kuntz.

Family Theaceae

Indian names Tea plant (English); *Caha, Syamaparni* (Sanskrit); *Chai* (Hindi); *Sapat* (Assamese); *Sapat* (Bengali); *Teyila* (Malayalam); *Teyilai* (Tamil); *Teyaku* (Telugu).

Description   A large evergreen shrub, 9–15 m in height; leaves simple, alternate, elliptic-ovate or lanceolate with serrate margin, usually glabrous, leathery; flowers pedicellate, fruits depressed, capsules bi- or triloculate; 3-seeded.

Flowering and fruiting   September–March

Distribution   Found abundant naturally in Assam and upper regions of Burma, West Bengal and Nilgiris. Cultivated throughout the North-Eastern regions of the world.

Parts used   Leaves

Chemical constituents   It contains ascorbic, malic and oxalic acids, theophylline, theobromine, xanthine, hypoxanthine, adenine, gums, dextrine and inositol.

Uses   It is the "Queen of beverage" in most of the world. Green tea is rich in antioxidants, which rejuvenate the cells, and also serve as anti-cancer agents. Abortif used in cuts, injuries and to prevent foul smell in mouth.

Canarium strictum Roxb. [Vulnerable]

Family   Burseraceae

Indian names   The black dammar tree (English); *Mandadhupa, Raladhupa* (Sanskrit); *Kaladamar* (Hindi); *Dong-Khreng* (Garo); *Hale maddu, Mada dhup, Kari dhopa* (Kannada); *Thelli, Pantham, Kunthirikkam* (Malayalam); *Sambrani, Karunkungilium* (Tamil); *Nalla rojanamu* (Telugu); *Dhup, Raldhup* (Marathi).

Description   Lofty trees up to 40 m high, buttressed at base; bark usually greyish or blackish brown, warty, exfoliating in overlaping rectangular flakes; crown dense, umbrella-shaped; leaves 30–60 cm long; leaflets 5–13, oblong-lanceolate, oblanceolate, obovate-elliptic, acuminate, base cuneate, rounded, truncate, oblong, glabrous and shining above, pubescent beneath; panicle supra-axillary, 20–30 cm long, flowers brownish-yellow; calyx cup-shaped, 2–3 mm long, lobed rounded; petals oblong, twice or thrice as long as calyx; drupes ellipsoid, ca. 0.5 cm long.

Flowering and fruiting   July–February

Distribution   Found throughout the deciduous and evergreen forest of North-Eastern parts of India; rather rare in Meghalaya, occur along riverbanks and forest margins of evergreen forests, associated with *Ficus, Cordia,* etc.

Propagation   By seeds

Parts used   Resin

Chemical constituents   It contains junenol, canarone, epi-khusinol, keto- α-amyrin β-amyrin.

Uses   Decoction or powder of the resin is given orally as a remedy for rheumatism, cough, fever, epilepsy, asthma, various poisons and haemorrhage. It is also used as an insect repellant.

Cannabis sativa L.

Family   Cannabinaceae

Indian names   True hemp, Soft hemp, Indian hemp (English); *Bhanga, Vijaya* (Sanskrit); *Bhang, Ganja, Charas* (Hindi); *Kanchavu* (Malayalam); *Ganja, Ghamgi* (Tamil); *Ganjayi* (Telugu); *Ganja* (Garo); *Kunja* (Khasi).

Description   Annual undershrubs up to 2 m high, strong smelling of variable height; branches angled or sulcate; leaves stipulate, simple, palmatisect, leaflets 4–20 × 0.5–2.5 cm (exterior ones smaller), linear–lanceolate, oblong-lanceolate, long acuminate, base narrowed, sharply serrate; cymes erect, up to 5 cm long; flowers greenish yellow, drooping, 0.3–0.45 cm long; achenes small, angular, brownish yellow when ripe; male flowers in cymes, sepals 5; female in compact cymes, bracteate, perianth entire, reduced bicarpellary ovary.

Flowering and fruiting   July–November

Distribution   Central Asia, cultivated in many parts; run wild in Meghalaya at higher elevations forming gregarious patches, particularly in moist and shady localities. Most common in valleys through North-East Region.

Propagation   By seeds

Parts used   Seeds and leaves

Chemical constituents   The plant contains cannabidiol, cannabinol, cannabicitran, stereo isomers of cannabitriol, cannagiglendol, etc.

Uses   All parts of plant are intoxicating (narcotic), stomachic, antispasmodic, analgesic, stimulant, aphrodisiac and sedative. Its habit leads to indigestion, body waste, melancholia and impotence. In large doses it produces, mental exultation, intoxication, a sense of double consciousness and finally loss of memory, gloominess, etc.

Canthium dicoccum (Gaertn.) Merrill.

Family   Ribiaceae

Indian names   *Bol-thing* (Garo).

Description   Trees or shrubs; crown lax with spreading branches; bark warty and vertically fissured, dark brown; leaves 5–15 × 3–6 cm. Ovate or oblong-lanceolate, acuminate, base rounded or cuneate, glabrous, glossy; flowers 0.7–1 cm long, white, fragrant; corolla lobes reflexed; fruits blackish purple, pyrenes convex on one side.

Flowering and fruiting   July–March

Distribution   Indo-Malaya; North-east India and Andamans. Very rare in the tropical deciduous forests of Meghalaya.

Propagation   By seeds

Parts used   Roots and leaves

Chemical constituents   Alkaloids and mannitol

Uses   The roots and leaves are astringent, diuretic, anthelmintic and tonic. They are useful in fever, general debility, and dysentry.

Canthium parviflorum Lam.

Family   Rubiaceae

Indian names   Carray cheddie, Wild jessamine (English); *Gangeruki, Chayatinisah, Naga-vala* (Sanskrit); *Kirma, Kadbar* (Hindi); *Cherukara, Kantankara, Karamullu, Kara* (Malayalam); *Karaycheddi, Nallakkarai* (Tamil); *Balusu* (Telugu); *Shiah-sohngiam* (Khasi); *Kakegida* (Kannada).

Description   A thorny subscandent shrub with spreading branchlets tomentose; leaves 1–4 × 0.7–3 cm, simple, small, opposite with interpetiolar stipules and axillary spines, obovate, obovate-oblong or orbicular, rounded, pubescent particularly along nerves beneath, stipules ovate-lanceolate; spines straight and sharp; flowers yellowish white, 4-merous, small, in axillary cymes; corolla tube short; fruit oblong-ellipsoid or compressed drupes, yellow when ripe.

Flowering and fruiting   March–November

Distribution   Indo-Malaya; confined to eastern India; frequently in Meghalaya at lower elevations, particularly in open sandy areas.

Propagation   By seeds

Parts used   Leaves and roots

Chemical constituents   Alkaloids and mannitol

Uses   The roots and leaves are astringent, sweet, thermogenic, diuretic, febrifuge, constipating, anthelmintic and tonic. They are useful in vitiated conditions of Kapha, diarrhoea, strangury, fever, leucorrhoea, intestinal worms and general debility.

Careya arborea Roxb.

Family   Lecythidaceae

Indian names   Patana oak, Ceylon oak (English); *Bhadrendrani, Girikarnika* (Sanskrit); *Khumbi* (Hindi); *Gavvahannu, Gavvele* (Kannada); *Pezha, Pezhu* (Malayalam); *Pela* (Tamil); *Kumbhi, Govadi* (Telugu); *Bol dimbel, Gambel* (Garo); *Godhajam, Kumbhi* (Assamese).

Description   A fairly large deciduous tree, bark dark grey; leaves alternate, obovate, oblanceolate or shortly acuminate; flowers sessile usually a few together in terminal panicles; stamens numerous in several series; ovary tetralocular, rarely pentalocular.

Distribution   North-East Region, and particularly in Meghalaya.

Propagation   By seeds

Parts used   Bark, flowers and fruits

Uses   Bark and fruit are astringent and demulcent.

Carica papaya L.

Family   Caricaceae

Indian names   Pawpaw tree, Papaya, Papain (English); *Brahmairandah, Erandakarkati* (Sanskrit); *Pappaya, Pappita* (Hindi); *Pappaya, Peragi, Piranji* (Kannada); *Pappaya, Karmmust, Pappali, Karmmati* (Malayalam); *Pappali* (Tamil); *Bappayi, Bobbasi* (Telugu); *Amita* (Assamese); *Madufal* (Garo).

Description   A small tree, soft wooded, large palmate long petioled alternate leaves; flower unisexual; female solitary large. Males in panicles, pentamerous; stamens 10 in 2 series; fruit a large one celled berry many seeded.

Flowering and fruiting   Throughout the year especially March–September

Distribution   Cultivated throughout the North-Eastern regions for its edible fruits

Propagation   By seeds

Parts Used   Fruits, latex, leaves and pulp

Chemical constituents   Papaverine, cryptoxanthin, mutatochrome, *cis*-violaxanthin, antheraxanthin, chrysanthemaxanthin and neoxanthin.

Uses   Leaves are used in bone fracture. Latex is used in ring worm, other skin diseases, dog bite and for tooth and gum ache. Fruit is digestive, cooling and is used for urinary bladder complications. Contains papain, a digestive enzyme and therefore useful in digestive disorders. The fruit and seeds are useful in bleeding piles and dyspepsia. It has emmenagogue properties. The seeds are carminative, abortifacient and vermifuge and inhibit fertility in male rats. The root is abortifacient and diuretic. It checks the irregular bleeding from the uterus. It is also used for piles. The root is applied in yaws, a bacterial disease on the soles of feet. The leaves are used in jaundice, gonorrhoea, fever and beri-beri. Leaves are also useful for asthma.

Carpesium nepalense Less.

Family   Asteraceae

Description   Stem and branches pubescent often more or less cottony; leaves sub sessile elliptic; heads with large leafy bracts at the base.

Distribution   Temperate Himalayas, Nilgiri hills and Khasi hills (Meghalaya).

Propagation   By seeds

Parts used   Whole plant

Used   Used as an astringent, diuretic and anthelmintic.

Carthamus tinctorius L.

Family   Asteraceae

Indian names   Safflower, American saffron, False saffron (English); *Cusumbha, Gramyakunkuma* (Sanskrit); *Karrah, Kasumba* (Hindi); *Kusambe* (Kannada); *Centurakam, Chendurakam* (Malayalam); *Chendurukam, Kusumba* (Tamil); *Agnisikha, Kushumba* (Telugu).

Description   A thistle-like branching herb.; leaves sessile, 0.5–2 × 0.2–0.6 in., lanceolate; usually spinosely serrate; heads large, terminal; flowers orange red; involucre bracts many seriate, outer foliaceous, green usually spinous, linear-lanceolate, ovate-oblong, acute; cypsela obovoid, 4-angled, truncate with 4 locules.

Distribution   Commonly cultivated in Combodia, China, India, Laos, Middle East, Vietnam, very common in Meghalaya.

Propagation   By seeds

Parts used Flowers

Uses Dysmenorrhoea, amenorrhoea, coronary heart diseases, angina pectoria, injuries, pains due to blood stasis.

Casearia vareca Roxb.

Family Flacourtiaceae

Indian names *Chhikramarg, Chhagladoi* (Assamese); *Dieng-soh-rang* (Jaintia).

Description Shrubs or small trees, bark greyish brown, branchlets, angular; leaves 7–14 × 3–4.5 cm, elliptic, oblong, acuminate, base cuneate, sparsely hairy, sharply crenate, serrate, veins prominent beneath; inflorescence in upper axils; flowers greenish-yellow, minute; capsules up to 1 cm long, ellipsoid, orange yellow, few seeded.

Flowering and fruiting August–May

Distribution Western Ghats and North-Eastern region of India to Malaya; common in tropical evergreen forests at lower elevations.

Propagation By seeds

Parts used Fruit

Uses Fruit juice is anthelmintic.

Cassia alata L.

Family   Caesalpiniaceae

Indian names   Ringworm shrub, Ringworm senna (English); *Dadrughna, Dvipagasti* (Sanskrit); *Dadmari, Dadmurdan* (Hindi); *Sheemigida, Dhavala gida* (Kannada); *Elakajam, Shimayakatti* (Malayalam); *Vandukolli, Simaiyagatti* (Tamil); *Mettatamara, Sheemaavisi* (Telugu).

Description   Shrubs up to 4 m height; bark greenish brown, with horizontal branches; leaves paripinnate, alternate; up to 60 cm long; leaflets 4–15 × 2.5–6 cm, larger towards the tip oblong, obovate-oblong, emarginate or rounded, mucronate at tip, unequal at base, glabrous, nerves much prominent beneath; Twigs and petioles usually reddish-brown. Infloresence in axillary and terminal erect spikes; florets dense; 3–4 cm across; pods compressed, winged at both sides, appearing 4-angled; seeds numerous, black.

Flowering and fruiting   September–April

Distribution   Throughout Central and South India, in most of the districts on dry stony hills and black cottony soil. Mostly in North-East India. Frequent in Meghalaya at lower elevations in water logged or damp localities associated with other species of *Cassia* and *Clerodendron*. Nangalbibra and Balphakram, Garo Hills.

Propagation   By seeds

Parts used   Leaves

Chemical constituents   The leaves contain anthraglucosides, chrysopanic acid and rhein.

Uses   The leaves and stem have antiseptic and laxative properties. They are prescribed for constipation, oedema, hepatitis and icterus in a tea-like infusion.

Cassia fistula L.

Family   Caesalpiniaceae

Indian names   Tanner's cassia, Tanner's senna (English); *Suvarnaka* (Sanskrit); *Amaltas, Kilvali, Kirala, Sinar* (Hindi); *Kokke* (Kannada); *Kanikonna, Kritamalam* (Malayalam); *Konnei, Alas* (Tamil); *Rela* (Telugu); *Sonaru* (Assamese); *Sonali, Bandarlati* (Bengali); *Sinari, Sonari* (Garo); *Bahava, Jampa* (Marathi).

Description   A moderate sized handsome deciduous tree, up to 12 m height; crown lax, with greenish grey smooth bark when young and rough when old, exfoliating in hard scales; Leaves up to 50 cm long, leaflets 5.5–12 × 3.7 cm, ovate, ovate-elliptic, acute or subacute, base cuneate, glaucous beneath; racemes up to 60 cm long; Flowers 4–5 cm across; calyx oblong, obtuse; petals obovate; pods 25–69 × 2.5–3 cm, cylindric, indehiscent, blackish; seeds up to 100, immersed in pulp, shining brown.

Flowering and fruiting   March–December

Distribution   Indo-Malaya, throughout India in deciduous forests, planted as well as wild in Meghalaya at lower elevations.

Propagation   By seeds

Parts used   Whole plant

Chemical constituents   The plant contains barbaloin, aloin, formic acid, butyric acid, their ethyl esters and oxalic acid, acetyl acid and thiocyanogen.

Uses   The roots are astrigent, cooling, purgative, febrifuge and tonic and are useful in skin diseases, tuberculosis, syphilis and burning sensation. The bark is used as laxative, emetic, diuretic and depurative and is useful in boils, pustules, leprosy, ringworm, constipation, fever, diabetes, etc.

Cassia occidentalis L.

Family   Caesalpiniaceae

Indian names   Negro coffee, Stinking weed, Coffee-senna (English); *Kasamardah* (Sanskrit); *Kasaumdi, Barikasaumdi* (Hindi); *Taw-eitniang thynthai* (Khasi); *Daddaagace* (Kannada); *Ponnavaram, Ponnariviram* (Malayalam); *Ponnavarai, Peravarai, Nattam takarai* (Tamil); *Kasinda* (Telugu).

Description   Undershrub with furrowed subglabrous branches; leaflets 3–5 pairs; flower yellow, in short peduncled with few florets, racemes fruits cylindrical or compressed, transversely septate, glabrous pods containing 20–30 seeds; seeds ovoid, compressed, hard, smooth and shiny dark, olive-green or pale brown.

Flowering and fruiting   March–December

Propagation   By seeds

Parts used   All parts of the plant

Chemical constituents   Chrysopanol, emodin, physcion, 7-rhamnoside, crude protein, crude fibre, ash, calcium phosphate, and ascorbic acid.

Uses   The plant is bitter, thermogenic, purgative, expectorant, and febrifuge. It is useful in cough, bronchitis, constipation, fever, epilepsy and convulsion. The roots are bitter, acrid, thermogenic, diuretic and inflammatory, digestive, stomachic, and tonic. They are useful in diabetes, strangury, ringworm infections, dyspepsia and scorpion sting. The leaves are bitter, sweet, thermogenic, vulnerary, anodyne, alexeteric, and aphrodisiac and are useful in vitiated conditions, leprosy, wounds, ulcers, cough, bronchitis, asthma, fever and hydrophobia. The seeds are bitter, acrid, diuretic, expectorant and purgative and febrifuge, they are useful in leprosy, ulcers, strangury, cough, bronchitis, flatulence, dyspepsia and fever.

Cassia tora L.

Family   Caesalpiniaceae

Indian names   Sickle senna, Sickle pod, Coffee weed, Tavara (English); *Dadamari* (Sanskrit); *Chakunda, Panevar* (Hindi); *Chakunda, Panevar* (Bengali); *Soneru* (Garo); *Sonaru* (Assamese); *Tantimu* (Tamil); *Tagarai* (Telugu).

Description   Shrubs attaining about 30–90 cm in height, with pubescent young parts; leaves alternate, pinnate, with 3 pairs of obovate leaflets; leaf rachis 1.75 – 2.5 in., more or less puberulous with two glands between the two lowest pair of leaflets, grooved; stipules linear, 0.3–0.6 in. long; leaflets three pairs, gradually decreasing in size downwards 1–2.5 × 0.4–1.2 in. obovate, oblong, obtuse or subacute, minutely mucronate; base unequal, glabrous above, pubescent beneath, main lateral nerves 6–9 on either half; petiole 0–1 in long; inflorescence is axillary raceme, shorter than the leaf; flowers yellow pods 1–3 cylindric, linear, slender, very long, curved; calyx segments 0.2–0.3 in. long, ovate to elliptic, usually obtuse, ciliate and thinly pubescent in bud, glabrous with age. Petals 0.4–0.5 in. obovate, oblong, prominently veined; stamens 7, perfect, 3 reduced to staminodes; pod 5–8 in. long; each pod 5–8 in. imperfectly septate between the seeds. Seeds numerous, dark-brown, shining.

Flowering and fruiting   March–December

Distribution   Gregarious along forests, roads and margins and cleared areas. In Meghalaya and North-East India.

Propagation   By seeds

Parts used   Seeds

Chemical constituents   The whole plant contains anthraglucosides, chrysopanol and rhein. The seeds yield fatty oil consisting of oleic, linoleic, palmitic and lignoceric acids and sitosterol.

Uses   The raw seeds are utilized as a laxative. The torrified seeds are effective in insomnia, headache, constipation, cough, ophthalmia and amblyopia. The alcoholic or vinegar maceration of pounded fresh leaves is used externally to treat eczema and dermatomycosis.

Castanea sativa Mill.

Family   Fagaceae

Indian names   Congo stick, Sweet chestnut (English); *Mitha khanor* (Hindi); *Kaskottai* (Tamil); *Singuri* (Assamese); *Soh-oh-heh* (Khasi).

Description   Middle sized trees with a gradually spreading crown; bark blackish or dark brown, scaly, reticulately fissured; leaves broadly oblong, oblong-elliptic, acuminate, base rounded, subcordate, pubescent beneath, lateral nerves ending in spinous serrature; spikes greenish or creamy white; fruits 4–6 cm in diameter, densely spinous.

Flowering and fruiting   May–November

Distribution Cultivated at Khasi hills for their edible fruits

Propagation   By seeds

Parts used   Leaves and bark

Chemical constituents   Sweet chestnut contains tannins, plastoquinones and mucilage.

Uses   Leaf is useful for whooping cough, bronchitis, rheumatic conditions, to ease lower back pain, bronchial catarrh, and also to relieve joint pain.

Carallia brachiata (Lour.)

Family   Rhizophoraceae

Indian names   *Kierpa* (Hindi); *Andhimaragala, Andhimuriyana* (Kannada); *Kare-kandel, Varangu* (Malayalam); *Andimiriam* (Tamil); *Kaarvalli, Gijuru chettu* (Telugu); *Thekra-aga* (Garo).

Description   Trees up to 15 m high; crown compact; dense, oval; bark dark brown; younger branches reddish-brown with faint vertical fissures; leaves 6–18 × 3–7 cm, broadly elliptic, oblong-obovate acuminate, often obtuse or rounded, base narrowed, cuneate, entire or sub-entire at the tip, glabrous and shining above; cymes 1–3 cm across; flowers sessile or subsessile, greenish-yellow; berry, globular, pink or red.

Flowering and fruiting   November–March

Distribution   Indo-Malaya extending to Australia, North-East, Southern India and Andamans, common in Meghalaya in evergreen forest margins, particularly along streams and rivers.

Propagation   By seeds

Parts used   Bark and fruits

Uses   The pulp mixed with turmeric and rice flour is useful for sapraemia. The fruits are used in the treatment of contagious ulcers. The bark is used for itch and stomach ache and other complications.

Cayranthia pedata (Lam.) Juss. ex. Gagnep.

Family   Vitaceae

Indian names   *Suvaha*, *Godhapadi* (Sanskrit); *Goallialota* (Assamese); *Kaama pattige balli* (Kannada); *Velutta sori valli* (Malayalam); *Pannikkodi*, *Tiripatakam*, *Kattu pirandai* (Tamil); *Gummadi tige* (Telugu); *Gorpadvi* (Marathi).

Description   A large but weak climbing shrub completely hairless. Leaf tendrils opposed, branched, wiry, coiled; stem hirsute; petioles 5–15 cm long; leaflets 8–20 × 2.5–8 cm, elliptic, oblong-lanceolate, shortly acuminate, base rounded, sub-cordate; corymbs 8–20 cm across; flowers 0.3–0.4 cm across; bracts 0.7–1 cm in diameter, depressed globose.

Flowering and fruiting   April–October

Distribution   Semi-evergreen to evergreen forests. Indo-Burma, fairly common in plains

Propagation   By seeds and vegetative methods

Parts used   Leaves

Chemical constituents   Green waxy oil, sterol.

Uses   The whole plant (excluding roots) has less diuretic activity and has been a reputed remedy for cough, bronchitis, asthma, joint pain and to check uterine reflexes. The leaves are stringent and refrigerant and are used as a remedy for ulcers and diarrhoea.

Celastrus paniculatus Wild. [Vulnerable]

Family   Celastraceae

Indian names   Climbing staff tree, Black oil plant, Intellect tree (English); *Jyotimati, Pitataila* (Sanskrit); *Malkangani, Malkunki* (Hindi); *Kariganna* (Kannada); *Cheruppunna, Valulavam, Palulavum* (Malayalam); *Valuluvai, Siruvaluluvai* (Tamil); *Danti cettu, Gundumida* (Telugu); *Bhumlofa* (Assamese).

Description   Large woody climbers; bark grey, corky, lenticellate; leaves 5–12 × 4–7 cm, obovate, orbicular, oblong-obovate, acuminate, base rounded or acute, nerves raised on both surfaces; inflorescence up to 15 cm long, lax, panicled; flowers minute, 0.4 cm across, pale green; capsules 10–15 cm, globose, obscurely trigonous, orange-yellow. Fruits capsules, globose, 3-lobed, bright yellow when ripe, opening to expose the brown seeds, covered with orange red aril.

Flowering and fruiting   May–December

Propogation   By seeds

Distribution   Indo-Malaya, nearly throughout India; in Meghalaya in lower elevations both in primary and secondary forests.

Chemical constituents   Sesquiterpene alkaloids, celapanigine and celapanine which on hydrolysis gave polyalcohol A, polyalcohol C and polyalcohol D.

Uses   Bark is an abortifacient, depurative and a brain tonic. Leaves are emmenagogue and the leaf sap is a commonly used antidote for opium poisoning. Seeds are acrid, bitter, thermogenic, stimulant, digestive, laxative, febrifuge and tonic and are useful in vitiated conditions of vata and kapha, abdominal disorder, leprosy, skin diseases, paralysis, leucoderma, cardiac debility, inflammation, nephropathy, amenorrhoea, beri-beri and fever.

Celtis tetrandra Roxb.

Family   Ulmaceae

Indian names   *Garuke, Hadhuwa* (Kannada); *Karukka, Garuka* (Malayalam); *Ada, Kona* (Tamil); *Jabjabal* (Telugu); *Bol kerasu* (Garo); *Dieng chini* (Khasi); *Brumaj* (Marathi).

Description   Large trees 20–39 m high; young parts pubescent, bark dark grey, horizontally wrinkled; leaves 4–12 × 2–6 cm, ovate, ovate-elliptic or elliptic-lanceolate, caudate-acuminate, base rounded or truncate, oblique with tufts of hairs on nerve axils beneath; cymes 1–3 cm long (female ones smaller); Calyx 0.4 cm across, greenish; drupe 0.5–0.8 cm long, subglobose or broadly ovoid, orange red.

Flowering and fruiting   February–October

Distribution   Indo-Malaya; throughout India, frequent in deciduous forests at lower elevations in Meghalaya.

Parts used   Leaves and fruits

Chemical constituents   Contains tannins and mucilage

Uses   Leaves and fruits are taken to reduce heavy menstrual and intermenstrual uterine bleeding. The fruit and leaves may be used to astringe the mucous membranes in peptic ulcers, diarrhoea and dysentery.

Centella asiatica (L.) Urban.

Family   Apiaceae

Indian names   Indian pennywort (English); *Mandukparni* (Sanskrit); *Brihmi, Brahmmanduki, Budhbrahmani* (Hindi); *Manimuni* (Assamese); *Thankhuria* (Bengali); *Vondelaga* (Kannada); *Muttil, Kutannal, Kutakan* (Malayalam); *Brahmi* (Marathi); *Brahmi* (Gujrati); *Vallarai keerai* (Tamil); *Saraswalaku* (Telugu); *Brahmi buti* (Punjabi).

Description   A prostrate herb rooting at the nodes; stem slender, rooting at the joints. Leaves alternate or tufted at each node, orbicular, round or kidney-shaped, obviously crenate inflorescence in single umbel, bearing 1–5 flowers, white or reddish, without stalks. Fruit oblong, dull brown, laterally compressed, pericarp hard and thickened, woody, white.

Distribution   This species is very common throughout Meghalaya in damp soil associated situation with *Oxalis coniculata* ascending up to 1800 m.

Propagation   By seeds and vegetative method

Parts used   Whole plant

Chemical constituents   Brahmoside, bicycloelemene, centelloside, astatic acid, betulinic acid, centellic acid, centellose, triterpenoid and trisaccharides.

Uses   Alternative, tonic, diuretic and local stimulant and sedative. Plant paste is applied on boils and tumours. Leaf decoction is taken in cough, cold, fever, and stomach ache and as an anthelmintic. Also used as a cure for diarrhoea and dysentery. The plant is used for treatment of leprosy and skin diseases and also to improve memory. The plant is used as an antidote to cholera and also to treat hydrocele. The plant is used as a tonic and used in bronchitis, asthma, gastritis, bronchial catarrh, leucorrhoea, kidney trouble, urethritis and dropsy. A decoction of shoots is given for haemarrhoids. The plant is also used in peptic ulcer and tumours. In large doses the plant acts as narcotic, producing a state of temporary unconsciousness, vertigo and also leads to coma. The plant is considered as aphrodisiac and used in venereal diseases. The plant showed anti-inflammatory and anti-fertility activity in female mice.

Cheilanthes farinosa (Kaulf)

Family   Cheilanthaceae

Indian name   *Inchentong Rani-sinka* (Naga).

Description   A small tufted fern with elongated black glossy stipes, scaly when young; leaves lanceolate in outline, blade small, waxy, papillae usually 2 pinnate, and pinnatified Sori at first small globose in almost continuous line close to the margin, finally confluent thin margin simulating an indusium sometimes covering the sori.

Distribution   All over North-East Region, common in Meghalaya

Propagation   By seeds

Parts used   Leaves and flowers

Uses   Roots are useful in stomach ache and menstrual disorders. Leaves are useful for seasonal cold and fever.

Chonemorpha fragrans

Family   Apocynaceae

Indian names   *Morala, Murva* (Sanskrit); *Moorva* (Hindi); *Chandrahoovina balli* (Kannada); *Belutta-kaka-kodi* (Malayalam); *Jyemi-longwan* (Khasi).

Description   Latex copious; leaves obovate, orbicular, elliptic; cymes terminal, 15–30 cm long; flowers white, fragrant, 8–10 cm across; corolla tube narrow, 20–40 cm long, sometimes cohering by tips; seeds beaked.

Flowering and fruiting   May–January

Distribution   Throughout India, common in Meghalaya, particularly in lower elevation of Garo hills.

Propagation   By seeds

Parts used   Leaves

Uses   Leaves are useful for diarrhoea and dysentery.

Chenopodium ambrosioides

Family   Chenopodiaceae

Indian names   Jesuit tea, Wormseed (English); *Kshetravastuka* (Sanskrit); *Khatua* (Hindi); *Kaduvoma* (Kannada); *Chiski-bol* (Khasi); *Kattasambadan* (Tamil).

Description   An erect perennial herb attaining almost 1.5 m high; with sulcate glandular hairy and minutely pubescent stem; leaves smaller, upward; flowers minute, sessile, clustered on axillary and terminal simple or panicled spikes in the axils of foliaceous bracts, which increase in size downward; perianth segments usually 4–5, rarely 6, elliptic-acute. Ovary globose; stigma 3–5, very minute; seeds brown.

Flowering and fruiting   January–December

Distribution   Entire North-East Region, Khasi and Jaintia hills in Meghalaya

Propagation   By seeds

Parts used   Flowers

Chemical constituents   Seed contains volatile oil, triterpenoid and saponins.

Uses   The plant yields an essential oil, which is supposed to be a tonic and an antispasmodic. It has a reputation as a useful remedy for nervous disorders.

Cinchona officinalis L.

Family   *Rubiaceae*

Indian names   Crown bark, Peruvian bark (English); *Sinkona, Kunayanah* (Sanskrit); *Quinine* (Hindi); *Sinkona, Barkino* (Kannada); *Sinkona, Kayina* (Malayalam); *Sinkona, Koyina* (Tamil); *Cinkona, Jaddapatta* (Telugu); *Quinine* (Assamese).

Description   A slender tree, 6–9 mm in height. Leaves simple opposite, lanceolate or ovate, shining, petioles reddish, secondary nerves 8–10 pairs with hairy pits in their axils; flowers pink or purple coloured in short corymbiform terminal and axillary compound cymes; fruits ovoid-oblong capsules.

Distribution   Cultivated in India in Nilgiris at an elevation of 1800–2400 m.

Propagation   By seeds

Parts used   Bark

Chemical constituents   Chinchonidine, cinchonine, quinamine, quinidine, quinine.

Uses   The dried bark has the drug quinine. It is well known for its effective use in malarial fevers. The drug causes quick remission of fever and with repeated or regulated doses, check relapse of malarial fevers. Quinine also destroys certain bacterial infections and in certain preparations has been found useful in pneumonia, amoebic dysentery and used for eye lotions. Preparation of quinine is also useful as local application on certain rheumatic pains.

Cinnamomum bejolghota (Buch-Ham.)

Family   Lauraceae

Indian names   *Katkaula* (Hindi); *Chhamejong* (Garo); *Nagadalchini* (Assamese).

Description   Large evergreen tree, bark brownish white; leaves opposite, large petiole, elliptic-oblong, obtuse, acute or acuminate, glabrous, glaucous beneath, coriaceous, petiole thick; panicle large long peduncled, pubescent; flowers yellowish white or greenish white, perianth persistent; fruits long and ellipsoid.

Flowering and fruiting   January–July

Distribution   Central Himalayas and eastwards; very common in Meghalaya in higher elevation.

Propagation   By seeds

Parts used   Bark

Uses   Decoction is given in urinary troubles and to remove gall bladder stone. The bark used is in dyspepsia and liver complaints.

Cinnamomum camphora Nees.

Family   Lauraceae

Indian names   Camphor tree, Formosan wood (English); *Gandhadravya* (Sanskrit); *Kapur, Karpur* (Hindi); *Karppuravriksham* (Malayalam); *Karuppuramaram* (Tamil); *Himavaluka, Kappooramu* (Telugu); *Kafur* (Assamese).

Description   Evergreen tree with 15 m height; trunk barks thick and grooved; leaves alternate, coriaceous, long-petioled, shining on the upper side, 3-nerved at the base. Inflorescence in axillary panicle, shorter than the leaf; flowers small, greenish-yellow; berry globose, black when ripe.

Flowering and fruiting   February–October

Distribution   North-East India and neighbouring countries; fairly common in Meghalaya at lower elevation below 1000 m.

Propagation   By seeds

Parts used   Roots and bark

Chemical constituents   The stem wood and leaves contain an essential oil consisting of camphor, cineol, terpineol, caryophyllin, safrol, limonene, camphorene and azulene.

Uses   The camphor from the trunk wood possesses cardiac analeptic, antibacterial, demulcent and anodyne properties. Injections of camphor oil and sodium camphosulfonale are prescribed in case of cardiovascular collapse. The oral application of camphor is effective for fever, colic, sore throat and impotence. It is applied as an antiseptic. Camphor is used in external strains, rheumatic conditions and inflammation.

Cinnamomum glanduliferum wall.

Family   Lauraceae

Indian names   *Dieng-pingwail* (Khasi); *Gonsalu* (Garo).

Description   Large tree up to 25 m high with an ovoid, dense crown; bark dark grey or blackish, scaly, furrowed; leaves ovate or obovate elliptic, acuminate, base cuneate, glabrous, pale beneath; panicles with flowers greenish yellow; fruits obovoid, ellipsoid.

Flowering and fruiting   April–August

Distribution   Central Himalayas and Eastern parts; very common in Meghalaya.

Propagation   By seeds

Parts used   Bark and leaves

Chemical constituents   Eugenol, camphor, cineol and linalool.

Uses   The bark is used for cough, diarrhoea and dysentery. Oil extracted from the roots bark and leaves is applied externally for rheumatic complaints.

Cinnamomum pauciflorum Nees [Endemic/Rare]

Family   Lauraceae

Indian name   *Dieng-lorthea* (Khasi).

Description   A large shrub or a small tree, branched more or less quadrangular, glabrous; bark grey with streaks of brown on stem, green on branches; young shoots and leaf buds glabrous; leaves elliptic-ovate, lanceolate, acuminate, sometimes caudate-acuminate, firmly coriaceous, and finely reticulated beneath; base acute, rounded or sub-cordate, usually 3-nerved, rarely 5-nerved; petioles 0.2–0.4 in., long; pedicel with 3 flowers; pedicels 0.25 in. long, minutely pubescent; perianth 0.15 in. across; segments, pubescent on both surfaces; inner surface usually more dense; filaments hairy; fruit globose, about 0.3 in. across, often insect attacked, seated on the truncate toothed accrescent base of the perianth.

Flowering and fruiting   May–November

Distribution   North-East India; rather rare in Meghalaya at higher elevations above 1000 m.

Propagation   By seeds

Part used   Roots and woods

Chemical constituents   The stem wood and leaves contain an essential oil consisting of camphor cineol, terpineol, safrole, limonene and azulene.

Uses   The camphor from stem wood possesses cardiac, analeptic, antibacterial, and demulcent properties. Injection of camphor oil and sodium camphosulfonale is useful in case of cardiovascular collapse. Camphor is extensively used for external applications in the treatment of muscular strains, rheumatic condition in addition with menthol or phenol. It relieves skin itching; it is a stimulant and a carminative.

Cinnamomum tamala Nees and Eberm.

Family   Lauraceae

Indian names   Indian cassia lignea, Cassia cinnamon (English); *Tamalapatram* (Sanskrit); *Talispatri, Tejpatt* (Hindi); *Kadu lavanga patte* (Kannada); *Paccila, Ilavangam* (Malayalam); *Elavangapattiri, Talishappattiri* (Tamil); *Talisa patri, Adavi-lavanga patri* (Telugu); *Tejpatta* (Assamese).

Description   A medium sized branching tree, up to 15 m height with a compact ovoid crown and suberect branches, leaves 8–20 × 3–5 cm, oblong-lanceolate or ovate, oblong-elliptic, acute or acuminate, base narrowed, cuneate, glaucous beneath; oblong panicles as long as the leaves, greenish pubescent; Flowers 0.5–0.8 cm across, grey outside and yellow inside; pubescent perianth, silky; stamens villous; fruit 0.8–1.5 cm across, black when ripe, ovoid.

Flowering and fruiting   January–March

Distribution   Tropical and subtropical Himalaya; cultivated and wild in secondary forests, moist deciduous to shoal forests.

Propagation   By seeds

Parts used   Bark and leaves

Chemical constituents   Cinnamic acid, linalool, eugenol acetate, benzaldehyde, camphor, cadinene, α-terpineol, α- and β-pinene, geraniol, ocimene, benzyl cinnamata, benzyl acetate.

Uses   Decoction is taken for cough and catarrah. The oil is used as dentifrice and for tooth ache. It shows antibacterial activity. The leaves are carminative and used in colic diarrhoea and rheumatism, and also used in cough and cold. The leaf powder has hypoglycaemic action. It is also used in treating diabetes. The dried leaves act as an antioxidant to oils and fats. The bark has carminative effect and is given for gonorrhoea.

Cinnamomum zeylanicum Blume

Family   Lauraceae

Indian names   *Darushila* (Sanskrit); *Dalchini* (Hindi); *Dalsini* (Assamese); *Dalchini* (Bengali); *Lavangpatti* (Kannada); *Dalchini* (Marathi); *Karuvapatta, Ilavannapatta* (Malayalam) *Ilavangapattai, Karuvap-pattai,*(Tamil); *Dalchini* (Punjabi); *Dalchini* (Gujrati).

Description   It is an evergreen tree and it is about 6–8 m high; leaves large, ovate, thick, leathery, pointed at tip; shining green, lighter coloured beneath, main nerves prominent on leaves 3 or 5, running from base of leaf to middle or tip. Flowers minute, in large hairy clusters; fruit oblong or ovate, about 1.5–2 cm long, dark purple; one-seeded.

Distribution   This is distributed in southern parts of India up to an altitude of about 1500 m and is more common at lower altitudes.

Propagation   By seeds

Parts used   Tree or stem bark

Chemical constituents   Cinnamon, bark oil, cinnamic aldehyde, coumarin.

Uses   The stem bark is used in diarrhoea, nausea and vomiting. It is commonly used as a condiment. The oil obtained from leaves is used as a flavouring agent and preservative for sweets, soaps, etc. and for local application on certain rheumatic pains.

Cissampelos pareira L.

Family   Menispermaceae

Indian names   False pareira root (English); *Akaisika, Ambashtha* (Sanskrit); *Pardhi, Akanadi* (Hindi); *Padavali, Aamaradaavalli, Kaaduballi* (Kannada); *Kattuvalli, Patakkilangu* (Malayalam); *Paadakkizhangu* (Tamil); *Adivibankatige* (Telugu); *Jyrmi salla* (Khasi); *Tubukilota* (Assamese).

Description   Twining shrubs, or climbers, which annually form perennial root stocks; young parts usually tomentose or pubescent; leaves 3–6 × 5–10 cm, ovate-orbicular, peltate, obtuse or some times retuse, mucronate at the apex, membranous or subcoriaceous, pubescent, grey, tomentose beneath or both surface glabrate specially the upper, base cordate or truncate, 5–7 nerved, petiole as long as, or longer than the blade, tomentose or glabrate; male flowers in cyme, female flowers racemes and crowded in the axil of leaf like bract; drupes 0.4–0.6 cm across, globose and deep red.

Flowering and fruiting   August–January

Distribution   Tropics, throughout India; common in Meghalaya along margins

Propagation   By seeds and vegetative method

Parts used   Root, bark and leaves

Uses   Stomachic, tonic and diuretic.

Cissus quadrangula L.

Family   Vitaceae

Indian names   Edible stammed vein (English); *Asthisrnkhala, Vajravalli* (Sanskrit); *Hadjod, Hadjora* (Hindi); *Harjora* (Assamese); *Hurjarap* (Garo); *Mangarahalli* (Kannada); *Kannalamparanta, Peranta* (Malayalam); *Pirandai, Vajjiravalli* (Tamil); *Nulleratiga* (Telugu).

Description   A tendril climber with stout fleshy jointed quadrangular stems, tendrils simple, long, slender and leaf-opposed; leaves simple cordate or reniform, crenate, glabrous, shortly petiolate; flowers whitish, tetramerous in umbellate cymes; petals 4, at base connate afterwards, free. Stamens 4 filaments slender; disc adnate to the base of the ovary. Fruit succulent, red when ripe and very acrid.

Flowering and fruiting   June–November

Distribution   Throughout North-Eastern part of India

Propagation   By vegetative method

Parts used   Whole plant

Chemical constituents   The stem contains two unsymmetric tetracyclic triterpenoids, and two steroidal principles I and II. Presence of β-sitosterol, δ-amyrin and δ-amyrone is also reported.

Uses   The plant is bitter, sour, thermogenic, laxative, anthelmintic, carminative, digestive, stomachic, depurative, haemostatic, anodyne, ophthalmic and urine-promoting and is useful in helminthiasis, flatulence, skin diseases, leprosy, haemorrhages, convulsions, swellings, bone fracture, wounds, cuts and swelling, paste of the whole plant is applied on fractured bones since it is supposed to hasten the joining. The leaf extract shows antifungal activity. The stem is useful in piles and its juice is used in scurvy and irregular menstruation and in diseases of the ear and in nasal bleeding. A paste of the stem is given in asthma and is useful for muscular pains, burns and wounds, bites of poisonous insects and for saddle scars of horses and camels. The powder of dry shoots is given in digestive troubles and body pains.

Cissus javanica DC.

Family   Vitaceae

Indian names   *Mel-harjarap* (Khasi); *Kongngouyen laba* (Manipuri).

Description   Slender climber; branchlets subangular, red; leaves ovate-lanceolate, acute-acuminate, base cordate or sub-turnate, bristly, crenate, at the base glabrous with irregular white blotches on the upper surface, purple beneath, tendril forked; flowers tetramerous in umbellate cymes. Calyx fleshy; petals yellowish; fruit reddish purple to black.

Flowering and fruiting   May–December

Distribution   Throughout North-Eastern region, common in Meghalaya.

Propagation   By seeds

Parts used   Shoot

Uses   Plant is bitter, sour, thermogenic, laxative, anthelmintic, carminative, digestive, stomachic, depurative and aphrodisiac. Shoot decoction is given for urinary troubles and to dissolve stones in kidney and gall bladder. Roots or leaves are useful for itches.

Cissus repens Lamk.

Family   Vitaceae

Indian names   *Amlavetasah* (Sanskrit); *Kondage balli, Manda kumbala* (Kannada); *Nerinnampuli* (Malayalam); *Nal tenga* (Assamese); *Mei hurjarap* (Khasi).

Description   Climber; young shoots reddish; leaves ovate, acute, base cordate, serrate; pedicles 4–10 cm long; flowers greenish-red, fruit berries, globose, ellipsoid, 0.3–0.4 cm in diameter.

Flowering and fruiting   August–January

Distribution   Indo-Malaya; Himalaya, common in Meghalaya along with secondary forests and forest margins.

Propagation   By seeds

Parts used   Leaves and roots

Uses   The paste of the roots and leaves is applied as a suppurant. Leaf paste is useful in skin diseases.

Citrus limetta Risso

Family   Rutaceae

Indian names   Limetta, Sweetie (English); *Madhu-karkatika* (Sanskrit); *Mitha-amritphal* (Hindi); *Gaja nimbe* (Kannada); *Elumitchanarakam* (Malayalam); *Elumichai* (Tamil); *Gajanimma* (Telugu).

Description   A bushy shrub, young shoots glabrous, somewhat-angled spines ascending, straight; bark greenish; leaflets light green, elliptic, ovate or oblong rounded or with an obtuse apex; flowers waxy white usually tinged with red, sweet scented; petals 4–5, stamens 20–40, fruit yellow when ripe very aromatic with thin spongy rind and coherent colourless small vesicles filled with acid aromatic juice.

Flowering and fruiting   Whole year

Distribution   Throughout India, entire North-East Region

Parts used   Fruits and leaves

Uses   Fruit juice is taken for kidney and urinary trouble.

Citrus medica L.

Family   Rutaceae

Indian names   Citron (English); *Amlakeshara, Begapura* (Sanskrit); *Baranimbu, Bijaura* (Hindi); *Ganapati-naranga, Kaipanaragam* (Malayalam); *Gadarangai, Kadaranarattai* (Tamil); Madhipala-pandu (Telugu).

Description   Shrub with straggling thorny branches and smooth yellowish brown bark; leaves oblong or elliptic with acute or rounded apex, glabrous, dull green above; flowers white tinged with pink, scented, in axillary cymes; fruits large berries onlong or globose, fleshy, rind thick, rough, irregular or warted, yellow when ripe; seeds few, smooth.

Flowering and fruiting   April–December

Distribution   Throughout India, in forests along streams, in areas up to 1,200 m elevation.

Propagation   By seeds and vegetative method.

Parts used   Roots and fruits

Chemical constituents   Campesterol, stigmasterol, sitosterol, cholesterol.

Uses   The roots are laxative, anthelmintic and diuretic and are useful in constipation, renal and vesicular calculi, tumours, vomiting, helminthiasis and dental scaries. The buds and flowers are astringent, stimulant, antiemetic and appetiser and are useful in anorexia, abdominal disorders, asthma, cough and hiccups. The ripe fruits are sweet, sour, astringent, stimulant, refrigerant, digestive, cardiac stimulant, haemostatic, and tonic and also useful in cough, asthma, hyperdipsis, otalgia, anorexia, vomiting and skin diseases. The seeds are stimulant, anti-inflammatory, emmenagogue and tonic and are useful in inflammations, skin diseases, haemorrhoids. The leaves are anodyne.

Clausena excavata Burm. F.

Family   Rutaceae

Indian name   *Sam-sweng* (Garo).

Description   Large shrub and sometimes a small tree up to 6–7 m high, with unpleasant odour, young parts and inflorescence grey tomentose; leaves terete, tomentose; leaflets 15–30, rarely up to 35, shortly petioled, obliquely oblong to ovate, acute, with large marginal translucent, pubescent gland, membranous, pubescent beneath panicles with spreading branches; Flowers tetramerous, dull-white; calyx lobes hairy, broad, acute, petals oblong, glabrous; ovary villous, fruit oblong or obovate.

Flowering and fruiting   March–May

Distribution   Throughout North-Eastern region

Propagation   By seeds

Parts used   Leaves

Uses   Leaf juice is useful for muscular pain and fever, acidity and malaria. The plant is diuretic and it is useful for digestive disorders. An infusion of the root, flowers or leaves is given in colic fever, pounded roots are applied as poultice for sores.

Clausena heptaphylla Wt. and Arn.

Family   Rutaceae

Indian name   *Sam-sweng* (Garo).

Description   Small bushy shrub, branchlets thin, glabrous, terete; leaves glabrous or minutely pubescent; Leaflets 5–9, elliptic, lanceolate membranous or thinly coriaceous, glabrous, pale, translucent dots of unequal size, beneath, panicles terminal and axillary, 1–4 in, long, with short slender spreading racemose branches; calyx lobes 4–5, short; petals 4–5 white, fading to pale yellow, oblong glabrous; ovary 4 grooved, glabrous.

Flowering and fruiting   March–May

Distribution   Found throughout the North-Eastern regions of India and also commonly in Meghalaya.

Propagation   By seeds

Parts used   Leaves

Uses   The leaves are chewed with pan leaves and are also used for flavouring tobacco. Leaf juice is useful to reduce headache.

Clematis buchananiana DC. Syst.

Family   Ranunculaceae

Indian names   *Lagulia, Kanguli* (Hindi); *Mel-lieh* (Khasi).

Description   Woody climbers; branches grooved, pubescent, white young. Leaves pinnately 5–7 foliate; leaflets 5–16 × 3.5–10 broadly ovate or sub-orbicular, generally cordate, acuminate, base cordate or truncate, sparsely hairy, dentate, serrate, 5–7 nerved at base, petiole often tendrillar, panicles lax; flowers dull white, 1–1.5 cm across; sepals oblong, greenish white; stamens yellowish; style terminal, ciliate, persistent; achenes ovate-lanceolate, reddish-brown.

Flowering and fruiting   October–April

Distribution   Himalayas, frequent in wet evergreen forests of Meghalaya, associated with *Camellia* sp., *Daphne* sp., *Agapetes* sp., etc.

Propagation   By seeds and vegetative method

Parts used   Root and stem

Chemical constituents   Anemonin, clematosides

Uses   Juice taken for stomach ache, caused due to tumour, also for skin diseases, sores and tumours.

Clematis gouriana DC.

Family   Ranunculaceae

Indian name   *Bytengdok* (Jyrmi).

Description   Woody climbers, branches grooved; nodes swollen; leaves narrower, ovate-lanceolate, acuminate, base-rounded, subcordate, usually 3–5 nerved from base, pubescent, atleast along the nerves, panicles tomentose. Flowers white, sepals oblong, stamens basifixed, silky, tomentose; achenes with persistent silky hairy and styles.

Flowering and fruiting   August–December

Distribution   Himalayas and Western Ghats; frequent in Meghalaya, usually shady areas, Garampani in Jaintia hills.

Propagation   By seeds

Parts used   Roots

Uses   Roots are useful for rheumatic arthritis, acute tonsillitis, throat inflammation, and tooth ache.

Cleome viscosa Linn.

Family   Capparidaceae

Indian names   Wild mustard, Sticky cleome (English); *Dasugandha* (Sanskrit); *Hulhul, Hurhur* (Hindi); *Nayibelu, Kadu sasive* (Kannada); *Atunarivela, Vela, Aryavela, Naivela* (Malayalam); *Naivelai, Naikkaduku* (Tamil); *Kukkavaminta, Nallavaminta* (Telugu).

Description   An annual herb with strong penetrating odour and clothed with glandular and simple hairs; leaves sessile, palmately 3–5 foliate, gradually becoming shorter upwards; leaflets obovate, sessile, membranous, flowers yellow in lax raceme; fruits capsules, compressed, hairy throughout; seeds brownish-black when ripe.

Distribution   Throughout North-East Region, particularly wastelands and shifting cultivation areas, common in Meghalaya.

Propagation   By seeds and vegetative method

Parts used   Whole plant

Chemical constituent   The aerial parts contain a macrocyclic diterpene, and a bycyclic diterpene, cleomeolide. The seeds contain coumarino-lignins, cleomiscosin, A,B,C and D.

Uses   The leaves are rubefacient, vesicant and suborific. The leaf juice is useful in the treatment of inflammations of the middle ear and also used in wounds and ulcers. The roots are stimulant, antiscorbutic and vermifuge. The leaf juice is digestive and is good for otalgia. The seeds are useful for fever, diarrhoea, worm infestations, cardiac disorders and dyspepsia.

Clerodendrum colebrookianum Walp.

Family   Verbenaceae

Indian names   *Dieng-jakangum* (Khasi); *Kinchang* (Arunachali); *Dermaliong* (Naga); *Jaren* (Manipuri).

Description   Shrubs about 5 m in height with globose crown and disagreeable smell; leaves shining, green, 6–20 × 5–17 cm, ovate-orbicular, obtuse, base subcordate or truncate, glabrous; Flowers white in broad terminal compact corymbiform compound cymes; bracts caducous; fruit bluish green to deep green when fully ripe, glassy, compressed.

Flowering and fruiting   June–December

Distribution   Indo-Burma; confined to North-East India; frequent in secondary and deciduous forests.

Propagation   By seeds and vegetative method

Parts used   Roots, bark and leaves

Uses   Leaf paste is rubbed to alleviate rheumatism; young leaves are used as anthelmintic. Decoction is given to cure malarial fever.

Clerodendrum fragrans (Vent.) Willd.

Family   Verbenaceae

Indian name   Chinese glory tree (English); *Sanaki* (Garo).

Description   Small shrub, 0.5–1.5 m high; sparsely branched, branchlets, tomentose; young twigs quadrangular, pubescent, leaves opposite, long-petioled, cordate or flat at the base, tomentose on both sides, margins toothed or wavy; corymbs 5–10 cm across; bracts reddish or purplish; Inflorescence in terminal head dense, flower 1.5–2 cm across, white turning purplish pink; corolla usually more than 1-seriate; fruits purplish, ovoid.

Flowering and fruiting   February–December

Distribution   Native in China, grows wild, cultivated in Meghalaya at lower elevations, often along roadside.

Propagation   By seeds and vegetative method

Parts used   Roots

Uses   The roots have antibacterial and antiinflammatory activity. They are utilized in treating menstrual disorders, osteodynia, lumbago, and hypertension. The decoction serves as an external antiseptic for cleansing of infected wounds, burns and impetigo.

Clerodendrum infortunatum L.

Family Verbenaceae

Indian names *Barhibarha* (Sanskrit); *Ghato, Thunera* (Hindi); *Basavanapada, Ibbane* (Kannada); *Peruvelam, Vattaperuvelam* (Malayalam); *Karukanni, Perugilai* (Tamil); *Basavanapadu, Bogada* (Telugu); *Samsikhs* (Garo).

Description Shrub or undershrub, often gregarious forming dense thickets; sometimes covered with large, raised lenticular warts, young parts rusty villous; leaves broadly ovate, acuminate, denticulate, pubescent, lateral nerves on either half base rounded or shallow cordate, petiole pubescent; flowers white, tinged pink, fragrant, in large lax terminal pyramidal panicles of cymes. Bracts elliptic; acuminate; calyx deeply 5-partite, pubescent, corolla tube slender, pubescent lobe elliptic, extended. Drupe bluish black.

Flowering and fruiting February–July

Distribution Common throughout North-East Region

Propagation By seeds and vegetative method

Parts used Leaves and roots

Chemical constituents Leaves contain clerodendrin, aceacetin and mesoinositol

Uses It is useful in joint pain, numbness and paralysis, and occasionally for eczema. The leaf decotion is used as a tonic and is suppossed to be a vermifuge. Roots are used for fermenting liquor (Garo Hills).

Clerodendrum serratum (L.) Spreng.

Family Verbenaceae

Indian names *Bharngi, Kharasakah* (Sanskrit); *Barangi* (Hindi); *Samgengol* (Garo); *Gantabarangi* (Kannada); *Cheruthekku* (Malayalam); *Sirutekku* (Tamil); *Gantubarangi* (Telugu).

Description   A slightly woody shrub with bluntly quadrangular stems and branches; up to 2 m high; bark greyish; leaves 6–25 × 2–10 cm, oblong or oblanceolate, obovate, acute or acuminate, base narrowed, cuneate, glabrescent; panicle up to 30 cm long pyramidal; flowers 1.5–2.5 cm long (excluding stamens) bluish white or reddish white; drupes 0.7–0.9 cm across, globose, obovoid, black when ripe.

Flowering and fruiting   May–December

Distribution   Throughout India, in forests up to 1,500 m elevation, Indo-Malaya, common in Meghalaya as a forest undergrowth at all elevations, particularly in pine and sal forests.

Propagation   By seeds and vegetative method.

Part used   Roots and leaves

Uses   The roots are bitter, acrid, thermogenic, anti-inflammatory, digestive, carminative, stomachic, anthelmintic, depurative, expectorant, stimulant and febrifuge and are useful in inflammation, dyspepsia, anorexia, flatulence, cough, asthma, bronchitis, hiccup, tumours, chronic nasitis, skin disease, leucoderma, leprosy and fevers. The leaves are used as an external application for cephalagia and ophthalmia; seeds aspirant, used in dropsy.

Clerodendrum viscosum Wall.

Family   Verbenaceae

Indian names   *Bhandirah* (Sanskrit); *Titabhamt, Bhates* (Hindi); *Sam-Mokhi* (Garo); *Basavana pada, Ibbane* (Kannada); *Perutheku* (Malayalam); *Peruthekhai, Perukilai* (Tamil); *Gurrapu katliyaku* (Telugu).

Description   Shrubs or undershrubs up to 3 m height; young parts tomentose, lenticular, warty; leaves 8–20 × 7–15 cm, ovate-lanceolate, acute or acuminate, base rounded or subtruncate, dentate, serrate or sub entire, subpubescent; panicles corymbose, up to 20 cm across;

flowers 2.5–3 cm long, excluding stamens, white or tinged with pink; bracts reddish; drupes bluish-black, 0.7–1 cm across, shining.

Flowering and fruiting   February–August

Distribution   Throughout India, common in Meghalaya usually as undergrowth of sal forests and gregarious along roadsides.

Propagation   By seeds and vegetative method

Parts used   Leaves

Chemical constituents   The alcoholic extraction of the roots yields sterol and glycoside. The major sterols are campesterol, cholesterol, sitosterol, etc.

Uses   The leaves are bitter, acrid, thermogenic, laxative, cholegogue, antiseptic, demulcent, anti-inflammatory, depurative, vermifuge, expectorant, antipyretic and tonic, and are useful in helminthiasis, ascarides, abscesses, tumours, skin diseases, indolent ulcers, cough, bronchitis, intermittent fever, malarial fever, general debility and protoptosis.

Clerodendrum wallichii Merrill.

Family   Verbenaceae

Indian names   *Samapul* (Garo); *Kudeng* (Manipuri); *Uther* (Mizo); *Baphul-kung* (Tripuri).

Description   A shrub about 4 m in height, glabrous; bark reddish brown, leaves lanceolate, oblanceolate or elliptic, acuminate, entire, membranous, glabrous; flowers white in lax, pendulous, panicled cymes; Calyx greenish, enlarged in fruit, brick red; corolla tube white.

Flowering and fruiting   September–April (next year)

Distribution   Entire North-East Region especially in Meghalaya

Propagation   By seeds

Parts used   Leaves

Chemical constituents   As in *Clerodendrum colebrookianum*

Uses   Leaf juice is useful for headache and to clear nasal block and ease breathing, astringent, carminative and heart tonic.

Codonopsis javanica (BL) H.K. [Rare]

Family   Campanulaceae

Description   Glabrous twiners; leaves ovate, acute, base-cordate, entire; flowers greenish white with purple streaks; calyx foliaceous, persistent, green, oblong, ovate; berries subglobose, 1–2 cm across, red to dark purple on maturity.

Flowering and fruiting   May–December

Distribution   Confined to North-East India, frequent in Meghalaya

Propagation   By seeds and vegetative method

Parts used   Roots

Chemical constituents   Roots contain sugars, fatty substances, essential oils, the glucoside scutellarin and small amount of alkaloids.

Uses   It is useful for diarrhoea, anaemia, fatigue, cough, jaundice, dyspepsia, nephritis, albuminuria, oedema, metroptosis and lymphatic system diseases.

Coelogyne stricta (D. Don.) Schlechter

Family   Orchidaceae

Description   Pseudobulbs narrowly ovoid; leaves in pairs, acute, thinly coriaceous; Inflorescence raceme bearing 5–11 flowers; bracts deciduous, ovate; acute. Sepals oblong, acute; petals linear, reflexed lobes narrow, keel 2, crenulated, white, orange red at apex.

Flowering and fruiting   April–June

Distribution   Indo-Burma, entire North-East Region, particularly in Meghalaya.

Propagation   By vegetative method

Parts used   Pseudobulbs

Uses   Pseudobulbs are used for headache and fever.

Coffea arabica L.

Family   Rubiaceae

Indian names   Coffee, Arabian coffee (English); *Caphi, Pilu* (Hindi); *Kafi, Bunna* (Kannada); *Dieng-Kophi* (Khasi); *Kappi, Kappi-kuru* (Malayalam); *Kappi* (Tamil).

Description   Middle sized trees 10–12 m high or usually bushy shrubs; bark dark, smooth; branchlets swollen at nodes; leaves 15–30 × 8–16 cm, broadly elliptic or broadly oblanceolate shortly acuminate or acute, base rounded or obtuse, glabrous, dark green above, lateral nerves much prominent beneath; fascicles 5–7 cm across at each node; flowers pure white, fragrant; drupes 1.2–2 cm long, oblong, purple when ripe.

Flowering and fruiting   March–December

Distribution   Domesticated in the mountains of North-East Africa. Cultivated first in the mountains of subtropical Asia. Also grown in sunny slopes of Western Ghats. Indigenous to Abyssinia and Sudan, extensively planted in South India and rare in Meghalaya as a beverage plant.

Propagation   By seeds and vegetative method

Parts used   Seeds

Chemical constituents   Alkaloids, caffeine. Important furan derivatives as flavour components in coffee are furfural, 5-methyl-furfural, acetylfuran and furan-carboxylic acid, methyl ester. Coffee berries contain the enzyme β-galactosidase, which increases during fruit ripening. Japanese coffee beans contain tocopherols.

Uses   The seeds are bitter, stimulant, diuretic, antipyretic and aromatic, and are used as a medicinal beverage for the sick and convalescent. It stimulates the flow of digestive juices and intestinal peristalsis. The infusion of unripe seeds is good for migraine, fever and gout, whereas infusion of ripe roasted seeds is good for diarrhoea. Coffee is a palliative in spasmodic asthma, whooping cough, hysteria and cardiospasim. A strong coffee is a remedy for poisining by opium, alcohol and other narcotics.

Coffea bengalensis Roxb.

Family   Rubiaceae

Indian names   *Coffee* (Garo); *Mirherai* (Arunachali).

Description   Deciduous shrubs with slender spreading branches pale greyish. Leaves elliptic or broadly ovate, acute or acuminate, entire, membranous, glabrous. Flowers slender, white fragrant (appearing with the leaves or occasionally before appearance of new leaves) showy.

Flowering and fruiting   February–December

Distribution   Common in entire North-East Region

Propagation   By seeds and vegetative method

Parts used   Leaves and seeds

Uses   Infusion of leaves is used to bathe infants to reduce fever. Seeds are used as subtitute for coffee.

Coix lacryma Jabi L.

Family   Poaceae

Indian names   Job's tears (English); *Gvendhukah* (Sanskrit); *Gurlu, Samku* (Hindi); *Majutti* (Kannada); *Sohrim* (Khasi); *Chaming* (Mizo); *Kattugotampu, Kakkappalunku* (Malayalam); *Kattukuntumani* (Tamil); *Adavi guruginja* (Telugu).

Description   An erect perennial grass up to 1.5 m high with basal rooting nodes; leaves linear-lanceolate, narrowed up from a broad cordate base, midrib stout, ligule narrow, membranous; racemes nodding or drooping from long peduncles carrying male and female spikelets; lower basal glume in male spikelet narrowly winged enclosing hard bract of female spikelet; fruits subglobose or ellipsoid, bluish grey, smooth; tear-shaped grains.

Flowering and fruiting   February–December (next year)

Distribution   Throughout India, common in North-East Region

Propagation   By seeds and vegetative method

Parts used   Roots and leaves

Chemical constituents   The seeds contain trans-ferulyl stigmasterol and trans-ferulyl campesterol, which form part of an ovulation-inducing or fertility drug. The roots contain adenosine and phenolic compounds, 4-ketopinorsinol, theo-and erythro-1-C-syringylglycerol and 2,6-dimethoxy-p-hydroquinone.

Uses   They are used in treating menstrual disorders and strangury. The seeds are bitter, tonic, diuretic, cathartic and depurative. Leaves are used in diarrhoea, dysentery, fever, small pox and also as a tonic. Decoction is also given in urinary trouble. The fruits are used in medicine as tincture or as decoction, for catarrhal afflictions of the air passage and inflammation of the urinary tract.

Colocasia esculenta (L.) Achott.

Family Araceae

Indian names Taro, Cocoyam (English); *Alupam, Alukam, Hachu* (Sanskrit); *Arvi* (Hindi); *Kaci, Samagoddi* (Kannada); *Kochu, Dhoka* (Assamese); *Thareng* (Garo); *Bimitang* (Naga); *Chempu* (Malayalam); *Seppankizhangu* (Tamil); *Cema dumpa, Cema gadda* (Telugu).

Description A herb of moist and marshy land, leaves broad large, ovate with a broad triangular basal sinus, spathe oblong 2–4 times shorter than the narrow lanceolate limb.

Flowering and fruiting Through the year

Distribution Throughout India, everywhere on wet soil and along the margins of ponds

Propagation By vegetative method

Parts used Rhizome, corm

Uses Rhizome and corns are used in cuts and wounds. The juice of petiole is used as a styptic or astringent.

Colchicum luteum Baker.

Family Liliaceae

Indian names *Hiranyatuh* (Sanskrit); *Hirantutiya* (Hindi); *Suranjam-karvi* (Punjabi); *Irkim, Moond* (Kashmiri).

Description An annual herb; corms brownish in colour, almost conical in shape, with one side flated, the other roundish; leaves very narrow, but broader towards tip, increasing in size as the plant approaches fruiting stage, 15–30 cm long, 0.8–1.5 cm broad; Flowers large, 2.5–4 cm in diameter, 7–10 cm long, yellow. Fruits 2.5–4 cm long, their beaks recurved.

Distribution   In North-Western Himalayas from about 700–2800 m altitude, usually in open lands.

Propagation   By seeds

Parts used   Seeds and corms

Chemical constituents   Colchicine, an alkaloid

Uses   The extract of the bulb is used in inducing chromosomal duplication.

Combretum pilosum Roxb.

Family   Combretaceae

Indian name   *Bhoree loth, Thoonia loth* (Hindi).

Description   Scandent shrubs; young parts rusty or blackish-brown, spreading, hairy; leaves oblong-oblanceolate, acuminate, base cordate or retuse, glabrescent or pubescent along nerves; spikes panicled, densely blackish or rusty, villous; flowers pink; petals exceeding the length of sepals; ovary pilose; fruits elliptic or ovate in outline, 5 winged.

Flowering and fruiting   January–October

Ditribution   Bangladesh and North-Eastern India, frequent in Jaintia hills.

Propagation   By seeds

Parts used   Stem

Uses   It is useful for dysentery, diarrhoea and stomach trouble.

Combretum roxburghii Roxb.

Family   Combretaceae

Indian names   *Dhoba lota, Tita sali* (Assamese); *Dagi shing, Dugrak* (Garo); *Ther sali, Mei longkhasaw* (Khasi); *Keuti* (Oriya).

Description   Large woody climbers; young parts brown villous; bark reddish-brown with vertical splits; leaves 6–15 × 3–7 cm, oblong, obovate-oblong, abruptly acuminate, base rounded, often cuncate, pubescent along nerves beneath; spikes in panicles of 40–50 cm long; flowers 0.5–0.6 cm, across, brownish-red or purplish-brown; anther red; fruits 2.5–3 cm long, oblong-ovoid, 5 or rarely 4-winged.

Flowering and fruiting   November–March

Distribution   Indo-Malaya, Northern and North-Eastern India; very common in Meghalaya in tropical evergreen and deciduous forests; it also forms a dominant element in secondary forests.

Propagation   By seeds

Parts used   Leaves

Uses   When in bloom, this gives the forest canopy a white look due to the bracts. Young leaves are used in Garo Hills for stomach troubles.

Commelina paludosa Blume.

Family   Commelinaceae

Indian names   *Jatakanchura* (Bengali); *Wangden-khobi* (Manipuri); *Keni* (Marathi).

Description   Stem stout, 30–60 cm, branched, leaves lanceolate, spathes subsessile, solitary or crowded funnel shaped; capsule trigonous, obovoid; seeds ellipsoid, compressed.

Distribution   All over the North-Eastern parts

Propagation   By rhizome

Parts used   Whole plant, stem and leaves.

Uses   Paste mixed with tobacco leaves and ginger applied for insect bites.

Corchorus capsularis L.

Family   Tiliaceae

Indian names   Jute (English); *Ciki* (Garo); *Marapat* (Assamese); *Chanal* (Malayalam); *Sanal Naru* (Tamil).

Description   An annual shrub; leaves oblong, acuminate, coarsely toothed, base prolonged into taillike appendages; flowers small, yellow, bracteate; sepals 4–5, petals 4–5; stamens generally numerous on a short torus; Ovary 2–5 celled, ovules, many in each cell, style short; capsule oblong, not beaked, wrinkled, muricate 2–5 valved without transverse septa, seeds numerous.

Family   Boraginaceae

Indian name   *Soh-lambrang* (Khasi).

Description   Tree or shrubs; leaves alternate, often with a marginal nerve, petiolate; flowers polygamous, calyx campanulate or tubular, accrescent in fruit; segments short, irregular or obscure. Corolla tubular or funnel-shaped, lobes usually 4–8, white, recurved, imbricate; stamens 4–8, adnate to the tube, anthers exserted; ovary 4-celled, ovules 1 in each cell, stigma capitate, fruits drupaceous, ovoid or ellipsoid; seeds exalbuminous.

Distribution   Throughout North-Eastern region

Propagation   By seeds

Parts used   Leaves and seeds

Uses   Leaves are demulcent, bitter, tonic, stomachic, laxative carminative, refrigerant and diuretic.

Corchorus olitorius L.

Family   Tiliaceae

Indian names   Tossa jute, Jews mallow (English); *Brihatchanchu, Divyagandha* (Sanskrit); *Koshta, Chamghas* (Hindi); *Chunchali gida, Kadusenabu* (Kannada); *Pirattikkirai, Natikam* (Tamil); *Parintakura, Janumu* (Telugu).

Description   Shrubs; leaves glabrous, ovate-lanceolate, 3–5 nerved, serrate, the two lower prolonged into long sharp points; stippes shorter than petioles. Peduncles 1–3 flowered; Sepals shortly pointed; petals yellow, spathulate, longer than the sepals. Stamens generally numerous on a short torus; capsule cylindric, elongated, glabrous, 10 ribbed, beaked, cells transversely septate.

Distribution   Throughout North-Eastern regions of India

Propagation   By seeds

Used   Leaves and seeds

Uses   It is used in fever and dysentery.

Cordia caudatus Geisel.

Family   Boraginaceae

Indian name   *Soh-lambrang* (Khasi).

Description   Tree or shrubs; leaves alternate, often with a marginal nerve, petiolate; flowers polygamous, calyx campanulate or tubular, accrescent in fruit; segments short, irregular or obscure. Corolla tubular or funnel-shaped, lobes usually 4–8, white, recurved, imbricate; stamens 4–8, adnate to the tube, anthers exserted; ovary 4-celled, ovules 1 in each cell, stigma capitate, fruits drupaceous, ovoid or ellipsoid; seeds exalbuminous.

Flowering and fruiting   May–September

Distribution   Khasi and Jaintia hills

Propagation   By seeds

Parts used   Bark and fruits

Uses   Bark astringent, anthelmintic and used in fever, diarrhoea, skin diseases, fruit astringent, diuretic, expectorant and used in lung and spleen diseases.

Cordia dichotoma Forst. [Very rare]

Family   Boraginaceae

Indian names   Sebestin plum (English); *Slesmatakah, Bahuvarah* (Sanskrit); *Lasura, Lasopra* (Hindi); *Cikkacalli, Doducallu* (Kannada); *Naruvari, Naruviri* (Malayalam); *Naruvili* (Tamil); *Cinnanakkeru, Botuka* (Telugu); *Bol-mingmong* (Garo).

Description   Middle sized trees with straight drooping branches; bark ashy or brownish with longitudinal wrinkles; leaves 2.5–14 × 2–10 cm, simple, entire or slightly dentate, broadly ovate or elliptic, entire, sinuate or crenate, acute or bluntly acuminate, sparsely hairy, rough above; corymbs up to 10 cm across; flowers white, fragrant, 0.4–0.5 cm long; fruits drupes, yellowish brown, pink or nearly black; ovoid, glossy yellow when ripe 1–1.5 cm in diameter, with a viscid sweetish transparent pulp surrounding a central stony part.

Flowering and fruiting   March–September

Distribution   From Africa to Australia; throughout India; very rare in Meghalaya, occur in tropical deciduous forests below 800 m; Songsak and Rangrenggiri, Garo Hills.

Propagation   By seeds

Parts used   Bark, leaves and fruits

Uses   The bark is bitter, astringent, acrid after digestion, constipating, anthelmintic, cooling and depurative, and is useful in dyspepsia, fever, diarrhoea, burning sensation, leprosy and skin diseases. The leaves are aphrodisiac and are useful in gonorrhoea and ophthalmodynia. The fruits are sweet, emollient, purgative, vulnerary, diuretic, expectorant, aphrodisiac and febrifuge, and are useful in ulcers, skin diseases, bronchitis, dry cough, strangury and ring worm.

Cordia fragrantissima Kurz. [Very rare]

Family   Boraginaceae

Indian name   *Bahari* (Garo).

Description   Middle sized trees, 8–15 m high; bark brownish grey, vertically fissured; Leaves 5–20 × 4–15 cm, ovate or orbicular, obtuse, base rounded or cuneate, alternate, large, ovate, 3 nerved, mature scabrous above, grey tomentose beneath; corymb large terminal; glabrescent; berry ellipsoid, obtuse, entire, sinuate or crenate, acute or bluntly acuminate, sparsely hairy, rough above; corymbs up to 10 cm across; flowers white, fragrant, fruits ellipsoid, 0.7–10.9 cm in diameter.

This closely resembles *C. grandis*, Wall. in its area of distribution, inflorescence, tetramerous flowers, and berries; differing only in the adult leaves, being densely stellately villous beneath.

Flowering and fruiting   October–April

Distribution   Indo-Burma; confined to North-East India; very rare in Meghalaya, usually in dense tropical evergreen forests.

Propagation   By seeds and vegetative method

Part Used   Stem and leaves

Uses   A preparation made from the stem is used to promote ovulation. The plant is also used to treat rheumatic aches and swelling of muscles and joints. A preparation made from leaves is used to treat knee wounds or skin ulcers. The leaves are used as a remedy for bronchitis and asthma.

Corydalis longipes DC.

Family   Fumariaceae

Description   A much branched perennial leafy prostrate plant of pasturelands; leaves pinnately lobed, the uppermost usually with 3 obovate or oblanceolate segments, membranous; Inflorescence racemose, few flowered; sepals small. Petals yellow; stamens 6, ovary 1-celled; style short, filiform; stigmas 2; capsule narrowed to the pedicel; seeds few small shining.

Flowering and fruiting   May–August

Distribution   Khasi hills 1500–2000 m

Propagation   By seeds and vegetative method

Parts used   Whole plant

Uses   It is useful for pains, Menorrhagia, headache, abdominal pain and fatigue.

Costus specious (Koening) Sm. [Vulnerable]

Family   Costaceae

Indian names   Elegant costus, Ceylon calumaba root, Turmeric tree (English); *Kemuka* (Sanskrit); *Keu* (Hindi); *Keu* (Bengali); *Chengalvakoshtu* (Kannada); *Kuiravam* (Tamil); *Chengavakoshtu* (Telugu); *Jomlakhuti* (Assamese).

Description   A succulent herb with long leafy spirally twisted stems, 2–3 m in height with horizontal rhizome; leaves simple, spirally arranged, oblanceolate or oblong, glabrous above, silky pubescent beneath with broad leaf sheath. Flowers white, fragrant, in dense terminal spikes; spike very dense 2–4 in bracts ovate, bright red, 1–1.5 in, lip white, suborbicular, the margins uncurved; filament 1.5–2 in, including the oblong petaloid connective; Capsule 1–2.5 cm, globose red, and crowned with the persistent calyx.

Distribution   Throughout India in moist localities

Propagation   By vegetative method

Chemical constituents   Dioscin, prosapogenins-A and – B of dioscin, gracillin and α-sitosterol-β- D-glucoside.

Parts used   Rhizome and seeds

Uses   The rhizome is bitter, astringent, acrid, febrifuge, and useful for burning sensation, flatulence, constipation, helminthiasis, leprosy, skin diseases, asthma and anaemia.

Crassocephalum crepidioides (Benth.) Moore.

Family   Asteraceae

Indian names   Thickhead (English); *Terapibi* (Manipuri).

Description   Herbs, slender, erect up to 1.5 m high; Leaves 5–15 × 2–6 cm, obovate-elliptic, lanceolate, acute-acuminate, tapering to base, narrowed below, irregularly shaped, dentate, membranous, glabrous or nearly so; corymbs drooping when young some what lobed, heads 0.7–1.5 cm long, deep red at tip, involucral bracts oblong-linear, scarious margined; achenes minute, blackish.

Flowering and fruiting   June–March

Distribution   A native of Africa, now common in tropics; throughout India, very common in Meghalaya particularly new plantations and open areas and in fallows.

Propagation   By seeds

Parts used   Leaves

Uses   Juice is applied on cuts and wounds as haematic, cures stomach disorders and headache.

Crataeva nurvala Buch-Ham. [Very Rare]

Family   Capparidaceae

Indian names   *Barhapushpa* (Sanskrit); *Barna, Varvunna* (Hindi); *Bilpatri* (Kannada); *Niirvala, Varana* (Malayalam); *Kattumavilangai* (Tamil); *Bilvaram* (Telugu).

Description   Large trees, 20–30 m tall, crown oval, bark grey or greenish grey, smooth or nearly so; leaves trifoliate. Leaflets 9–18 × 4–9 cm, ovate- lanceolate or elliptic, acuminate, base rounded, often oblique, glabrous, shining, green above, pale beneath; flowers large, yellow or purplish 4–5 cm across, polygamous, white turning yellow; sepals 4, 0.5–0.7 cm long, cohering below with the convex lobed disk; petals 4, 1–2.5 cm long ovate, prominently nerved, narrowly clawed; filaments filiform, up to 4 cm long; gynophore 3–4 cm long; ovary oblong, on a slender stalk, 1-celled; stigma sessile, depressed; ovules many, on 2 parietal placentas; berry fleshy, fruits up to 5 cm across; seeds many, flat, ovoid; embedded in yellow fleshy pulp.

Flowering and fruiting   March–November

Distribution   Tropical and cosmopolitan. South-East Asia; North-East India; rather rare in Meghalaya in dense evergreen forests in all parts of India.

Propagation   By seeds and vegetative method

Parts Used   Roots, stem bark, leaves and flowers

Chemical constituents   From root and stem are isolated triterpenoids lupeol and varunol. The leaves yield flavonoids including rutin, quercetin and isoquercetin.

Uses   It is the drug of choice in all kapha disorders of the urinary tract, and in renal and bladder calculi. It also shows anthelmintic properties. It is used for intestinal and hepatic infections.

Croton caudatus Gies.

Family   Euphorbiaceae

Indian name   *Dumi-shak* (Garo).

Description   Shrubs, often scandent; young parts stellate tomentose, sticky, glandular; bark brown; leaves young and senescent leaves redish in colour; 5–12 × 4.5–12 cm (often broader) ovate, orbicular, acute or obtuse, base cordate, 5 to 7-nerved; recemes up to 30 cm long; flowers yellowish or greenish-yellow, 0.8–1.2 cm across; fruits oblong, 1.5–2.5 cm long, tomentose, obscurely trigonous.

Flowering and fruitig   March–October

Distribution   Philippines, Himalayas and Southern India; very common in Meghalaya as a forest weed in plantations and riversides.

Propagation   By seeds

Parts used   Leaves

Chemical constituents   The seeds yield a fatty oil composed of stearin, palmitin, glycerides of crotonic and tiglic acids; proteins 18%, the glucoside crotonoside; amino acid, lysine and alkaloids.

Uses   Leaf extract is used for malaria and cholera in Garo Hills. Leaves are employed in the treatment of abscesses, boils, impetigo, bloody stool, gastric and duodenal ulcer and dyspepsia.

Croton tiglium L. [Rare]

Family   Euphorbiaceae

Indian names   Purgative croton, True croton, Croton-oil plant (English); *Jayapala* (Sanskrit); *Jamalgota* (Hindi); *Jayapala* (Bengali); *Kohi-Bih* (Assamese); *Nepala* (Kannada); *Nervalam, Katalavanakku* (Malayalam); *Sevalamkottai* (Tamil).

Description   Medium sized tree, 3–6 cm in height; leaves alternate, toothed, glabrous, pinkish-violet when young; Bark grey, corky, lenticellate; leaves 5–15 × 2.5–8 cm, ovate or elliptic oblong, acute or acuminate, base oblique, rounded, serrate or subserrate, glabrous, base glanded, racemes up to 8 cm long; flowers unisexual, monoecious; female flowers without petals; capsule ovoid, obtusely 3-lobed, yellow; seeds small with hard brownish-yellow shell.

Flowering and fruiting   May–November

Distribution   Grows wild in hills and damp forests. Sino-Malaya; natural in North-East India rather rare in Meghalaya.

Propagation   By seeds and vegetative methods

Parts used   Seeds

Chemical constituents   The seeds yield a fatty oil composed of stearin, palmitin, glycerides of crotonic and tiglic acids; proteins 18%, the glucoside crotonoside; amino acid, lysine; alkaloids.

Uses   The processed seeds are used in treating flatulence, dyspepsia, constipation, oedema, dyspnoea and persistent cough. *Croton tiglium* poisoning is treated with a decoction of string beans.

Cryptocarya amygdalina Nees.

Family   Lauraceae

Indian name   *Dagappa* (Garo).

Description   Middle sized to large trees up to 20 m high with spreading crown, bark reticulately fissured; leaves 8–20 × 2.5–8 cm, broadly oblong, elliptic, elliptic-lanceolate, acuminate to subacute, base cuneate, tertiary nerves prominent beneath; panicles up to 15 cm long; flowers 0.3–0.4 cm across, silky tomentose, yellowish; fruits 2–2.5 cm long, ellipsoid, oblong, ribbed or smooth.

Flowering and fruiting   February–October

Distribution   Eastern Himalayas and Andamans; common in Meghalaya particularly in sal tracts.

Propagation   By seeds

Parts used   Leaves and bark

Cucurbita moschata Duch. Ex. Poir.

Family   Cucurbitaceae

Indian names   Pumpkin, Winter squash (English); *Kumrah, Kolaro vela* (Hindi); *Dangara balli, Dangara kaayi* (Kannada); *Mathan* (Malayalam); *Paranki pucani* (Tamil); *Potti gummadi* (Telugu); *Kumra, Lal kumra* (Assamese).

Description   A trailing annual herb with somewhat prickly or hairy stem and axillary tendril; leaves simple, nearly orbicular in outline, lobed with deep sinus at the base; flowers large, yellow, unisexual, solitary; male flowers solitary; sepals foliaceous; corolla pale yellow; stamens 5, filaments swollen at the base, anthers one, single-celled; female flowers, solitary; calyx and corolla same as in male flowers with inferior, ovary short style and three stigma; fruits fleshy; seeds ovoid or oblong.

Distribution   Cultivated throughout India, extensively cultivated in Meghalaya

Propagation   By seeds and vegetative method

Parts used   Fruits and whole plant

Chemical constituents   The leaves contain calcium, magnesium, iron, zinc and copper. An esterase was purified from the fruits.

Uses   The fruits are sweet, refrigerant, diuretic, sedative and tonic. They are useful for burns, asthma, bronchitis, headache, etc.

Curcuma amada Roxb.

Family   Zingiberaceae

Indian names   Mango-ginger (English); *Amrardrakam, Karpuraharidra* (Sanskrit); *Amahald* (Hindi); *Dike* (Garo); *Ambrasini* (Kannada); *Mangayinchi* (Malayalam); *Mangayingi* (Tamil); *Mamidi allamu* (Telugu).

Description   Perennial herbs, stem rhizomatous, pale yellow, leaves oblong-elliptic ending in short somewhat twisted spikes arising from the centre of the leaf tuft; sepals white to pale yellow; flowers white or pale yellow, lip semi-elliptic, yellow 3-lobed, the midlobe emarginated; flowers almost as long as bracts.

Distribution   Common in the outskirts of the forests of Meghalaya

Propagation   By rhizome

Parts used   Rhizome

Chemical constituents   Myrcene, linalool, citronellal, α-and β-terpineol, β- and δ-elements, bisabolene, α-zingiberene and a termerone.

Uses   The rhizome is bitter, sour, aromatic, cooling, appetiser, carminative, digestive, stomachic, febrifuge, laxative, diuretic, erxpectorant and anti-inflammatory. Infusion and paste are used.

Curcuma aromatica Salisb.

Family   Zingiberaceae

Indian names   Wild turmeric, Cochin turmeric, Yellow zedoary (English); *Aranyaharidra, Vanaharidra* (Sanskrit); *Ban-haridra, Jangli-haldi* (Hindi); *Dike* (Garo); *Ban-haldi* (Assamese); *Kasturi-arsina* (Kannada); *Kasturimanjal, Kattumanna* (Malayalam); *Kattumanjal, Kasturimanjal* (Tamil); *Kasturi pasupu* (Telugu).

Description   A perennial tuberous herb with annulate aromatic yellow rhizome which is internally orange-red in colour; leaves elliptic or lanceolate-oblong, caudate-acuminate, 30–60 cm long; petioles long or even longer, bracts ovate, recurved, more or less tinged with red or pink; flowers pink, lip yellow, obovate, deflexed, subentire or obscurely three lobed; fruits dehiscent, globose, three valved capsules.

Flowering and fruiting   March–October

Distribution   Commonly seen widely in India and throughout North-Eastern regions.

Propagation   By vegetative method

Parts used   Rhizome

Chemical constituents   D-camphene, D-camphore, sesquiterpenes, sesquiterpene alcohols, a sample from essential oil, showing high amount of α-curcumene, β-curcumene and xanthorrhizol. Presence of 2,4-methylphenol is also reported.

Uses   Rhizome is bitter, carminative, appetiser and tonic, and used in combination with astringents and aromatics for bruises, sprains, hiccup, bronchitis, cough, leucoderma and skin problems.

Curcuma longa Salish. Parod.

Family   Zingiberaceae

Indian names   Turmeric (English); *Gandhapalashika* (Sanskrit); *Halad, Haldi* (Hindi); *Arasina, Haldi* (Kannada); *Manjal* (Malayalam); *Manjal* (Tamil); *Haridra* (Telugu).

Description   A perennial herb, root stock large, ovoid, sessile, rhizome thick; leaves sub-sessile; flowers orange white.

Distribution   Throughout North-Eastern regions

Propagation   By vegetative method

Parts used   Rhizome

Chemical constituents   The rhizome contains the pigment curcumin, an essential oil of sesquiterpenes, zingiberene, D-phellandrene, turmerone, dehydroturmerone, α-alantolactone, curcumene and cineol.

Uses   Aromatic, stimulant, tonic, carminative and anthelmintic. The rhizome is useful in gastric ulcer, anti-inflammatory and cholegogic properties. It is prescribed in the therapy of gastric and duodenal ulcer and also for skin ailments.

Cynodon dactylon (L.) Pers.

Family   Poaceae

Indian names   Dhub grass, Barmuda or Bahama grass, Conch grass (English); *Niladurva, Durva* (Sanskrit); *Dub, Durba* (Hindi); *Samchi* (Garo); *Garika hallu* (Kannada); *Karuka* (Malayalam); *Arukampillu* (Tamil); *Garicagaddi, Gerike* (Telugu).

Description   A prostrate extensively creeping glabrous, highly branched perennial grass, rooting at every nodes, forming matted tufts; leaves narrow, linear, soft, smooth, distichous at the base, ligule with a very fine ciliate rim; inflorescence a terminal spike, green or purplish axis slender, involucral glumes acute to subulate mucronulate, floral glume obliquely oblong to semi-ovate; fruits are grains, oblong, laterally compressed about 1 mm long.

Distribution   Throughout India and common in Meghalaya

Propagation   By roots and vegetative methods

Parts used   Whole plant

Chemical constituents   β-ionone, 2-propionic, 4-hydroxybenzoic, 2-propionic and 3-methoxy-4-hydroxybenzoic, phytol, phytone, glycosides, saponins, tannins, flavonoids and carbohydrates.

Uses   Plant is an astringent, sweet, cooling, depurative, vulnerary, constipating, diuretic and tonic and is useful in burning sensation, haemorrhages, wounds, haemarrhoids, conjunctivitis, skin diseases, vomiting, diarrhoea, dysentery and general debility.

D

Dalbergia pinnata (Lour.) Prain.

Family   Fabaceae

Indian names   *Ketti* (Assamese); *Dat bigli* (Garo); *Dukhentri khot* (Khasi); *Hruitengtere* (Mizoram).

Description   Scandent, bushy shrubs or small trees with a spreading crown; bark blackish brown, lenticellate; leaves 7–14 cm long; leaflets 0.8–1.8 × 0.4–0.8 cm, rounded and slightly emerginate, base oblique, hairy beneath; panicles tomentose, up to 5 cm long; flowers greenish white, 0.6–0.8 cm long; bracts obovate-orbicular; pods 4–7 × 0.7–1.5 cm, strap-shaped, 1–3-seeded.

Flowering and fruiting   February–October

Distribution   Indo-Malaya, North-East and southern India, common in Meghalaya in deciduous forests, associated with *Holarrhena antidysentrica*, *Grwia microcos*, *Bridelia* sp. *Careya arborea*, etc.

Propagation   By seeds and vegetative methods.

Parts used   Roots, leaves, bark and hardwood.

Chemical constituents   The root contains antifungal essential oils isoflavones, biochanin-A and tiobioside. It also contains (S)-4-methoxydalbergione, (R)-latifolin and dalbergin.

Uses   The roots are astringent and constipating and are useful in diarrhoea and dysentery. The leaves are bitter, ophthalmic, styptic, digestive, constipating, anthelmintic, diuretic and stimulant and are used to treat gonorrhoea and burning sensation. The bark and hardwood are astringent, acrid, bitter, thermogenic, anthelmintic, depurative, vulnerary, anti-inflammatory, expectorant, aphrodisiac, antipyretic and appetiser and also useful in hyperdipsia, burning sensation, vomiting, skin diseases, scabies, ulcers, dysentery, scalding of urine, syphilis, gastropathy and intermittent fevers.

Dalbergia sisso Roxb-ex. DC.

Family Fabaceae

Indian names Sissoo (English); *Simsapa* (Sanskrit); *Sisam*, *Sissu* (Hindi); *Agaru*, *Biridi* (Kannada); *Irupul*, *Sisam*, *Iruvil* (Malayalam); *Sisu*, *Itti*, *Tesimaram* (Tamil); *Errasissu* (Telugu); *Sissu* (Garo); *Sissu* (Assamese).

Description Large trees up to 25 m high; crown oval, branches slender; bark light brown, vertically fissured, peeling off in long strips; leaves 8–15 cm long, leaflets 3–7 cm, ovate, orbicular, broadly elliptic, acuminate or caudate, base rounded, obtuse, glabrous at length, tomentose when young; panicle terminal and axillary; flowers 0.5–0.8 cm long, yellowish-white, standard petal long clawed; pods 5–7 cm long; flat, compressed, oblong elliptic, rounded at tip, usually 1-seeded, rarely 2–3-seeded.

Flowering and fruiting February–December

Distribution Throughout Northern India, Afganistan and in eastern parts of Meghalaya planted in lower elevation for their highly valuable timbers.

Propagation Seeds and vegetative method.

Parts used Roots, leaves, bark and hardwood.

Chemical constituents The root contains antifungal essential oil isoflavones, biochanin-A, and tiobioside. It also contains (S)-4-methoxydalbergione, (R)-latifolin and dalbergin. The fatty acid composition of the seed oil is palmitic, 16.2%; stearic, 7.0%; oleic, 14.6%; linoleic, 52.5; and linolenic, 8.0%.

Uses The roots are astringent and constipating, and are useful in diarrhoea and dysentery. The leaves are bitter, ophthalmic, styptic, digestive, constipating, anthelmintic, diuretic and stimulant and are useful in gonorrhoea, vomiting, and burning sensation. The bark and hardwood are astringent, acrid, bitter, thermogenic, anthelmintic, depurative, vulnerary and expectorant and are useful in hyperdipsia, vomiting, skin diseases, leprosy, scabies, ulcers, dyspepsia, dysentery, scalding of urine, syphilis, gastropathy, intermittent fevers, etc. The mucilage of leaves mixed with sesamum oil is a good application in excoriation. The decoction of leaves is given in the acute stage of gonorrhoea (Unani).

Dalbergia volubilis Roxb.

Family   Fabaceae

Indian names   *Sirisika* (Sanskrit); *Bankhara* (Hindi); *Mryti* (Malayalam); *Punalippu* (Tamil); *Baddugaruttuginja* (Telugu); *Budu galwang* (Garo).

Description   Lianas; stems ash-coloured, deeply fluted; leaves up to 18 cm long; leaflets 2.5–5 × 1–3.5 cm, obovate, obtuse, truncate or emarginate, base cuneate, adpressed, pubescent beneath; panicles rusty pubescent, up to 12 cm long; flowers 0.5–0.7 cm long, blush or light mauve, bracteate and bracteolate; pods 4–7 × 1.5–2.5 cm, glabrous, usually 1-seeded.

Flowering and fruiting   December–June

Distribution   Burma, Bangladesh, Sri Lanka, India, rather uncommon in Meghalaya, in deciduous forests; associated with *Acacia* sp., *Bridelia* sp., *Combretum* sp., etc.

Propagation   Seeds and vegetative methods

Parts used   Roots, leaves and bark

Uses   The roots are astringent and are useful in diarrhoea and dysentery. The leaves are bitter, digestive, constipating, diuretic and stimulant and are useful in gonorrhoea and vomiting. The bark is astringent, acrid, bitter, and is useful in vomiting, skin diseases, leprosy, scabies, dysentery, syphilis and gastropathy.

Dalhousiea bracteata Grah.

Family   Fabaceae

Indian names   *Mel-dieng* (Khasi); *Laungy-aurgthu* (Arunachali).

Description   Large, woody climber; bark greyish-white, branchlets hairy; It is an erect or scandent evergreen shrub with rigid ovate elliptic or oblong leaves; base rounded or subcordate; racemes, bracts ovate, bracteoles larger, ovate-orbicular; flowers 0.6–0.8 cm long; pods 7–10 × 3–4 cm, turgid, compressed at both ends.

Flowering and fruiting   May–January

Distribution   Occurs in tropical Eastern Himalayas, India, Burma and Bangladesh. Reported as very common but now rare in Meghalaya, in lower elevations.

Parts used   Leaves

Uses   Leaves are pounded into paste and applied on wounds.

Daphne papyracea Wall.

Family   Thymelaceae

Indian names   *Sher Shing* (Arunachali); *Dieng-thulu-thyrmia* (Jaintia).

Description   Shrub about 3 m high, leaves lanceolate narrow; flowers pale yellow to purple or white, scented, crowded in head-like clusters on very short peduncles.

Distribution   Eastern Himalayas from Nepal to Sikkim and Bhutan, Khasi hills and Meghalaya.

Propagation   By seeds and vegetative methods

Parts used   Whole plant

Chemical constituents   Plant contains diterpenes, daphnetoxin, tannins and mucilage. Daphnetoxin is considered to be highly toxic and also shows the presence of anti-cancerous agents which are utilized in the treatment of cancer.

Uses   It contains an acrid poison and is an emetic and purgative. The barks of some are used for making paper.

Datura metel L.

Family   Solanaceae

Indian names   *Dhatturah* (Sanskrit); *Dhattura, Kaladhattura* (Hindi); *Dhattura* (Kannada); *Ummam, Ummata* (Malayalam); *Adukkumattai, Ummattai* (Tamil); *Tellavummetta, Ummetha* (Telugu).

Description   A sub-glabrous herb; leaves triangular ovate in outline, unequal at the base; flowers, often double or triple, white, reddish purple or white; fruit globose, tuberculate or muricate.

Distribution   Throughout India

Propagation   By seeds and vegetative method

Parts used   Leaves and flowers

Chemical constituents   Leaves and flowers contain alkaloids. Scopolamine, hyoscyamine, atropine and norhyoscyamine as well as vitamin C.

Uses   Root is pounded and taken with warm water to cure dysentery. Leaves possess antispasmodic, sedative and anodyne properties; they are used for treating cough, asthma, rheumatism, gastric ulcers, seasickness and airsickness.

Datura stramonium L.

Family   Solanaceae

Indian names   Thornapple (English); *Dhatura* (Sanskrit); *Dhatura* (Hindi); *Ummam* (Malayalam); *Ummathai* (Tamil); *Datturamu* (Telugu); *Sada dhatura* (Bengali); *Dhatara* (Garo); *Datura* (Assamese); *Dholo dhatura* (Gujrati).

Description   Long tubular, robust annual growing to 1 m has lobed oval leaves, long white or violet trumpet-shaped flowers and spiny fruit capsules similar to those of the horse chestnut.

Flowering and fruiting   Throughout the year

Distribution   America, Europe, Asia and North Africa. It is cultivated for medicinal uses and it is common in Meghalaya.

Propagation   By seeds and vegetative method

Parts used   Leaves, flowers and seeds

Chemical constituents   Thornapple contains 0.2–0.45% tropane alkaloids, flavonoids, withanolides, coumarins and tannins. The tropane alkaloids are similar to those found in deadly nightshade acting to reduce secretion and relax smooth muscle.

Uses   Thornapple is a common remedy for asthma, whooping cough, muscle spasm and the symptoms of Parkinsonism. It also relaxes the muscles of the gastro-intestine, bronchial and urinary tracts, and reduces digestive and mucous secretions.

Dillenia indica L.

Family   Dilleniaceae

Indian names   Elephant apple (English); *Bhavya, Ruvya* (Sanskrit); *Girnar, Kanigala* (Hindi); *Bettakanigala, Kadkanagula* (Kannada); *Punna, Vazchapunna* (Malayalam); *Uvattekku, Kattarali, Kurukati* (Tamil); *Uppu ponna, Peda kalinga* (Telugu).

Description   A large evergreen tree up to 25 m tall, with a large crown, with oval, compact, dense foliage; bark greyish brown; leaves 14–30 × 5–12 cm, oblanceolate or narrow elliptic, acute, base rounded or cuneate, dentate, glabrous above, strigose along nerves beneath, petiole narrowly winged; flowers large white, solitary, pendant; peduncles 3–8 cm long, stout; sepals 5, 4.5 cm thick, fleshy, green, pale along margin; petals 7–9 cm long, caducous, white; stamens up to 2 cm long, numerous, white, anther tips often incurved; carpels many, styles 10-many, green, deflexed; ovules many; fruits green, large, 3–5 in diameter, seeds small, immersed in pinkish pulp, compressed.

Flowering and fruiting   May–February

Distribution   Throughout India, except North-West India, common in North-Eastern region, all along river sides and evergreen patches.

Propagation   By seeds

Parts used   Fruit, leaves and bark

Uses   Fruit is taken as tonic, also used for chest complications, cholera dysentery, fever and sores. Juice of the fruit mixed with sugar and water is used as a cooling beverage for fever and as a cough syrup. Bark and leaves have astringent property.

Dillenia pentagyna Roxb.

Family   Dilleniaceae

Indian names   Dog teak, Nepalese elephant apple (English); *Aksikiphal, Punnaga* (Sanskrit); *Karmal* (Hindi); *Kaadu kanigalu* (Kannada); *Dieng soh bar* (Khasi); *Kutapunna, Pattippunna, Vaazhappunna* (Malayalam); *Nay-t-tekku, Punnai vakai* (Tamil); *Chinna kalinga, Revada* (Telugu).

Description   A large deciduous tree with straight cylindrical poles 9–21 m high, 2.5 m in girth with ascending branches bearing a rounded crown of large leaves; leaves oblanceolate, very narrow on the base, often stem clasping, serrate, coriaceous and hard, glabrous above and glabrescent beneath; flowers fragrant, yellow, 3 cm across, in umbels; fruit orange yellow, succulent and edible.

Flowering and fruiting   March–September

Distribution   It is common in sal forests of Meghalaya and Bengal and is scattered throughout ravines and hill slopes in the North-Eastern region.

Propagation   By seeds

Parts used   Fruits and leaves

Uses   Fruit is taken as tonic, also used for chest complications, cholera dysentery, fever and sores. Juice of the fruit mixed with sugar and water is used as a cooling beverage for fever and as a cough syrup. Bark and leaves have astringent property.

Dendrobium nobile Lindl.

Family   Orchidaceae

Indian names   *Palhutura* (Garo); *Yerumlei* (Manipuri).

Description   An epiphyte, with erect compressed stem; flowers 2–4 in raceme; sepals and petals purple above, white below; leaves oblong obliquely notched; petals much broader than linear, oblong obtuse; sepals short, wide, labellum almost sessile, wide ovate, oblong, hairy, base shortly convolute, margins curved.

Distribution   Throughout India, common in Khasi hills, 1500 m elevation

Propagation   By seeds

Parts used   Seeds

Uses   Seed powder is applied for cuts and wounds for quick healing.

Dendrocnide sinuata (Bl.) Chew.

Family   Urticaceae

Indian names   *Gilmat-Jakma* (Garo); *Torash* (Assamese); *Dieng synrem* (Khasi); *Otta-pilava* (Tamil).

Description   Shrubs, branches greenish white, clothed with stinging hairs; leaves 15–35 × 7–15 cm, ovate, broadly elliptic, acuminate, base rounded or cordate, upper half usually crenate or serrate; panicles 3–5 cm across; utricles white; calyx 0.3 cm across, inflexed.

Flowering and fruiting   July–December

Distribution   Indo-Malaya, Himalayas and South India; common in Meghalaya in shady areas and undergrowth in evergreen forests.

Propagation   By seeds and vegetative method

Parts used   Bark and leaves

Uses   Bark is used in treating symptoms of general illness, as a pain in the lungs with vomiting of blood (tuberculosis). Liquid squeezed from the leaves is given to cure fits in children, sickness after birth, and to aid expulsion of the afterbirth. Tea made from the leaves of stinging needles is used in the treatment of venereal diseases. The bark is used to treat pain in bones and joints and also for post-natal depression.

D *Dendrophthoe falcata* (L.f.) Etting.

Family   Loranthaceae

Indian names   Mistletoe (English); *Vrksadani* (Sanskrit); *Banda* (Hindi); *Badanike* (Kannada); *Ittilkkanni* (Malayalam); *Pulluri, Pulluruvi* (Tamil); *Badanika, Jiddu* (Telugu); *Tuthekmi* (Garo).

Description   A stem parasite with smooth, grey bark; leaves usually opposite, thick, variable in shape from ovate to linear, midrib prominent, secondary nerves obscure; branches thick dark or greyish-brown; leaves variable, ovate, elliptic, oblong, obovate, obtuse or rounded, base rounded or cuneate; racemes, erect, flowers, orange yellow, pinkish or reddish green, berry, oblong black.

Flowering and fruiting   February–November

Distribution   Throughout India, very common in Meghalaya particularly in drier places, parasitic on *Lannea* sp., *Callicarpa* sp. and *Quercus* sp.

Propagation   By seeds and vegetative method

Parts used   Whole plant

Chemical constituents   Cytokinins, casein, kaempferal, quercetin, myricetin, meratin and rutin.

Uses   The plant is cooling, bitter, antrigent, narcotic, diuretic and is useful in pulmonary tuberculosis, asthma, menstrual disorders, swelling, wounds, ulcers, renal and vesical calculi.

Derris ferruginea (Roxb.) Benth.

Family   Fabaceae

Indian names   *Bol-tara* (Garo); *Ruphang-doukha* (Assamese); *Kho* (Manipuri).

Description   Woody climber; bark blackish brown; young parts ferruginous pubescent; leaves up to 25 cm long; leaflets elliptic or obovate or lanceolate, bluntly acuminate, rounded at base, glabrous above, hairy beneath; racemes panicled, up to 30 cm long.

Flowering and fruiting   April–October

Distribution   Indo-Malaya; frequent in Meghalaya forests and forest margins

Propagation   By seeds

Parts used   Stem bark and leaves

Uses   Bark is used for dysentery and diarrhoea. Leaves are used as fodder.

Derris trifoliate Lour.

Family   Fabaceae

Indian names   Common derris (English); *Angaar valli* (Sanskrit); *Panlata* (Hindi); *Angarvalli, Nalla tiga* (Telugu); *Kelia* (Oriya); *Kakharu* (Garo).

Description   Large climber, branchlets with large white lenticels; leaves up to 25 cm long; leaflets 4.5 × 2.5–6.5 cm, oblong, oblanceolate ovate, acute or caudate, acuminate, base obtuse or rounded, glabrous above, pubescent along nerves beneath; racemes up to 15 cm long; flowers reddish-purple; calyx teeth obscure; pods oblong, flat, rigid, 1–2-seeded.

Flowering and fruiting   April–November

Distribution   Tropics, throughout India; frequent in Meghalaya in evergreen forests, usually in shady and wet localities.

Propagation   By seeds

Parts used   Leaves

Desmodium gangeticum (L.) DC.

Family   Fabaceae

Indian names   *Prsnipami, Prthakparni* (Sanskrit); *Salpan, Salvan* (Hindi); *Nabiyalabune* (Kannada); *Orila* (Malayalam); *Orilai* (Tamil); *Gitanaram* (Telugu); *Achhak* (Garo).

Description   An erect, diffusely branched undershrub, 90–120 cm in height with short woody stem and numerous branches provided with soft grey hairs; leaves unifoliate, ovate to ovate-lanceolate, membranous and mottled with grey patches; flowers white, purple or lilac in elongate lax, terminal or axillary racemes; fruits moniliform, 6–8 jointed, glabrescent pods, joints of pods sparsely pubescent with hooked hairs, points separating when ripe into indehiscent one-seeded segments; seeds compressed, reniform.

Flowering and fruiting   October–February

Distribution   Indo-Malaya to Philippines; mostly throughout India, common in Meghalaya at lower elevations in open areas, grassland and secondary forests.

Propagation   By seeds and vegetative method

Parts used   Roots

Uses   The roots are bitter, thermogenic, nervine tonic, carminative, diuretic, febrifuge, cardiotonic, anti-inflammatory, expectorant properties which are useful in anorexia, dyspepsis, haemarrhoids, dysentery, strangury, fever, inflammations, cough, asthma, bronchitis and cardiac diseases.

Desmodium heterocarpon (L.) DC. [Critically Rare]

Family   Fabaceae

Indian names   *Salaparni* (Sanskrit); *Adavi vehinta* (Telugu); *Samphleng* (Garo).

Description   Shrubs or herbs, up to 1.5 m height; branches angular, sulcate, adpressed hairy; leaflets 2–9 × 1–4 cm, oblong, elliptic, obovate, rounded or notched, base obtuse or rounded, adpressed, hairy; racemes up to 10 cm long. Leaves tri-folate or simple, stipulate; flowers small, 0.5–0.8 cm long, purplish pink, red, in copious usually dense racemes; Calyx campanulate or turbinate, teeth longer or shorter than the tube, the two upper often subconnate; corolla exerted, standard petal broad; wing petal more or less adhering to the obtuse keel petal; upper stamen entirely or partially free, the other 9 united; ovary sessile or stipulate, few or many ovules; style incurved, stigma minute, capitate; pods 1–2 × 0.2–0.3 cm, ciliate; usually composed of several one seeded indehiscent, pieces compressed, never muricated, the upper suture finally splits open.

Flowering and fruiting   June–November

Distribution   Widespread in the tropical and in the cape and North America. Indo-Malaya, Himalayas and North-East India.

Propagation   By seeds and vegetative methods

Parts used   Leaves

Chemical constituents   The leaves contain tannin, flavonoids, organic acids and alkaloids.

Uses   It is used in bacillary dysentery and diarrhoea. Snake bite is treated with finely pounded fresh leaves, the juice being extracted for oral administration and the residue used for poultice.

Desmodium microphyllum (Thunb.) DC.

Family   Fabaceae.

Indian names   *Sunsuni* (Hindi); *Silli-wrong* (Arunachali).

Description   Stems densely caespitose and branched; stipules linear, setaceous, persistent; leaflets oblong or obovate, pubescent below; racemes copious, terminal 6–10 flowered; bracts deciduous; calyx densely pubescent.

Distribution   In Khasi hills

Propagation   By seeds

Parts used   Whole plant

Uses   Decoction is given in cold, cough and asthma.

Desmodium triangulare (Retz.) Merr.

Family   Fabaceae

Indian names   *Naaga thagare* (Kannada); *Sambraphong* (Garo); *Porongkhok* (Mizo).

Description   Small tree, 2–3 cm high, branched; leaves alternate, leaflets 3, ovate to lanceolate, the terminal largest, covered with a dense silvery coat, pubescent beneath; inflorescence in short axillary racemes; flowers white. Pod minutely constricted between the seeds; seeds reniform.

Flowering and fruiting   July–March

Distribution   Indo-Malaya extending to Philippines, throughout India; common in wastelands of Meghalaya, roadsides and secondary forests.

Propagation   By seeds and the vegetative method

Parts used   Leaves

Chemical constituents   The leaves contain tannin, flavonoids, organic acids and alkaloids

Uses   The leaves possess antibacterial and anti-inflammatory properties. They are indicated in the treatment of bacillary dysentery and diarrhoea. Snake bite is treated with finely pounded fresh leaves, the juice being extracted for oral administration and the residue used for a poultice.

Desmodium triflorum (L.) DC.

Family   Fabaceae

Indian names   Matty desmodium (English); *Hamsapadi, Tripadi* (Sanskrit); *Kudaliya* (Hindi); *Kaadu pullampurasi, Nelaparande* (Kannada); *Nilamparanta* (Malayalam); *Sirupullati* (Tamil); *Muntamandu* (Telugu).

Description   A small perennial herb, much branched trailing with slender stems rooted at the nodes; leaves trifoliate, leaflets membranous, obovate, cuneate, truncate, emarginate or rarely rounded recticulately veined; flowers pink or white, 1–5 fascicled in the axils of leaves; fruits pods, with straight upper edge and indented lower edge.

Flowering and fruiting   June–February

Distribution   Throughout India as a weed up to 900 m elevations, common in Meghalaya

Propagation   By seeds and vegetative method

Parts used   Whole plant

Uses   The plant is acrid, sweet, cooling, expectorant and and galactagogue and is useful in cough, bronchitis, wounds, abscess, sores, dysentery, flatulence and burning sensation.

Desmodium triquetrum (L.) D.

Family   Fabaceae

Indian names   *Doddaotta, Molada gida* (Kannada); *Kattarali* (Malayalam); *Dammidi* (Telugu); *Ulucha* (Assamese); *Etang* (Garo).

Description   Undershrubs up to 2 m high; leaves 3–12 × 1–3 cm, ovate, lanceolate or oblong, acute, base rounded to truncate, sparsely hairy; racemes 3.5–12 cm long, usually simple; flowers purplish pink, 0.3–0.6 cm long; pods 2.5–4 × 0.4–0.6 cm, densely hairy.

Flowering and fruiting   July–March

Distribution   Throughout India, common in Meghalaya in wastelands and roadsides

Propagation   By seeds and vegetative method

Parts used   Leaves

Chemical constituents   Same as *Desmodium triflorum*

Uses   Leaves are useful in treating dysentery, diarrhoea and snake bite.

Dioscorea bulbifera L.

Family   Dioscoreaceae

Indian names   Air potato, Aerial yam, Bulb-bearing yam (English); *Badarakachha, Bilvamula* (Sanskrit); *Bhirvolikanda* (Hindi); *Kaimoode gedde, Kunta genasu* (Kannada); *Kattukachil* (Malayalam); *Kattukkilangu, Kayvalli* (Tamil); *Malakakayapendalamu* (Telugu).

Description   A perennial herb; tubers variable, 3–10 cm in diameter; bulbils numerous, irregular in shape, 1 cm or more across, brown, warty; stem twining to the left; leaves alternate, ovate-cordate, 7–14 × 6–13 cm wide; apex acuminate, base deeply or broadly cordate, margin entire, 7–11 nerved, glabrous; flowers unisexual, dioecious; capsules 1.8–2.2 cm long, oblong; seeds winged at the base.

Flowering and fruiting   July–October

Distribution   In China, Japan and throughout India

Propagation   By seeds and vegetative methods

Parts used   Tubers

Uses   Haemoptysis, epistaxis, pharynagitis, goitre, pyogenic infections, scrofula, orchitis, sprains and injuries.

Dioscorea pentaphylla L. (15) [Endemic]

Family Dioscoriaceae

Indian names *Kantakalu* (Sanskrit); *Bursa, Gajaria* (Hindi); *Kaadugumbala, Adavigummathiga* (Kannada); *Nurankizhannu* (Malayalam); *Kattukkilangu* (Tamil); *Karuchemba* (Telugu).

Description Tubers oblong 12–15 cm; stem slender, prickly at the base, rarely above, leaves as in *D. tomentosa*, but never softly tomentose, obovate, acuminate or cuspidate. Male panicles and flowers glabrous, hispidly pubescent or villous, spikes lax or dense, flowers sessile or pedicelled, 1/16–1/12 in diameter, fragrant; filaments and staminodes very short; capsule 0.75–2.5 cm, rounded at both ends and base cordate and tip apiculate, glabrous or pubescent; seeds 1/3–0.5 in., wing broader than the nucelus.

Flowering and fruiting May–November

Distribution China, Nepal, Burma, Bhutan and North-East India

Propagation By seeds and vegetative method

Parts used Tubers

Uses Haemoptysis, goitre, infections, sprains and injuries.

Diospyros lancaefolia Roxb.

Family Ebenaceae

Indian names *Dieng-sohle* (Khasi); *Puja* (Naga).

Description Middle sized trees, 6–15 m high with a spreading crown and pungent smelling branches; bark blackish, rough, parallel fissured; leaves, oblong, lanceolate or elliptic, acuminate, base narrowed, acute, coriaceous; flowers 4–5-merous, sessile, in clusters or solitary, yellowish, rusty silky; fruits subglobose, ovoid, rusty, villous, supported by the persistent calyx.

Flowering and fruiting   April–February

Distribution   Indo-Burma, confined to North-East India; common in Meghalaya at lower elevations evergreen forests usually along riverbank.

Propagation   By seeds and vegetative method.

Parts used   Seeds

Uses   Paste applied for skin diseases.

Diospyros malabarica (Desr.) Kost.

Family   Ebenaceae

Indian names   Gaub persimmon, Indian persimmon (English); *Tindukah, Tinduki* (Sanskrit); *Gabh, Tedu* (Hindi); *Holitupara, Kusarta, Bandadamara* (Kannada); *Paniccai, Panicika* (Tamil); *Panachi, Pananci* (Malayalam); *Tinduki, Gabu* (Telugu).

Description   A medium sized spreading evergreen tree about 15 m in height with dark grey or black bark, exfoliating in rectangular scales and young parts clothed with grey or twiny tomentum; leaves oblong elliptic or lanceolate, obtuse or acute, base rounded or subacute, glabrous, glossy, green above, pale beneath, coriaceous; male flowers in short rusty pubescent cymes, female flowers considered longer than males, solitary, subsessile; fruit ovoid or globose, yellow when ripe; seeds compressed 2–8, oblong, shiny.

Flowering and fruiting   November–August

Distribution   Nearly throughout India; frequently found in Meghalaya at higher elevations

Propagation   By seeds and vegetative methods

Parts used   Bark, leaves and fruits

Chemical constituents   The ethyl acetate extract showed anti-stress and anti-ulcerogenic activity.

Uses   The bark is astringent, acrid, cooling, anti-inflammatory, constipating, diurative and febrifuge. It is useful for burning sensation, diarrhoea, dysentery, skin diseases, burns and diabetes. The leaves are diuretic and aphrodisiac, and are useful in leucorrhoea, nyctalopia, anaemia and scabies. The fruits are bitter, acrid, cooling, digestive and carminative.

Diospyros montana Roxb. Hiern.

Family   Ebenaceae

Indian names   Bombay ebony (English); *Vayasapiluka* (Sanskrit); *Bistendu, Dasaundu* (Hindi); *Jagalugante, Kadubalekayi* (Kannada); *Malayakathi* (Malayalam); *Vakkanai, Vellaittuvarai* (Tamil); *Eddayagata, Gatugata* (Telugu).

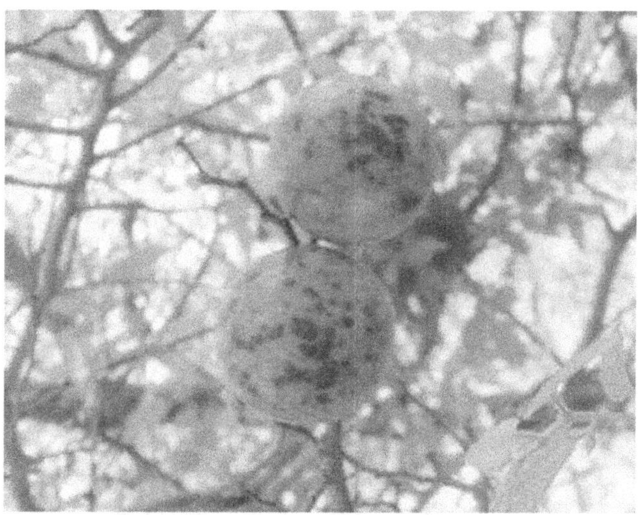

Description   Small trees or 2–10 m in height, often spiny; bark dark grey; leaves 5–10 × 2.5– 4 cm, ovate, obovate-oblong, acuminate, base acute or cuneate; male flowers in few flowered cymes, yellowish; berries globose, 2–2.5 cm in diameter, yellow, glabrous.

Flowering and fruiting   November–August

Distribution   Indo-Malaya to Australia; nearly throughout India, common in Meghalaya

Propagation   By seeds and vegetative methods

Parts used   Roots and bark

Chemical constituents   The bark contains β-sitosterol, and diospyrin, while wood contains isodiospyrin and a trinaphthaquinine xylopyrin. The roots also contain diospyrin, besides betulin and betulinic acid.

Uses   A decoction of root bark is used to cure joint pain and swelling.

Diospyros pilosula (DC.) Hiem. [Extremely Rare]

Family   Ebenaceae

Indian name   *Dieng-sohsit-lankait* (Khasi).

Description   Small tree, bark blackish, smooth, reddish brown, inside, fibrous, rather hard, somewhat granular, thick; leaves bifarious 1.5 × 0.7–1.5 in., elliptic ovate, oblanceolate or linear, sub-coriaceous, hairy along midrib, often shining underneath, lateral nerves very conspicuous, oblique, base cuneate; petiole 0.15– 0.2 long; flowers solitary, below the leaves, petals shorter than the sepal lobes; fruit 1.5–2 cm in diameter, globose, reddish-brown.

Flowering and fruiting   February–August

Distribution   Burma and India; confined to North-East and Andaman, extremely rare in Meghalaya at lower elevation.

Propagation   By seeds and vegetative method.

Parts used   Bark, leaves, flowers and fruits.

Chemical constituents   The ethyl acetate extract showed anti-stress and anti-ulcerogenic activity similar to ginseng.

Uses   The bark is an astringent, acrid, cooling, anti-inflammatory, constipating, depurative and febrifuge, and is useful in burning sensation, inflammations, diarrhoea, dysentery, skin diseases, pruritus, haemarrhages, diabetes, fever and vaginal disorder. All parts of the plant are medicinal used for stomach disorder, piles, kidney stone, etc.

Disporum cantoniense (Lour) Merr.

Family   Liliaceae

Description   A stout herb dichotomously branching above, lower stem stouter and twice larger than above. Sepals succate at the bases. Flowers in umbels.

Distribution   Sub-tropics and temperate regions, frequently in Khasi hills (Meghalaya)

Propagation   By seeds

Parts used   Stem

Uses   Extract of stem is useful for rheumatism and back pain.

Drymaria diandra Bl.

Family   Caryophyllaceae

Indian name   *Pipita-batong* (Naga).

Description   Sub-erect dichotomously branching herbs; leaves opposite, stipule small, often fugacious; flowers axillary or terminal, solitary or cymes; sepals 5, herbaceous; petals 5, stamens 5, rarely fewer, slightly perigynous; ovary 1-celled; style trifid, ovules few or numerous; capsule trivalvate; seeds globose, reniform or compressed, hilum lateral, embryo curved.

Distribution   Entire North-Eastern India.

Parts used   Whole plant.

Uses   Plant juice is useful for blisters in the mouth and tongue.

Dysoxylum binectariferum Hk.f.et. Bedd.

Family   Meliaceae

Indian names   *Kadugadda, Kadugandha* (Kannada); *Akuni, Akunivakil* (Tamil); *Banderdima, Bandardima* (Assamese); *Bol-narang, Masispel, Bandardima* (Garo).

Description   Large tree up to 30 m high; crown oval; bark grey or nearly smooth, vertically fissured in old tree; leaves 30–75 cm long; leaflets 7–17 × 3–8 cm, obliquely ovate, oblong, shortly acuminate, cuneate at base, entire, obscurely dentate, glabrous; panicles 20–30 cm long, flowers 3–5 cm long, pale white; calyx cup-shaped, 5-partite; petals valvate; staminal tube nearly disc, exceeding the ovary; capsule 2.5–3 m, obovoid, 4-celled, pale-yellow turning orange; seeds with large, yellow hileum and white aril.

Flowering and fruiting   March–January

Distribution   Throughout India; common in Meghalaya at lower elevations

Propagation   By seeds and vegetative methods

Parts used   Stem, bark and fruits

Chemical constituents   The stem bark contains an alkaloid, rohitukine possessing anti-inflammatory and immunomodulatory property. The fruits contain a tetranotriterpene, dysobinin which showed mild anti-inflammatory activity.

Uses   Wood is used to cure rheumatism; Wood oil is used in treating ear and eye disorders.

E

Eclipta alba L.

Family   Asteraceae

Indian names   Trailing eclipta (English); *Bhrangarajah, Tekarajah* (Sanskrit); *Ghamgra, Mocakand, Babri* (Hindi); *Gatagada soppu* (Kannada); *Kanjunni, Kayyonni* (Malayalam); *Karishalanganni* (Tamil); *Galagara* (Telegu); *Kehoraji* (Assamese).

Description   Perennial herb, erect, 30–40 cm high, stems green or purple, bristly, thickened at the nodes; leaves opposite, subsessile, lanceolate, oblong, denticulate, hirsute on both sides; flowers white in axillary or terminal head; female radiated, bisexual, in the centre; achene 3-angled, slightly flattened.

Distribution   Throughout India, at all elevations in waste lands and on roadsides

Propagation   By seeds and vegetative methods

Parts used   Whole plant

Chemical constituents   The roots are very rich in thiopene acetylenes. They contain the dithiopene derivatives, 5-seneocioyl oxymethylene 2-dithiopene and 5 triglyoxymethylene 2-dithiopene in addition to 2,5-thiopene.

Uses   The plant is bitter, acrid, thermogenic, alternative, anti-inflammatory, anthelmintic, anodyne, vulnerary, digestive, carminative, diuretic, aphrodisiac and febrifuge, and is useful in gastropathy, anorexia, skin disease, wounds, ulcers, debility, hypertension, fever, jaundice. The seeds are good for increasing sexual vigour.

Eclipta prostrata L.

Family   Asteraceae

Indian names   *Kesaranjan, Kesharaj* (Sanskrit); *Bhangra, Bharangraj* (Hindi); *Bhringaraja, Kaeshavardhana* (Kannada); *Karishanganni, Kayyonni* (Malayalam); *Karishalanganni* (Tamil); *Guntagalijeru, Guntakalaagara* (Telugu).

Description   Small herbs; bracts in one or two series; leaves alternate, simple, exstipulate; inflorescence capitulum, surrounded by involucral bracts; florets of the capitulum sessile, bisexual, sepals modified into pappus. Petals 5, tubular, stamens-5, syngenecious, epipetalous, carpels-2, ovary inferior, 1-celled with one ascending, anatropous ovule borne on basal placenta stigma bifid; fruit- cypsela.

Distribution   Wet places in the fields in Korea, Japan and China, throughout India

Propagation   By seeds

Parts used   Whole plant

Chemical constituents   Ecliptine, wedelactones, thiopene-derivatives

Uses   Antidote for snake venom, antihepatotoxic, anti-inflammatory; used in eruption leucorrhoea, enterohaemorrhage. The juice of leaves is used for drawing factor.

Elaeocarpus sphaericus (Gaertn.) K. Schum. [Vulnerable]

Family   Elaeocarpaceae

Indian names   Utrasum bead tree (English); *Rudraksah* (Sanskrit); *Rudraki*, *Rudrake* (Hindi); *Hakkarike* (Kannada); *Rudraksham* (Malayalam); *Ruttiracam* (Tamil); *Hastikaraka* (Telugu); *Udrok* (Garo).

Description   A medium sized evergreen tree with a spreading handsome crown, up to 20 m high, often buttressed at base; bark greyish brown with light reticulation and white blotches; leaves simple, 8–15 × 2.5–5 cm, oblong lanceolate or oblanceolate, acute or acuminate, narrowed, sub acute at base, minutely crenate, serrate, glabrous or nearly so; flowers usually hermophrodite, rarely polygamous, in axillary racemes; sepals 5, distinct; petals 5, usually laciniate at the apex, rarely entire, springing from the outside of a cushion-shaped, often 5-lobed torus; stamens usually indefinite, never less than 10, arising from the inside of the torus and more or less aggregated with the glands of the torus; anthers innate, linear, opening by terminal pores; ovary sessile, 2–5 celled, ovoid; style columnar. Drupe with a single bony stone, which is 3–5, or by abortion 1-celled; seeds pendulous, albumen fleshy, cotyledons flat.

Flowering and fruiting   May–December

Distribution   About 50 species, most abundant in the hotter parts of India and the Indian archipelago. A few species are found in some of the South Sea Islands, New Zealand and Australia.

Propagation   By seed and vegetative method

Parts used   Fruits and seed kernels

Uses   The fruits are sedative and appetiser; they are useful in cough, bronchitis, epileptic fits, brain disorders and vitiated conditions of kapha and vata. Seed (kernel) is sweet, cooling, emollient, cerebral sedative. It is useful in epileptic fits, melancholia, mental disorders, convulsions, hypertension, bronchitis and fever.

Elephantopus scaber L.

Family   Asteraceae

Indian names   Prickly leaves elephant's foot (English); *Hastipadi, Gojiliva* (Sanskrit); *Gobhi, Gojivha* (Hindi); *Hakkarike* (Kannada); *Aanayadiyan* (Malayalam); *Yanaichuvadi* (Tamil); *Hastikaraka* (Telugu); *Gobhi* (Garo).

Description   A stiff rigid herb, creeping in habit, dichotomously branched; radical leaves, oblanceolate, obovate, cuneate; amplexicaul, alternate, sessile, 1–3 × 0.5–1 in. heads homogamous; 2–5 flowered in dense cluster, each cluster, being supported by usually 3, cordate, rigid, ovate, foliaceous bracts, which are dry, flat or conduplicate; corolla purplish, limb 4–5 toothed; style arms subulate; fruit truncate, 15 in. long, brown ribbed, hairy pappus of 5 bristles, dialated at base.

Flowering and fruiting   Throughout the year

Distribution   Widespread in India, especially in dry localities

Propagation   By seeds and vegetative method

Parts used   Roots, leaves and flowers

Chemical constituents   The plant contains hydroxylated germacrianolides, molephantin and molephantinin having cytotoxic and antitumour properties.

Uses   The plant is bitter, acrid, mucilagenous, cardiac tonic, astringent, mucilagenous and febrifuge. Decoction of leaf and root is given in urinary trouble, dysurea and other urethral discharges or complaints. It is also administered in diarrhoea, dysentery and swelling or pains in the stomach. The root is given to arrest vomiting and used in tooth ache. Leaves are applied to ulcers and eczema.

Elsholtzia blanda Benth.

Family   Lamiaceae

Indian names   *Tesak-moletong* (Naga); *Ban tulsi* (Assamese); *Panet* (Garo).

Description   Under shrubs or shrubs up to 2.5 cm high; branches angular, greenish, aromatic; leaves elliptic, lanceolate, acuminate, base narrowed, crenate, serrate, gland punctate, puberulous above; spikes close, erect, panicled, up to 30 cm long, axillary and terminal, flowers greenish-white, 0.2–0.25 cm long; calyx urceolate, glandular, pubescent, short, mouth contracted; corolla tube short, sparingly pubescent.

Flowering and fruiting   September–March

Distribution   Indo-Malaya; confined to Himalayas; very common in Meghalaya at higher elevations above 1200 mm, particularly in pine forests.

Propagation   By seeds and vegetative methods

Parts used   Whole plant

Chemical constituents   The plant contains an essential oil consisting of ketones.

Uses   The is useful for influenza, headache, sore throat, measles, ostealgia and furunculosis. The paste of the whole plant is applied on cuts and wounds. Paste of the flower is applied on rashes and skin diseases. The leaves are used as mosquito repellent.

Elsholtzia fruticosa (D. Don.) Rehder

Family   Lamiaceae

Description   Shrub, bark fibrous; branchlets obscurely quadrangular; leaves elliptic, lanceolate, acuminate, serrate or crenate, thinly coriaceous, pubescent, gland dotted beneath; spikes long, flowers minute white, pubescent.

Flowering and fruiting   November–February

Distribution   Khashi hills in Meghalaya, 1500–3000 m

Propagation   By seeds and vegetative methods

Parts used   Flowering top

Uses   It is used as carminative for diarrhoea, fever, headache and oedema.

Embelia ribes Burn F. [Critically Endemic]

Family   Myrsinaceae

Indian names   Embelia (English); *Vidangah, Vellah* (Sanskrit); *Vayyidamg, Bhabhiramg* (Hindi); *Vayuvidangu* (Kannada); *Vizhal* (Malayalam); *Vayu-vilamga* (Tamil); *Vayuvidangamu* (Telegu); *Balungra* (Garo).

Description   A large scandent shrub with long, slender stem, up to 10 m high; flexible branches and bark studded with lenticels; stem warty, branches densely lenticellate; leaves simple, alternate, elliptic lanceolate, short and obtusely acuminate, entire, coriaceous, glabrous on both sides, shiny above, slivery beneath, gland dotted, glandular pits near the midrib beneath.

Flowers white or greenish white in terminal or axillary, lax panicled racemes; fruits globular, dull red to nearly black berries, longitudinally striated with short, slender, persistent pedicels; seeds single globose, hollowed at the base, white spotted, testa, membranous, albumen pitted.

Flowering and fruiting   January–April

Distribution   Throughout India, in areas up to 1500 m elevation in hilly regions, Indo-Malaya, common in Meghalaya as undergrowth in evergreen forests.

Propagation   By seeds and vegetative method

Parts used   Roots, leaves and fruits

Chemical constituents   Aflatoxin B1, embellin, 2,5- isobutylamine salts

Uses   The roots are useful in vitated conditions of vata, odontalgia and dyspepsia. The leaves are astringent, thermogenic, demulcent are useful in pruritus, skin diseases and leprosy. The fruits are acrid, thermogenic, anthelmintic, brain tonic, digestive, stomachic, diuretic, stimulant, laxative and febrifuge. The tonic extracted from the leaves is useful in treating skin diseases, nervous disorders, cardiac disorders, constipation, tumour, asthma, etc.

Emblica officinalis Gaertn.

Family   Euphorbiaceae

Indian names   Emli (English); *Amlika* (Sanskrit); *Amla* (Hindi); *Nellikka, Amlakam* (Malayalam); *Nellikanni* (Tamil); *Ambara sogin* (Garo); *Chukna, Amlaki* (Assamese), *Amlaki* (Bengali); *Amran* (Gujrati); *Amla* (Marathi); *Aora* (Oriya).

Description   A small deciduous tree with smooth, greenish grey, exfoliating bark; leaves feathery with small narrowly oblong, pinnately arranged leaflets; fruits depressed, globose, 0.5–1 cm in diameter, fleshy and obscurely 6-lobed, containing 6 trigonous seeds.

Flowering and fruiting   July–February

Distribution   Entire North-East Region in wasteland and moist grassland

Propagation   By seeds and vegetative method

Parts used   Root, bark, flowers and fruits

Chemical constituents   Aflatoxin B1, Embellin, 2,5-isobutylamine salts

Uses   The root bark is astringent; flowers are cooling and act as a refrigerant. Seeds are used in the treatment of asthma, and bronchitis. Fruit is an astringent, a natural source of vitamin C and used in treatment of human scurvy. It is acrid, cooling, refrigerant, diuretic and laxative. Dried fruit is useful in haemorrhage, diarrhoea and dysentery. It is used with iron in anaemia, jaundice, cough and dyspepsia. It is also used as laxative and in headache, constipation, piles, enlarged liver and ascites. The exudates of fruit are useful in inflammation of the eye.

Engelhardtia spicata Leschn. ex. Bl.

Family   Juglandaceae

Indian names   *Silapoma, Gadh mauha* (Hindi); *Dieng lamba* (Khasi); *Heijuga-manbi* (Manipuri); *Hnum* (Mizoram); *Bol-sne* (Garo).

Description   Lofty trees up to 25–35 cm high; crown ovoid, umbrella-shaped; bark grey or greyish–brown, vertically split and scaly; leaves 13–30 cm long; leaftets 4–15 × 1.5–5 cm, oblong, elliptic, oblanceolate, acute or shortly acuminate, base rounded, unequal, ultimately glabrate beneath; male spikes slender, 5–20 cm long; female spikes pendulous; fruits pale brown when ripe, median wing; nut bristly hairy.

Flowering and fruiting   November–June

Distribution   Confined to North-East Region, very common throughout Meghalaya, often gregarious forming tree canopy.

Propagation   By seeds

Parts used   Fruits

Uses   Fruits have pesticidal property.

Entada pursaetha DC.

Family   Mimosaceae

Indian names   Matchbox bean (English); *Shuri* (Garo); *Gitaphal, Saruni* (Assamese); *Naap* (Manipuri).

Description   A large woody climber, stem angled and twisted; leaves compound and cluster of 2-pinnate, tendril bifid, stipules bristle-like, pinnae 2 pairs, leaflets 2–4 pairs obovate or oblong elliptic, emarginate; flowers in huge, spikes or panicles, small, pale yellow; pod jointed, seeds flat, containing black, glossy seeds; pods 1.5 m in length, making them the largest-growing bean in the world.

Flowering and fruiting   March–May (next year)

Distribution   Native to Australia and tropical regions of Asia and Africa. Most common climbers in the territory at the outskirts of forests in hills and hill cuttings up to 1000 m in Meghalaya.

Propagation   By seeds and vegetative method

Parts used   Seeds, leaves and pods

Chemical constituents   Matchbox bean contains significant amount of saponins

Uses   Australia aborigins use the seeds to treat female sterility and indigestion, and as a painkiller. Seeds are irritant, emetic and a food poison. Seeds contain a turbid oil and saponin; glucoside and alkaloid. Leaf decoction is used in cuts and wounds, dropsy, epilepsy and liver complications. Juice of the pod is used for curing dandruff.

Erigeron bellidioides Benth.

Family   Asteraceae

Indian names   Daisy fleabane, Hairy fleabane, Erigeron (English); *Min-gum* (Arunachali).

Description   Perennial plant, glabrous, stem very slender, grooved, sparingly branched; radical leaves, lanceolate acutely serrate, cauline, sessile, oblong or linear, entire or crenate; heads 3 cm in diameter, long-peduncled, ligules thrice as long as the pappus; achenes sub-silky.

Distribution   Central to western Himalayas, from Kashmir to Nepal. It is common in North-Eastern region.

Propagation   By seeds and vegetative method

Parts used   Whole plant

Chemical constituents   It contains volatile oil, flavonoids, terpenes, acids and tannins.

Uses   The plant is generally used for blood purification.

Eryngium foetidum L.

Family   Apiaceae

Indian names   *Kaadu kothambari* (Kannada); *Awaphadigo* (Mizo); *Jongali-memedo, Podomosolla* (Assamese).

Description   Perennial herb, erect, aromatic, glabrous, diffuse with spathulate, spinous, marginal leaves; flowers in heads; sepals teeth ridged, acute; petals white, fruit ellipsoid.

Distribution   Entire North-Eastern region in moist areas and common in Meghalaya.

Propagation   By seeds

Parts used   Whole plant

Uses   Leaves are useful in reducing fever and also used as tonic. The root is used for stomach disorders.

Erythrina stricta Roxb.

Family   Fabaceae

Indian names   *Mura, Paribhadra* (Sanskrit); *Dhol-dak* (Hindi); *Hemmuruku mara, Muruku mara* (Kannada); *Murukku, Mullumurukku* (Malayalam); *Murukkai* (Tamil); *Mullumoduga, Baarjapu* (Telugu); *Dieng-songkhar* (Khasi); *Bol-madal* (Garo).

Description   Large tree, up to 4 m high; crown oval, branches horizontally spreading; bark yellowish grey, vertically fissured with conical thorn; leaves up to 50 cm long; leaflets, often broader than long, rhomboid-orbicular or ovate, lateral ones oblique, acuminate, base broadly cuneate, glaucous beneath; racemes with flowers, deep red, usually one sided; pods long, 2–3 seeded; seeds black.

Flowering and fruiting   February–June

Distribution   Burma, South and North-East India, fairly common in lower elevations in deciduous forests in Meghalaya.

Propagation   By seeds and vegetative method

Parts used   Leaves, stem and bark

Chemical constituents   The leaves and stem show the presence of an alkaloid erythrinaline. The seeds yield the alkaloid hypaphorine and a saponin called migarrhin.

Uses   The bark and leaves are sedative, carminative, digestive, anthelmintic, expectorant and diuretic; it is useful for relief of insomnia and anxiety. Crushed fresh leaves are used externally as poultice in haemarrhoids and metroptosis. Leaf is anthelmintic, diuretic, stomachic, used in rheumatism, itching, burning, fever, fainting, asthma, leprosy and epilepsy. Powdered leaves are topically applied for wound and ulcers. The stem bark is used against rheumatism in the form of a decoction.

Erythroxylum kunthianum Wall. ex kung [Endemic]

Family   Erythroxylaceae

Indian names   *Dieng-painkhar* (Khasi); *Dieng-sugsi* (Jaintia).

Description   Bushy shrubs or small trees; bark brown, reddish brown; leaves 3–7 × 1–2.5 cm, obovate, lanceolate, elliptic, subacute or rounded at apex, base gradually tapering to a short petiole; flowers white, sepals 0.4 cm across; fruits 1–1.5 × 0.4–0.5 cm, falcate, erect, red when ripe.

Flowering and fruiting   April–September

Distribution   Indo-Burma; largely confined in North-East India; in Meghalaya this species forms common undergrowth in sacred forests and wet evergreen forests.

Propagation   By seeds and vegetative method

Parts used   Leaves

Chemical constituents   It contains various alkaloids, a volatile oil, flavonoids, Vitamins A and B_2 and minerals.

Uses   The leaves are used in nausea, vomiting and asthma and have been used to speed convalescence.

Euonymus lawsonii Cl and Pr. [Endemic, Rare]

Family   Celastraceae

Indian name   *Soh-dadin* (Khasi)

Description   Small tree or shrubs; bark grey, reticulately fissured, often scandent, lithophytic or epiphytic; leaves elliptic, oblong lanceolate, acuminate, base narrow, cuneate, finely crenate; inflorescence 1, or 2 or 3-chotomous cymes; flowers bisexual, tetramerous or pentamerous; sepals reflexed in fruit; disc fleshy, angled or lobed; stamens on or outside the disc; ovary partly immersed in the disc; ovules 2 in each ovary; capsules 3–5 angled, 6-seeded.

Flowering and fruiting   February–December

Distribution   Endemic to Meghalaya; frequent in shady slopes and along the forest margins.

Propagation   By seeds and vegetative method

Parts used   Stem bark, root bark and seeds

Chemical constituents   Bark contains cardenolides similar to digitoxin, asparagines, sterols and tannins.

Uses   Bark is considered as a gall bladder remedy with laxative and diuretic properties. Bark in syphilis, indigestion, liver disorder; seed oils for lice.

Eupatorium adenophorum Spreng.

Family   Asteraceae

Indian names   *Bakura* (Hindi); *Bat-iong* (Khasi); *Phlang-burma* (Jaintia).

Description   Undershrubs or herbs up to 2 m high, much branched; leaves ovate, triangular, rhomboid, acute or acuminate, base cuneate or truncate, pubescent or glandular along nerves; heads 0.4–0.6 cm long in dense corymbs, in leafy panicles; involucral bracts oblong or lanceolate; florets white; achene 5-angled, black; pappus hairs twice as long as achenes.

Flowering and fruiting   March–May

Distribution   Native of Central America; abundant in Meghalaya at higher elevations, particularly under pine forests.

Propagation   By seeds and vegetative method

Parts used   Aerial parts

Chemical constituents   It contains sesquiterpenes, lactones, polysaccharides, flavonoids, diterpenes, sterols and volatile oil.

Uses   The plant stimulates resistance to viral and bacterial infections, and reduces fever by encouraging sweating.

Eupatorium odoratum L.

Family   Asteraceae

Indian names   Germany bone (English); *Matmati* (Assamese); *Dura-sumak* (Garo); *Tlangsam* (Mizoram).

Description   Erect or straggling undershrub or shrub up to 5 m high; leaves 5–12 × 2–6 cm, ovate, lanceolate or triangular, acute or acuminate, base cuneate, irregularly dentate, serrate or entire, pubescent, particularly along nerve beneath; heads 0.7–1.5 cm long in terminal and upper axillary corymbs; involucral bracts ovate, lanceolate; achenes angular; Calyx 0.4 cm long, blackish; pappus hairs stiff, white.

Flowering and fruiting   November–March

Distribution   Native of Central America, throughout India, very common in Meghalaya at lower elevations in forest-cleared areas, secondary forests and in plantations, usually gregarious in habit.

Propagation   By seeds

Parts used   Leaves

Chemical constituents   Germany bone contains alkaloids, flavanoids, sesquiterpenes, lactones, volatile oils and polysaccharides P-cymene is antiviral while eupatoriopicrin has anti-tumour properties which inhibit cell enlargement.

Uses   Leaves crushed and used for wounds and leech bite; also a noxious weed in forestry operation.

Eupatorium cannabinum L.

Family   Asteraceae

Indian names   Hemp agrimony (English); *Dura-samak* (Garo).

Description   Perennial, growing to a height of 1.5 m, stem red, downy leaves usually opposite, sometimes alternate, lanceolate, ovate, coarsely serrate, acuminates acabrid above pubescent along nerve beneath; heads 0.4–0.6 cm long, narrow, in dense terminal corymbs; involucral bracts oblong, lanceolate with scarious margins; corolla regular, tubular, 5 lobed, stamens syngenesious; minute pappus much longer; achene black, .

Flowering and fruiting   August–December

Distribution   Entire North-East Region in moist localities ascending up to 1800 m height.

Propagation   By seeds

Parts used   Roots and leaves

Chemical constituents   Germany bone contains alkaloids, flavanoides, sesquiterpenes, lactones, volatile oils and polysaccharides. p-cymene is antiviral while eupatoriopicrin has anti-tumour properties which inhibit cell enlargement.

Uses   Root are useful for fever, cold, flu and other acute viral conditions and also stimulates the removal of waste products via the kidneys. Leaves are kept with children suffering from small pox probably to reduce virulence of infection. The plant is diuretic, antiscorbutic, cathartic and emetic. It is used for fomenting sores and ulcers.

Euphorbia antiquorum L.

Family   Euphorbiaceae

Indian names   Triangular spruge (English); *Vajrakantakah, Tridhara* (Sanskrit); *Tidhara* (Hindi); *Mudumula, Katakkalli* (Kannada); *Chaturakkalli* (Malayalam); *Saturakkalli, Tiruvargalli* (Tamil); *Bommajemudu, Pedda jemmudu* (Telugu); *Hiju-arang* (Arunachali).

Description   Large cactus like shrub or tree with whorled fleshy branches, branchlets thick and broad, 3–5 winged, having sharp stipular spines; leaves few, deciduous; involucres ternate with short peduncled cymes hemispheric, yellow, the low lateral ones on thick pedicels, the central one sessile; female, bracteate many fimbriate.

Flowering and fruiting   January–March

Distribution   Cultivated throughout North-East Region for fencing.

Propagation   By vegetative method.

Parts used   Roots and juice

Chemical constituents   The stem contains friedelin, diacetate, friedelanacetate, friedelanol, lupeol cinnamate, mortenone, lupenone, taraxerone, β-amyrin, euphol, β-taraxerol and Ψ-taraxastane-β-antiquorin, etc.

Uses   The roots are bitter, acrid, thermogenic, anodyne, purgative, emetic, stomachic and digestive and useful in vitiated conditions of flatulence, constipition, dyspepsia, wounds and ulcers. The juice of the plant is used as an emollic in rheumatism. It is also used for fish poisoning.

Euphorbia hirta L.

Family   Euphorbiaceae

Indian names   *Nagarjuni, Pusiton* (Sanskrit); *Lal dudhi* (Hindi); *Nelapalai* (Malayalam); *Amampatchai arisi* (Tamil); *Hiju-arang* (Arunachali); *Barakeru* (Bengali); *Dudheli* (Gujrati); *Moti dudhi* (Marathi).

Description   An annual hispid herb; 15–30 cm high, long yellowish, crisp hairs, branches ascending, quadrangular; leaves opposite, elliptic oblong, acute, dentate or serrulate; base obliquely cordate, shortly petiolate; inflorescence of many male florets surrounding a solitary female floret enclosed within involucres. Involucres; in axillary and terminal dense flowered, the sessile or stalked cymes.

Distribution   Native to India and Australia, throughout North-East Region, common in Meghalaya in wastelands and grasslands in tropical and sub-tropical regions.

Propagation   By seeds and vegetative method.

Parts used   Whole plant

Chemical constituents   It contains flavonoids, terpenoids, alkanes, phenolic acids, shikimic acid and choline.

Uses   The plant is useful for a specific treatment bronchial asthma, causing spurge which relaxes the bronchial tubes and eases breathing, mildly sedative and expectorant. It is also taken for bronchitis and other respiratory tract conditions. The juice is employed in earache and in ophthalmia and in asthma.

Euphorbia ligularia Roxb.

Family   Euphorbiaceae

Indian names   Common milk hedge (English); *Snuhi* (Sanskrit); *S-hund*, *Thundar* (Hindi); *El-kalli* (Kannada); *llakkalli* (Malayalam); *Ilaikkalli* (Tamil); *Akujemudu* (Telugu); *Hiju-arang* (Arunachali).

Description   Small tree or shrub; leaves crowded at the end of the branches; deciduous, obovate, spathulate or oblong, base tapering, cymes solitary or 2, central flowers male; lateral ones bisexual stamens many; be inbundles.

Flowering and fruiting   March–May

Propagation   By seeds and vegetative method

Distribution   Entire North-East Region, particularly in Meghalaya

Parts used   Stem, root and latex

Chemical constituents   Anti-coagulant activity. The triterpenoids, euphol 24-methylenecyloartenol, euphorbol hexacosonate, glutenone, glutenyl acetate, taraxerol, friedelan, have been reported from the plant.

Uses   The juice obtained from latex is purgative and expectorant. Root is antispasmodic. The latex is acrid, rubefacient, purgative and expectorant. It is useful for dermatitis. The juice is applied for ear ache, opthalmia and is used in asthma.

Euphorbia uniflora Roxb.

Family   Euphorbiaceae

Description   Annual plant, sparsely hairy; stem many, prostrate and spreading from the root; leaves opposite, involucres subsolitary axillary, glabrous; style short; seeds smooth.

Distribution   Throughout India and entire North-Eastern region, common in Meghalaya.

Propagation   By seeds and vegetative method

Parts used   Latex

Uses   Latex is very useful in ring worm and skin diseases.

Eurya acuminata DC.

Family   Theaceae

Indian names   *Bakryal* (Hindi); *Dieng-lapyrshit* (Jaintia); *Almaset* (Naga); *Lim bung-pyrchit* (Manipuri).

Description   Trees or large shrubs; bark brown; branches usually horizontal or drooping; leaves 4–9, oblong lanceolate or linear, acuminate, base cuneate, rounded or obtuse, hairy beneath; flowers axillary leaf axils, fascicled, fragrant, yellow, 2–4 mm across; sepals ovate, outer smaller; petals yellow, ovate; stamens yellow, many; ovary pubescent; styles united, capsules up to 5 mm across, crowned by the persistent style.

Flowering and fruiting July–February

Distribution Indo-Malaya extending to Fiji Islands; very common in varied type of forests in Meghalaya.

Propagation By seeds

Parts used Leaves

Uses Juice of young leaves mixed with salt are useful as remedy for cholera, dysentery and stomach disorders.

Eurya japonica Thumb.

Family Theaceae

Indian names *Baunra, Gonte* (Hindi); *Huluni, Hulayane* (Kannada); *Attathuvarei* (Malayalam); *Huluni* (Tamil); *Dieng-pyrshit* (Khasi); *Chhamist* (Garo).

Description Shrubs; bark brown; leaves oblanceolate, oblong elliptic, acute, base cuneate, glabrous, crenate, often narrowly revolute, midrib and lateral nerves depressed above; flowers small, dull white, dioecious, axillary, fascicled or solitary on upper axils; calyx 5 pubescent outside; corolla 5 united at the base; stamens many shorter than the corolla; ovary 3 celled, styles 3, united persistent in fruit; fruit 3–5 mm across, glabrous, purple.

Flowering and fruiting June–December

Propagation By seeds

Parts used   Leaves

Uses   Leaves are used for poulticing skin eruption.

Eusteralis stellata (Lour) Panig.

Family   Lamiaceae

Description   Stem creeping below, multiple branches; leaves broad; bracteoles, and spikes, clavate or liliform, sepals, copular or conical, very variable; fruit is enlarged, petals tube, lobes hirsute, very short; nutlets very small shining.

Distribution   Common in Meghalaya

Propapagation   By seeds

Parts used   Leaves

Uses   The mixture of leaf powder and coconut oil is very useful on rashes and skin diseases.

F

Fagopyrum dibotrys (D.Don) Hara.

Family   Polygonaceae

Indian names   *Ban ogal, Kanjolya* (Hindi); *Jarain* (Khasi); *Tem* (Arunachali); *Ja-rain* (Manipuri).

Description   Undershrubs or herbs up to 2 m high, puberulous, laxly branched; branches spreading; leaves 7–18 × 6–12 cm, ovate–deltoid, acuminate, base cordate, puberulous, panicles of cymes, branched second; flowers 0.25–0.5 cm long, white or pinkish; nuts 0.3–0.5 cm long.

Flowering and fruiting   July–December

Distribution   Indo-China; temperate Himalayas; common in Meghalaya at higher elevations along pine forests and secondary forests and also wastelands.

Propagation   By seeds and vegetative method

Parts used   Whole plant

Chemical constituents   More or less same as *F. esculentum*.

Uses Used for a wide range of circulatory problems; grain is best taken as a tea or tablet, accompanied by vitamin C or lemon juice to aid absorption. It is useful for arthritis and is vermicidal. The grains are used in diarrhoea, and for abdominal obstructions.

Fagopyrum esculentum Moench.

Family Polygonaceae

Indian names Buckwheat (English); *Kotu, Phaphra* (Hindi); *Kaadu godhi* (Kannada); *Kotul doron, Lappasak* (Assamese).

Description Annual herbs; glabrous, leaves triangular-cordate; acute; long petioled at the base, gradually shortening upwards and then at the top becoming sessile; flowers white or pinkish, in axillary and terminal peduncled sub-capitate many-fid cymes; nut ovate, angles acute.

Distribution Cultivated in Khasi hills; throughout the Himalayas and western Tibet at elevations of 2000 to 12,000 feet, and in Nilgiri hills.

Propagation By seeds and vegetative method

Parts used Root and seeds

Chemical constituents Buckwheat contains bioflavonoids, especially rutin, which is an antioxidant. It strengthens the inner lining of blood vessels.

Uses The root is useful for lung disorders, rheumatism, typhoid, urinary complications, hypertension, haemorrhage and haemophilia. The flour of seeds is used in sprain as an emollient.

Flacourtia jangomas (Lour.) Raeusch.

Family   Flacourtiaceae

Indian names   *Pracinamalaka, Sruvavrksah* (Sanskrit); *Paniyamalak, Talisapatri* (Hindi); *Shamper, Chanchali mara* (Kannada); *Vayyamkaitha* (Malayalam); *Mullumukanchi, Talicam* (Tamil); *Mullumaana, Thaaleesapathramu* (Telugu); *Dorichik* (Garo); *Dieng sohmluh* (Khasi).

Description   Middle sized trees up to 15 m high; crown oval, young shoots often reddish, dense bark brown, nearly smooth, armed with compound spines at basal portion; leaves ovate-lanceolate, acuminate, broadly cuneate at base, crenate-serrate, glabrous; flowers greenish-yellow, or white, fragrant in short simple or branched tomentose racemes; calyx orbicular, ciliate, stamens in male flowers filiform; female flowers spreading, stigma 4–6; fruits globose, dark purple drupes with juicy pulp and hard endocarp; seeds 8–16.

Flowering and fruiting   March–October

Distribution   Entire North-East Region common in Meghalaya, usually along the river bank or water courses and forest margins.

Propagation   By seeds and vegetative method.

Parts used   Roots, leaves and fruits.

Chemical constituents   Arabinose, mannose, ester, steroid, ramontoside, etc.

Uses   The roots are sweet, refrigerant, depurative, alexipharmic and diuretic. Leaves are used in pruritis and scabies; fruits are useful in strangury, jaundice, gastropathy and splenomegaly.

Ficus bengalensis L.

Family   Moraceae

Indian names   Banyan (English); *Nyagrdhah, Vatah* (Sanskrit); *Bat, Bargad* (Hindi); *Ala* (Kannada); *Alamaram* (Tamil); *Paddammarri* (Telegu); *Vadavriksham* (Malayalam); *Gonok* (Garo); *Ahat gas* (Assamese).

Description   Spreading evergreen tree, epiphytic in early life; bark smooth, greyish; young parts softly pubescent. Leaves 4–8 × 2–5 in., ovate, elliptic, entire, obtuse or rounded, coriaceous, green and glossy above, pubescent beneath; lateral nerves 4–7 on either half, looped within the margin, prominent beneath; base 3–7 nerved upper pair stout, usually rounded or sub-cordate; petiole, stipulate, deltoid, acute, coriaceous. Male and female flowers are in the same receptacle; perianth segments 4, stamens 1; receptacle 0.5–0.75 in., across, globose, pubescent, sessile, axillary, supported by 3 rounded coriaceous spreading bracts, scarlet when ripe.

Flowering and fruiting   September–July

Distribution   Throughout India; both planted and wild in Meghalaya

Propagation   By seeds and vegetative method

Parts used   Whole plant

Chemical constituents   Bengalenoside and the flavonoid glycosides, leucocyanidin 3-0-α-D-galactosyl cellobioside 3,5-dimethyl ether and leucopelargonidin rhamnoside, 5,7-dimethyl ether.

Uses Whole plant is astringent, acrid, sweet, refrigerant, anodyne, vulnerary, depurative, anti-inflammatory, ophthalmic, styptic, antiarthritic, diaphoretic, antiemetic and tonic. The leaves are good for ulcers, leprosy, allergic conditions of skin, burning sensation and abscesses; buds are useful in diarrhoea and dysentery. The latex is useful in neuralgia, rheumatism, lumbago, bruises, nasitis, ulitis, haemarrhoids, gonorrhoea, and cracks of the sole and skin diseases.

Ficus fistulosa Reinw. Ex. Bl.

Family Moraceae

Indian names *Ka-lapang* (Khasi); *Towak* (Garo); *Maudimaru* (Assamese).

Description Small tree up to 10 m high, crown spreading; bark brown, smooth with annular horizontal ridges at intervals greyish-brown, smooth; inside light-brown; leaves alternate or opposite, oblanceolate, obovate-oblong or elliptic, stipules scarious, ovate-lanceolate; male flowers few; sepals concave, imbricate; stamens 1; filaments long, thick; female flowers perianth similar to gall flowers, style lateral; stigma cylindric. Receptacles dimorphous, axillary and short peduncled, in some undivided in others in dense bunches on stem and branches and long peduncles containing only fertile female flowers.

Flowerin and fruiting March–July

Distribution Burma and North-East Region, it is frequent in Meghalaya in tropical deciduous forests.

Propagation By seeds and vegetative method

Parts used   Whole plant

Uses   Bark, fruit and seeds are emetic and purgative and used in anaemia, jaundice, fever and ulcer.

Ficus heterophylla L. f.

Family   Moraceae

Indian names   *Trayantika* (Sanskrit); *Adavibende, Bili athi* (Kannada); *Valliteregam* (Malayalam); *Kotiyatti, Avanam* (Tamil); *Buroni, Kuvvu juvvi* (Telugu); *Purai* (Garo); *Buna* (Mizo); *Pui* (Arunachali).

Description   Trailling shrubs; leaves ovate in outline, 1–5 lobed, base rounded, lobes elliptic-lanceolate, acuminate, dentate, serrate, stipulate, glabrous; male flowers perianth gamophyllous, monandrous; female flowers-perianth gamophyllous; receptacle axillary, solitary, pyriform or globose, hispid when young, orange yellowish and smooth when ripe; achene subglobose, minutely tubercled viscid, basal bracts midway on peduncles.

Flowering and fruiting   February–July

Distribution   Warmer parts of Indo-Malaya, frequent in Meghalaya, on river banks of most of the localities.

Propagation   By seeds and vegetative method

Parts used   Whole plant

Uses   The fruit sugary substance present in the fruit, i.e., within the fig has a pronounced but gentle laxative effect; syrup of figs is still a remedy for mild constipation. The fruit pulp helps to relieve pain and inflammation, and it has been used to treat tumours, swelling and gum abscesses the fruit often being roasted before application.

Ficus hispida Vahl.

Family   Moraceae

Indian names   *Chitrabhashaja, Dhvankshanamni* (Sanskrit); *Gobla, Katgularia* (Hindi); *Adaviyatti, Kadatti* (Kannada); *Erumanakku, Parakam* (Malayalam); *Ottannalam, Peyatti* (Tamil); *Adaviatti, Boddamatti* (Telugu); *Panthap* (Garo); *Dumoru* (Assamese).

Description   Small tree or shrub with hollow branchlets; bark is greenish grey or brownish, often warty, otherwise smooth, often with horizontal wrinkles; leaves usually opposite, obovate, oblong, elliptic, acute or shortly acuminate, serrate or somewhat dentate, rarely entire, subcoriaceous, hairy and scabrid above, hispid, pubescent beneath; staples ovate lanceolate, pubescent externally, numerous in one set of receptacles, 1–1.5 cm across, ovoid, globose, faintly ribbed and often with scattered bracts, greenish yellow, tomentols, peduncled, axillary, in pairs or in clusters.

Flowering and fruiting   February–December

Distribution   Common throughout Meghalaya

Propagation   By seeds and fruits

Parts used   Bark, leaves and fruits

Uses   Bark, leaves and twigs are elephant fodder; green fruits are good for liver.

Ficus religiosa L.

Family   Moraceae

Indian names   Peepal tree, Sacred fig (English); *Pippalah, Asvatthah* (Sanskrit); *Pippal, Pipli, Pipar* (Hindi); *Aswatha* (Kannada); *Arayal* (Malayalam); *Arasu, Asvatham* (Tamil); *Ravi* (Telugu).

Description   Large tree with a spreading crown, bark pale brown; leaves bright green, the apex produced into a linear, lanceolate tail about half long as the main portion of the blade, ovate-orbicular, long, caudate, acuminate, base truncate, subcordate, entire or undulate, glabrous; stipule 0.5–1 cm long, ovate-acute; receptacles occurring in pairs, axillary, depressed, globose, smooth, purplish when ripe, sessile, basal bracts broad.

Flowering and fruiting   February–July

Distribution   Mostly in India and neighbouring countries; mostly, planted occasionally in Meghalaya at lower elevation.

Propagation   By seeds and vegetative method

Parts used Whole plant

Chemical constituents Arabinose, mannose, glucose, phenolic glucoside ester, flacourtia, steroid, ramontoside, β-sitosterol and its α-D-glucopyranoside.

Uses The bark is astringent, sweet, cooling and aphrodisiac, and an aqueous extract of it shows antibacterial activity against *Staphylococcus aureus* and *Escherichia coli*. Leaves and tender shoots have purgative properties and are also recommended for wounds and skin diseases. Fruits are laxative and digestive; the dried fruits pulverised and taken along with water to cure asthma.

Ficus rumphii Bl.

Family Moraceae

Indian names *Nandivrksa, Parkataki, Plaksa* (Sanskrit); *Gajiun, Gajna* (Hindi); *Bettaarali, Kadarali* (Kannada); *Phrap rakseng* (Garo); *Pakhri bor* (Assamese).

Description   Large deciduous tree, often epiphytic; bark greyish, smooth, inside light reddish-white with irregular streaks of white and faint purple. Leaves 7–15 × 4–8 cm, ovate, oblong, acuminate, base rounded or truncate, sometimes narrowed, cuneate, entire, glabrous, subcoriaceous, upper surface dotted and shining; stipules 1–2.5 cm long; Male flowers few near the mouth of the receptacle, perianth segments spathulate, stamen 1, female flowers, perianth segments 3, lanceolate, ovary ovoid, smooth, style elongated, stigma clavate; receptacle sessile in axillary pairs, globose, black, when ripe.

Flowering and fruiting   September–July

Distribution   Throughout India, frequent in Meghalaya in lower elevations

Propagation   By seeds and vegetative method

Parts used   Bark

Uses   Lac insects are reared on it; leaves and twigs are used as cattle and elephant fodder. Bark is emetic; it is useful for snake bite and asthma.

Flemingia macrophylla (Willd.) Prain ex. Merr.

Family   Fabaceae

Indian names   *Salaparni* (Sanskrit); *Kandran regu, Antinta* (Telugu); *Charariaw* (Manipuri); *Samnaskhat* (Assamese); *Samnaskhat* (Garo).

Description   Shrubs up to 5 m high, sparingly branched, triquetrous, hairy, leaflets 4–18 × 3–11 cm, oblong to ovate, acute or acuminate, base rounded, glabrescent above; cymes 3–4-flowered; flowers in racemes up to 20 cm long, purple, calyx velvety; pods 0.7–1.2 × 0.6–0.7 cm, oblong, tomentose, 2-seeded.

Flowering and fruiting   February–October

Distribution   Nearly throughout India, frequent in Meghalaya in secondary forests and forest margins and also in ravines in hills.

Propagation   By seeds

Parts used   Whole plant

Uses   Root paste is useful for muscular pain, blindness, cholera, dysentery, fever, fistula, spleen complications, ulcers and swelling.

Fragaria indica Andr.

Family   Rosaceae

Indian names   *Strawberry, Bhumla* (Hindi); *Bukger* (Arunachali); *Et-thi* (Garo).

Description   Low-growing perennial herb spreading by runners; leaves compound, digitately trifoliate or three-lobed; flowers yellow, solitary on long axillary peduncles; sepals many on a convex receptacle; fruits numerous, minute and black.

Distribution   Very common in North-Eastern Region

Propagation   By vegetative method

Parts used   Fruit, leaves and flowers

Chemical constituents   The leaves contain flavonoids, tannins and a volatile oil. Fruit contains methyl salicylate and borneol.

Uses   Fruit is useful for ulcer, leaves for cuts, bruises and flowers for eye diseases.

G

Garcinia cowa Roxb. Ex DC.

Family   Clusiaceae

Indian names   *Dvipakharjuri, Paravata* (Sanskrit); *Chengkek* (Mizoram); *Kau* (Manipuri); *Kauthekera, Kangach* (Assamese); *Rengran* (Garo).

Description   Small or middle sized tree up to 20 m high, bark brown, greyish outside, nearly smooth. Leaves acute, acuminate, base cuneate, young leaves reddish; male flowers about 1 cm across, yellow, calyx ovate, corolla double the length of calyx; stamens on fleshy receptacle; female flowers slightly larger than the males; ovary subglobose, stigma lobed, staminodes in a ring, fruits are numerous, minute black. Flowers and fruits throughout the year.

Flowering and fruiting   March–September

Distribution   Indo-Malaya; North-East India; frequent in evergreen and mixed forests at low elevations in Meghalaya.

Propagation   By seeds and vegetative method.

Parts used   Fruits

Chemical constituents   The fruit contains a polyisoprenylated phenolic pigment, garcinol and its isomer isogarcinol, along with hydroxycitric acid, cyanidin 3-glucoside and cyanidin 3-sambubioside.

Uses   Plant cathartic; leaves and fruits acidic and used to cure dysentery, stomach trouble.

Garcinia morella (Gaertn.) Desr.

Family   Clusiaceae

Indian names   *Kankustha, Tamala* (Sanskrit); *Tamal* (Hindi); *Devana huli, Pon puli, Arsina gurgi* (Kannada); *Chigiri, Karukam puli, Pinnar puli* (Malayalam); *Tamala* (Marathi); *Makki, Solai puli* (Tamil); *Reval chinni* (Telugu); *Thizru* (Garo).

Description   Tree up to 15 m tall, with spreading branches and dense crown; wood yellow, mottled hard; bark brownish grey, thin about 5 mm thick, smooth, exuding brilliant, dark yellow, sticky, thick latex. Leaves ovate-lanceolate or broadly elliptic, sub-acute or sub-acuminate, base cuneate, glabrous, coriaceous, margin sub-entire, midrib prominent below, pale beneath; flowers axillary usually in fallen axils, solitary, sessile or sub-sessile; male flowers 0.5–0.8 cm across; calyx orbicular, concave, corolla longer than calyx; stamens on a short androphore; female flowers terminal and axillary, slightly larger than male flowers; berries 3–5 × 1.5–2.5 cm, oblong, flattened, purplish brown when ripe, ending into a short, thick style with a broad imbricate stigma which is about 8 mm across. Seeds 1 or 2 about 3 × 1.5 cm, brown, smooth and shiny.

Flowering and fruiting   December–June

Distribution   Western ghats and North-East India; confined to evergreen forests in Meghalaya.

Propagation   By seeds and vegetative method

Parts used   Fruits

Gaultheria fragrantissima Wallich.

Family   Ericaceae

Indian names   *Carmapatra, Hemantaharita* (Sanskrit); *Gandapuro* (Hindi); *Gandhapoora mara, Haemanth harithaa* (Kannada); *Kolakkai* (Tamil); *Jirhap* (Khasi).

Description   Evergreen shrubs up to 4 m high with brownish bark; leaves 5, 2 × 1–2.5, aromatic due to the presence of methyl salicylate (oil of winter green) when bruised, oblong, lanceolate, elliptic or rhomboid, acute or acuminate, serrate or serrulate, serratures often ending in minute deciduous; tips petiole channelled, often tinged with red; flowers white. Usually untimely; in pubescent axillary, racemes; bracts, broad, ovate or oblong, sub-acute or acute; pedicles usually exceeding the bracts, bracteoles small, opposite towards the top of the pedicel, concave, ovate-acute or obtuse; calyx white, deeply 5-lobed, ovate, lanceolate, acute sometimes sub-acute, ciliate; corolla ovoid, urceolate, teeth very small, reflexed; stamens 10, filaments adnate to the base of the corolla; pilose; anther red when young; ovary deeply 5-grooved, 5-celled, pubescent at the top; base glabrate and glandular; ovules many in each cell, deep pink; capsule 5-celled.

Flowering and fruiting   March–August

Distribution   Indo-Malaya, North-East and South India; common at higher elevations in Shillong, particularly under pine forests.

Propagation   By seeds and vegetative method

Parts used   Leaves

Chemical constituents   The winter green oil is the volatile oil obtained from fresh plants of the species plants also containing benzoic acid, gaultheria, gaultherin, arbutin, and salicylic acid.

Uses   Plant is a stimulant, carminative and used in rheumatism, neuralgia and as an antiseptic wormicidal against hookworm; root used in menstrual disorder. The oil is used in the preparation of products used for killing or repelling mosquitoes, flies and other insects.

Gaultheria nummularioides Don.

Family   Ericaceae

Indian names   *Bhwinla* (Hindi); *Riligkon* (Arunachali).

Description   Erect undershrub with densely hirsute stems; leaves broad, ovate acute, glabrous above, setulose beneath and on the margins, base of the bristles glandular, nerves rather indistinct, 2–4 on either half, arched and forming loops near the margin; base rounded or cordate; petiole minute; flowers pink or white, axillary, solitary; pedicels, densely clothed with small glabrous ovate or oblong, bracts; calyx-teeth lanceolate. Corolla tubular, urceolate, teeth minute, recurved; capsule enclosed by blue black succulent calyx.

Flowering and fruiting   March–October

Distribution   Throughout North-East India and common in Meghalaya.

Propagation   By seeds and vegetative method

Parts used   Leaves

Uses   Leaf is useful for neuralgia of head and face, skin smarting and burning with intense erythema, which becomes worse by cold washing.

Gleichenia linearis Clarke.

Family   Gleicheniaceae

Description   The plant has creeping rhizome which gives a zigzag appearance, repeatedly forked, stipes bearing pinnae in pairs.

Distribution   Tropical and subtropical regions and entire North-Eastern Region.

Propagation   By vegetative method

Parts used   Whole plant

Uses   Rhizome, root and fruit are used as an anthelmintic. The fluid of fronds shows antibacterial properties.

Glochidion lanceolarium (Roxb.) Voight.

Family   Euphorbiaceae

Indian names   *Armlochan* (Assamese); *Lamgi-heikru* (Manipuri); *Bol-chiring* (Garo).

Description   Tree or shrubs; with spreading crown; bark greyish or greyish-brown, closely and reticulately fissured. Leaves 8–20 × 3–9 cm, elliptic-lanceolate or elliptic-oblong, acuminate, base acute, glabrous, pale beneath; male flowers yellow or pale yellow; pedicelled; female flowers green, sessile; capsules 8–12 lobed, depressed globose.

Flowering and fruiting   September–April

Distribution   Throughout India, North-East Region and Khasi hills, Meghalaya.

Propagation   By seeds

Parts Used   Bark

Uses   Bark is useful in stomach ailments.

Gloriosa superba L. [Vulnerable]

Family   Liliaceae

Indian names   Flame lily, Climbing lily (English); *Agnijvala, Agnimukhi* (Sanskrit); *Carrihari, Kalihari* (Hindi); *Aginsikhe, Akkatangaballi* (Kannada); *Kandal, Mendoni* (Malayalam); *Akkinichilam, Anaravam* (Tamil); *Adavi-nabhi, Adaviabhi* (Telugu).

Description   Plant is a herbaceous climber; rootstock is a chain of fleshy arched tuberous, cylindrical structures, 15–25 × 2.5–4 cm, pointed at both ends, 'V' or 'L'- shaped, fleshy, white, and roots fibrous; branchlets hairless; stem 3–6.5 m, terete herbaceous; leaves alternate, rarely opposite or in whorls, egg-shaped lanceolate, stalkless, about 5–15 × 2–4 cm, base heart-shaped, apex modified into tendrils which are highly coiled, margin entire, hairless, papery; lateral nerves parallel, 13–18 cm, sessile or shortly petioled, variable in breadth, many nerved; flowers solitary or subcorymbose towards the ends of the branches from the nearness of the leaves; pedicles 5–13 cm, tip deflexed; perianth 6–10 cm in diameter; segments linear lanceolate, bright red above and golden yellow below. Filaments stout, golden yellow, connective green; capsules ellipsoid-oblong, 3 × 1–2.5 cm, splitting by 3-valves; seeds numerous, globose, dorsally compressed, with warty projections, straw-coloured.

Flowering and fruiting   November–May

Distribution   Throughout tropical India; from the North-West Himalayas to Assam, Burma, Malaya and Ceylon, ascending to 1500 m.

Propagation   By seeds and vegetative method

Chemical constituents   The tubers contain colchicines, methylcolchine, 1,2-dimethylcolchicine, 2,3-dimethylcolchicine, *N*-formyl *N*-diacetyl colchicine and colchicoside. The seeds are considered a rich source of colchicines and gloriosine.

Uses   The tuberous roots are useful in curing inflammations, ulcers, scrofula, bleeding piles, white discharge, indigestion, snake bite and blindness. Seeds are used for relieving rheumatic pain and as a muscle relaxant.

Gmelina arborea Roxb.

Family   Verbenaceae

Indian names   Gamar (English); *Bhadraparni, Gambhari* (Sanskrit); *Gambari, Gambhara* (Hindi); *Bachanige, Kumbalamara* (Kannada); *Cumbulu, Kumbil* (Malayalam); *Perumkumbil, Perungumil* (Tamil); *Adavigummudu* (Telugu); *Dieng-lophiong* (Khasi); *Bol-Gippak* (Garo).

Description   Tree deciduous, bark grey or ashy, warty with lenticular tubercles, exfoliate in irregular plates which leaves shallow depressions; branchlet, nearly pubescent, quadrangular; leaves broadly ovate-acuminate, glabrescent above, fairly tomentose or almost glaucous beneath the lateral nerves; flowers 2–5 cm across, brownish yellow, in decussate cymes arranged on terminal tomentose panicles, bracts linear-lanceolate; sepals copular or funnel-shaped, pubescent, persistent in fruit teeth 5, acute, petals pubescent with crenulated margin; stamens 4, didynamous; ovary 4-celled; style slender stigma unequally bifid, ovules solitary in each cell; drupe succulent, ovoid or pyriform, about 2–4 cm long, glossy and yellow when ripe; pulp aromatic; endocarp horny; seeds 1–3, lenticular, exalbuminous, cotyledons fleshy.

Flowering and fruiting   February–July

Distribution   Common throughout Meghalaya

Propagation   By seeds

Parts used   Root, leaves and flowers

Uses   The juice of leaves is given in case of gonorrhoea, cough and ulcers. Parts of leaves are applied on the head of the patient to cure high fever. Flowers are prescribed for blood diseases. Roots are demulcent, alternative, astringent and aromatic, used for rheumatism, gonorrhoea and bladder troubles. Roots are one of the major constituents of Dasmool given after childbirth.

Gomphostemma parviflora Wall.

Family   Lamiaceae

Indian name   *Bhedaitita* (Assamese).

Description   Undershrubs unbranched, up to 2 m high; leaves broadly elliptic, lanceolate, acuminate or acute, base cuneate, serrate, white, stellate, woolly beneath; cymes axillary, often in leafless axils, fascicled, 4–8 cm long, woolly; flowers yellow or yellowish-white.

Flowering and fruiting   August–March

Distribution   Entire North-East Region

Propagation   By seeds

Parts used   Root and leaves

Uses   Root decoction is given after confinement; leaf paste with camphor is useful for swelling in the joints, and headache.

Gonatanthus pumilus (Don.)

Family   Araceae

Indian name   *Ban pindalu, Jal papri* (Hindi).

Description   Herbs with a small crown, stem with slender branches with small bulbil; leaves ovate, cordate, peltate; fruit berries coloured yellow.

Distribution   Very common in Khasi hills of North-East India.

Propagation   By seeds

Parts used   Stem and leaves

Uses   Application of leaf juice is helpful for wounds; stem paste is used to cattle as a germicide.

Gouania tiliaefolia Lamk.

Family   Rhamnaceae

Indian names   *Dumigong* (Garo); *Mei teiniong* (Khasi); *Penki tiga* (Telugu); *Dumigong, Dugithang* (Manipuri); *Tungcheong-monrik* (Sikkimis).

Description   An unarmed climbing shrub with grey bark bearing tendrils at the ends of the branchlets; branches glabrous; young parts and inflorescence pubescent; leaves alternate, 2–4 × 1.5–2.5 in, ovate, acuminate, crenate, glandular, subcoriaceous, sparsely rusty, pubescent beneath when young; lateral nerves 5–6 on either side of the midrib, generally opposite, the lowest pair sub-basal and laterally branched base rounded or subcordate; inflorescence simple racemes, appearing as terminal panicles. Flower greenish yellow, 3–4 mm across, disc 5-angled; ovary immersed in disc.

Flowering and fruiting   July–November

Distribution   Commonly found up to 9000 feet in Garo, Khasi hills and North Cachar hills.

Propagation   By seeds

Parts used   Bark, roots and leaves

Uses   Bathing in water in which crushed leaves are added and mixed, helps in curing fever. Leaf paste is used for skin diseases and poultices for sores. The bark and root which contain saponin are used for washing hair to destroy vermin.

Grewia multiflora Juss.

Family   Tiliaceae

Description   Small trees; bark smooth, dark-brown, branchlets appressed, hairy; leaves elliptic lanceolate, oblanceolate, smaller ones orbicular, acuminate, crenate, serrate, base obtuse or sub-truncate, sparsely stellate, hairy beneath, glabrescent above, 3-nerved from base, petiole 0.3–0.5 cm long; stipules linear-lanceolate, subpersistent; inflorescence axillary, 30 flowered cymes; bracts linear; flowers white, calyx 1 cm long, acute, hairy without; corolla 3–4 mm long, ovate, glandular, up to half the oblong length, fringed with hairs; stamens shorter than calyx, filaments yellow; ovary ovoid, hairy, style short, stigma bilobed; drupe 2–4 lobed, ovoid, purplish-black, glabrous and fleshy.

Flowering and fruiting   September–December

Distribution   Eastern Himalayas; Khasi hills, common in Meghalaya

Propagation   By seeds and vegetative method

Parts used   Bark and fruits

Uses   Bark astringent, expectorant and used in cough, skin disease and for removing hairs from the root.

Gynocardia odorata R. Br.

Family   Flacourtiaceae

Indian names   *Alasakapaha* (Sanskrit); *Chaulmogra, Chaulmugra* (Hindi); *Surantaeil* (Kannada); *Dieng-sohliang* (Khasi); *Masribu, Thithopha-bipha* (Garo).

Description   Small or middle-sized glabrous tree with slender branches; bark grey or greenish grey and generally warty outside, uniformly pale brown inside; leaves bifarious oblong, abruptly acuminate, rounded at base, entire, thinly coriaceous lateral nerves on either side of the midrib including the sub-basal pair; flowers dioecious, pale-yellow, sweet scented, in few flowered axillary fascicles or large from the trunk; female flowers comparatively larger, buds globose; peduncles, long with minute bracts at the base; Sepals 5, saucer-shaped, leathery; petals oblong or ovate, fleshy; male flowers have stamens about 100 filaments woolly, female flowers have staminodes 10 or 15, villous ovary 1-celled styles 5, fruit 3–5, globose; seeds about 2.5 cm long, obovoid or oblong, circular in cross section, embryo somewhat eccentric, cotyledons thick, set in thick oily albumen.

Flowering and fruiting   March–January

Distribution   Fairly common in evergreen forests throughout Meghalaya.

Propagation   By seeds

Parts used   Seeds

Uses   Seeds are used as a local medicine for dysentery and diarrhoea.

H

H

Hedychium acuminatum Roscoe.

Family   Zingiberaceae

Indian name   *Aithur* (Mizoram).

Description   A robust rhizomatous tall herbaceous plant, growing 1–3 m high; leaves are glabrous with white ascending flowers borne in dense terminal spikes; the leaves are woven in mats; flowers bisexual, pure white or tinged with yellow, fragrant, with broad round to inverted heart shaped lip that is 2-lobed at apex; seeds many, small, black, covered with red coloured aril; the dried fruit is added to meat and pulses during cooking to soften them and hasten cooking.

Distribution   Entire North-Eastern Region

Propagation   By seeds and vegetative method

Parts used   Flowers and rhizomes

Chemical constituents   The flowers have the following main constituents, linalool, methyl benzoate, eugenol, lactone, and methyl jasmonate; rhizome contains cytotoxic principles such as diterpenes, coronarin, ethyl ethers of coronarin-D, isocoronarin-D, etc.

Uses   The leaf juice is used for stomach ailments, and in eye diseases. The rhizome is stomachic, carminative, stimulant and a tonic. It is used in dyspepsia. Rhizome powder is used for hair growth.

Hedyotis scandens Roxb.

Family   Rubiaceae

Indian names   *Bhedeli, Bhedelilota* (Assamese); *Mikrisim, Samreting* (Garo); *Jermiskie, Meid* (Khasi); *Bonhrathrin* (Nepali).

Description   Stems often purplish tinged; leaves 5–12 × 1–4 cm, elliptic lanceolate or oblanceolate, acuminate, base narrowed, cuneate, shining above, pale beneath, entire; flowers white or turning cream in axillary and terminal compact compound trichomatous sub-corymbose puberulous cymes; sepals, tube campanulate, lobes equalling in length to the tube; petals wider at mouth, lobes recurved, capsule globose or ovoid.

Flowering and fruiting   August–April

Propagation   By seeds and vegetative method

Parts used   Leaves and roots

Chemical constituents   The whole plant contains alkaloids, saponins and tannins

Uses   The stem is an effective remedy for gastralgia, gastric ulcer and heartburn during the colic pains or before meals. A tonic prepared with hedyotis stem liquid extract and honey is used for ulcerous stomatitis. Leaf paste is applied to skin diseases. Roots are used in many diseases such as fever, stomach ache, dysentery, tooth ache and tapeworm.

Hedyotis verticillata (L) Lamk. [Rare]

Family   Rubiaceae

Description   An erect glabrous bushy annual into diffused branches and leaves sessile elliptic or linear-lanceolate, acute or acuminate 1-nerved, margin recurved, stipules with long cilia, flowers 2–6 in axillary clusters, calyx-teeth triangular about equalling the hispid ovoid capsule; corolla tube slender; fruits subglobose capsules. Capsule valves elongate, seeds ellipsoid.

Flowering and fruiting   July–March

Distribution   Common in Meghalaya, particularly in Khasi hills

Propagation   By seeds and vegetative method

Parts used   Leaves

Uses   The leaf paste is useful to reduce temperature and also in skin diseases.

Helicia excelsa Bl. [Rare]

Family   Proteaceae

Indian name   *S-ialhma, Siathma* (Mizoram).

Description   Medium sized or large tree; young shoots tomentose, yellowish brown, bark greyish, warty but plain, thick, inside grandular; leaves obovate or oblanceolate, acuminate; entire or coarsely serrate, coriaceous, glabrous, lateral nerves 7–10 on either half, impressed above, prominent beneath, base acute or connate, petiole 0.3–1.4 in. long; racemes rusty tomentose, nearly as long as the leaves; ovary tomentose; fruit subglobose smooth.

Flowering and fruiting   June–February

Distribution   Indo-Burma, confined to North-East India, rather rare in Meghalaya.

Propagation   By seeds

Parts used   Bark

Uses   Bark used for colic.

Hemidesmus indicus L. Br. Schult.

Family   Asclepiadaceae

Indian names   *Aasphota, Ananta* (Sanskrit); *Anantamul, Hindi-salsa* (Hindi); *Dudavali, Haleberuballi* (Kannada); *Kizhanna, Nannari, Naru-nindi* (Malayalam); *Anantvel, Makur* (Marathi); *Arakkam, Aritinviyachi* (Tamil); *Adavippatitige, Gadisugandhi* (Telugu).

Description   A perennial, slender, laticiferous, twining or prostrate wiry shrub with woody foot stock and numerous slender, terete stems having thickened nodes, hairless; leaves vary greatly in shape and size, 5–10 cm long, but their breadth varies from 0.5–4 cm, dark green, often with whitish blotches, pale or whitish, hairy on lower surface. Flowers very small, greenish, in small, compact clusters; fruits 10–15 cm long, green, narrow, cylindric, pointed at tip, in pairs; seeds small, black, with a tuft of white hairs at top. All parts of the plant have white milky juice.

Distribution   The plant occurs almost throughout India, particularly in North-East India.

Propagation   By seeds and vegetative method

Parts used   Root, stem and leaves

Chemical constituents   Air-dried roots yield essential oil containing *p*-methoxy salicylic aldehyde as the major constituents. The aroma of the drug is attributed to the aldehyde. Other constituents present in the roots are β-sitosterol, α- and β-amyrin, lupeol, tetracyclic triterpene, alcohols, small amount of resin, fatty acids, tannins, saponins, a glycoside and a ketone.

Uses   The drugs are useful in fever, skin diseases, loss of appetite, syphilis, leucorrhoea and other urinary complaints. The diuretic action of the drug has been shown experimentally. The drug is largely used as a blood purified and in rheumatism. Test in hospitals have confirmed the utility of Indian Sarsaparilla as a substitute for true sarsapailla, which is obtained from plants of the genus *Smilax*.

Hemiphragma heterophyllum Wall.

Family   Scrophulariaceae

Description   A pubescent herb; stem slender, loosely tufted, prostrate; leaves dimorphic; stem leaves opposite orbicular, cordate, short petioled; bracts linear, less acute, sessile; flowers pinkish, axillary, solitary, sepals 5-partite, petals tube, slender, short; lobes 5 spreading, broad, nearly equal; stamens 4, equal, fruit capsular, 2-valved, ultimately septicidal, black, shining.

Flowering and fruiting   August–December

Distribution   Entire North-East Region, especially in Khasi Hills, Meghalaya.

Propagation   By seeds

Parts used   Whole plant

Uses   Paste is used in syphilitic wounds, cuts and bruises.

Hibiscus rosa-sinensis L.

Family   Malvaceae

Indian names   Shoe flower, Chinese hibiscus (English); *Japa, Ondrapushpi* (Sanskrit); *Jasum, Jasuth, Java* (Hindi); *Dasavala* (Kannada); *Chembaruthi* (Tamil); *Chemparathi* (Malayalam); *Dasana, Mandara* (Telugu).

Description   Small tree or large shrubs; leaves ovate, acuminate, more or less serrate, glabrous with few scattered hairs on the veins; flowers solitary, axillary, bellshaped, with pistil and stamens projecting from the centre; capsules roundish, many seeded.

Distribution   Cultivated as an ornamental plant throughout India.

Propagation   By seeds and vegetative method

Parts used   Leaves and flowers

Chemical constituents   Taraxeryl acetate, beta-sitosterol, campesterol, stigmasterol, cholesterol, ergosterol, lipids, citric, tartaric and oxalic acids, flavonoids, hibiscetin, cyanidin and cyanin glucosides and alkanes.

Uses   To induce abortion caused by menstrual cramps and to help in childbirth; to treat headaches; the leaves are used to treat post-partum relapse sickness; to treat bottle, sores and inflammations, good for hairs; flowers are used as an antifertility drug, in childbirth, impotency, menstruation and urinary complications; bark is useful in hypotension; leaves are emollient, anadyne and operient or laxative.

Hibiscus surattensis L.

Family   Malvaceae

Indian names   *Ran bhindi* (Hindi); *Mullu gogu* (Kannada); *Sehnap* (Mizoram); *Kashlikirai, Kattuppuliccai* (Tamil); *Mullugogu* (Telugu).

Description   Undershrubs, stems armed with recurved, transparent spines leaves 5–8 × 5–10 cm, palmately 3–5 lobed, lobes acute at apex, serrate, truncate or sub-cordate at base 7–9 nerved at base, petioles 2–3 cm long, retrorsely spiny, stipule lanceolate, serrate, with a prominent midrib; bracteoles more than 10, armed with bristles, divided above; sepals purple tinged, accrescent in fruit, not spreading, petals 5, bright yellow with a purple centre, stigma purple; carpels ovoid, bristly hairy; seeds smooth.

Flowering and fruiting   August–December

Distribution   Throughout in India. In Meghalaya it occurs in disturbed forests or forest cleared areas.

Propagation   By seeds and vegetative method

Parts used   Leaves and flowers

Chemical constituents   Same as *H. rosa-sinensis.*

Uses   The leaves and the flowers possess antibacterial, demulcent and diuretic properties. They are useful for the treatment of boils, particularly on the chin, in the form of poultice. Powdered dried leaves and flowers mixed with concentrated tea infusion, makes the boils burst earlier and less painful.

Hiptage benghalensis (L.) Kurz.

Family   Malpighiaceae

Indian names   *Atimuktaka, Bhadralata* (Sanskrit); *Madhabilata, Jutimukta* (Hindi); *Adimurtte, Adirganti* (Kannada); *Sitampu* (Malayalam); *Adigandi, Karungodi* (Tamil); *Atimutamu, Kuruvenda* (Telugu).

Description Climbing or suberect shrubs; leaves opposite, quite entire, coriaceous, eglandular, or with a row of remote intermarginal glands beneath, stipules 0; racemes terminal or axillary, simple or compound; peduncles erect, bracteate, articulate with the 2-bracteolate pedicels. Flowers white, fragrant, the 5th corolla discoloured. Calyx 5-partite, glands large, adnate to the pedicel; corolla 5, clawed, unequal, silky; stamens 10, declinate, all fertile, one much larger, filaments connate at the base. Ovary 3-lobed, lobes appendiculate; style 1–2 circinate; seeds subglobose; cotyledons thick, unequal.

Flowering and fruiting February–June

Distribution Indo-Malaya, throughout India, common in Meghalaya in deciduous and secondary forests, often along river courses.

Propagation By seeds and vegetative method

Parts used Bark, leaves and flowers

Chemical constituents The bark is an aromatic, bitter and it contains a crystalline glucoside, hiptagin and tannin.

Uses The bark, leaves and flowers are acrid, astringent, refrigerant, vulnerary, expectorant, cardiotonic, anti-inflammatory and insecticidal. They are useful in burning sensation, wounds, ulcers, cough, asthma, cardiac, debility, inflammation, skin diseases, leprosy, scabies and rheumatism.

Holarrhena antidysenterica (L.) Wall [Vulnerable]

Family Apocynaceae

Indian names *Girimallika, Indravrksa* (Sanskrit); *Karchi, Karvaindarjau* (Hindi); *Kodamurike, Korasigina-gida* (Kannada); *Kotakappala, Venpala* (Malayalam); *Erukkalaipalai, Kudagappalai* (Tamil); *Amkudu, Amkudu-vittulu* (Telugu).

Description Small trees, often shrubby, reaching about 3–10 m in height or more; branches usually from near base; bark pale brown, wrinkled and warty, lenticellate, young twigs tomentose; leaves shiny above, dull and hairy below with conspicuous nerves. Leaves 5–30 × 2.5–10 cm, opposite, ovate-lanceolate, obovate, acuminate, base rounded or acute, glabrous above, glabrescent beneath; cymes up to 15 cm across; subsessile. Leaf stalk very small, flowers white in terminal branches; fruit, slender and paired, cylindrical, dark grey with white streaks all over; flowers white in axillary or terminal corymbiferous cyme; follicles cylindrical in pairs, long narrow curved; seeds numerous, brownish, crowned with a tuft of long hairs.

Flowering and fruiting April–December

Distribution Indo-China and Malaysia, from Pakistan to Malacca, throughout India particularly in drier parts; very common in Meghalaya at lower elevations. It is abundant in lower forests of Darjeeling, Tarai, Sikkim hills, Garo hills in Meghalaya and Assam. It is found here and there along with the lower valley (300–900 m) and up to the altitude of 1000 m in wild stages.

Propagation   By seeds and vegetative method

Part used   Bark and seeds

Chemical constituents   The trunk bark contains alkaloids: conessine, nor-conessine, conessimine, isoconessimine, curchine, conimine, holarrhine, holarrhenine and holarrhimine; gums, tannin, β-sitosterol.

Uses   In Ayurveda the drug is used in acute and chronic amoebic dysentery; hence its botanical name *antidysenterica*. *Conessine* is steroidal alkaloid from bark experimentally proved to be toxic to *Entamoeba histolytica*. The seeds find their traditional application in flushing out intestinal parasites and worms that invade the digestive tract and also in the treatment of digestive disorders, biliousness, diarrhoea, dysentery, loss of appetite, etc. In folk medicine, a paste made of the dried bark is rubbed over the body in cases of dropsy. It is also traditionally used in the treatment of leprosy. It has tonic and febrifuge properties. Traditionally the ailment and its prescription are:

1.   Mix ¼ to ½ teaspoon of the seed powder in warm water and drink, in case of blood in urine.

2.   Grind half teaspoon of the seeds of Kutaja and Vidanga, add 1 cup water and drink, in case of cholera.

3.   Mix ½ teaspoon each of the powdered bark of Siora and Kutaja, add 1 cup water and drink, in case of dropsy.

4.   In case of dysentery, loss of appetite mix ¼ teaspoon bark powder in ½ teaspoon honey.

5.   In case of haemorrhoids, boil ¼ teaspoon seed powder in milk and drink. Dutkhuri has a masked role in setting right ecological imbalance. It also played a great role in several afforestation programmes as a pioneer. A bath containing a decoction of bark or leaves cures scabies. Bark is an astringent, anthelmintic, stomachic, antipyretic, tonic and antidysenteric used in amoebic dysentery and diarrhoea.

Although *Holarrhena* has numerous medicinal properties, it may have many dangerous side effects if taken indiscriminately. It is also reported to raise the blood pressure in small doses; however; in larger doses it tends to be hypotensive. A Kutaja of dried stem bark collected from trees over 10 years old is reported to contain a higher concentration of alkaloid. The native physicians commence their collection during the post-monsoon months, i.e., July to September.

Holmskioldia sanguinea Retz.

Family   Verbenaceae

Indian names   Chinese hat plant, Cup-and-saucer plant (English); *Ukapni, Kaphlagula* (Hindi); *Jermi snamkhmut* (Khasi); *Misi nasil* (Garo); *Kharam leithong* (Manipuri).

Description   Large struggling shrubs; branchlets arching, subquadrangular; bark grey, peeling off in long strips; leaves ovate, lanceolate, long, acuminate, base truncate, rounded or subcordate, crenate-serrate, glaucous beneath; cymes corolla red or orange or yellow; stamens didynamous, fruits 4-lobed, supported by the calyx; ovary 2-celled.

Flowering and fruiting   September–February

Distribution   Subtropical Himalayas and North-East India; very common in Meghalaya at lower elevations.

Propagation   By seeds

Parts used   Root

Uses   Root juice is useful to relieve fever and headache.

Holboellia latifolia Wall.

Family   Berberidaceae

Indian names   *Gomphal, Gukhnial* (Hindi); *Mirang ksa, Soh lyngkait* (Khasi); *Soh tymbra* (Jaintia).

Description   Large climbing shrubs; stem corky when old; leaves alternate, digitate, 3–09 foliate; leaflets very variable in size and shape, 3–5 by 1.5–2 in. broadly ovate, oblong or lanceolate, acuminate, coriaceous, glabrous, somewhat shining above, pale beneath, usually 3-nerved at the base petiole 3–7 in.; flowers monoecious, green or purplish green, sweet-scented, in axillary fascicles or racemes, pedicels, slender; calyx 6 in 2 series, corolla 6, stamens 6, ripe carpels berry like, seeds black, generally embedded in yellow edible pulp.

Flowering and fruiting   February–November

Distribution   In Khasi and Jaintia hills

Parts used   Leaves

Uses   Crushed leaf is applied on burns.

Holoptelea integrifolia (Roxb). Planch. [Rare]

Family   Ulmaceae

Indian names   *Chirabilva, Karanja* (Sanskrit); *Banchilla, Bisenda* (Hindi); *Kaladri, Nilavahi* (Kannada); *Nettavil, Aaval* (Malayalam); *Avali* (Tamil); *Pedanevili, Tapasi* (Telugu).

Description   Tree with a compact ovoid crown; Bark greyish or ashy brown, longitudinally fissured; leaves 5–15 × 3–6 cm, ovate elliptic, oblong, acuminate, base rounded or sub-cordate, coriaceous, pubescent when young, flowers greenish yellow, male and hermaphrodite mixed in short racemes of fascicles near the leaf scars; fruits suborbicular samara, with membranous reticulately veined wings.

Flowering and fruiting   February–June

Distribution   Throughout India, very rare in Meghalaya at lower elevations.

Propagation   By seeds and vegetative method.

Parts used   Bark and leaves.

Chemical constituents   The seed cake contains N –10.3%, lysine –3.3%, glutamic acid –13.0%, and histidine –1.3%.

Uses   The bark and leaves are bitter, astringent, acrid, thermogenic, digestive, carminative, laxative, and depurative. They are useful in inflammations, flatulence, vomiting, skin disease, diabetes and rheumatism.

Homonoia riparia Lour.

Family   Euphorbiaceae

Indian names   *Jalavetasah, Kshudrapashanabheda* (Sanskrit); *Jalbent, Serni* (Hindi); *Niraganugulu* (Kannada); *Puzhavanchi, Nirnochil* (Malayalam); *Kattalari, Peruncuntari* (Tamil); *Cheppunjerinjal, Siridamanu* (Telugu).

Description   A rigid evergreen gregarious shrub or small tree, young parts pubescent; bark dark grey or brown, rough; leaves linear, oblong or lanceolate, acute, or somewhat glandular, toothed towards the apex, glabrescent above, papillose on both surfaces, clothed with scattered and numerous round scales beneath; lateral nerves many, about 10–30 on either half, prominent beneath, base acute, subulate; flowers 0.5–0.7 cm across; anthers red; capsules 0.3–0.5 cm across, globose.

Flowering and fruiting   February–July

Distribution   Throughout India, common in Meghalaya in Rocky River beds at lower elevations.

Propagation   By seeds and vegetative method

Parts used   Roots, leaves and fruits

Uses   The roots are laxative, diuretic, and refrigerant, depurative and emetic. They are useful in gonorrhoea, syphilis, vesical calculi, strangury and haemarrhoids.

Houttuynia cordata Thumb.

Family   Sauraceae

Indian names   *Muchandari* (Assamese); *Simdalu* (Hindi); *Toningkhok* (Manipuri); *Vaithinthang* (Mizoram).

Description   Glabrous perennial herb with creeping rhizome, 20–50 cm high; leaves alternate, ovate 3–8 cm long; 3–6 cm wide, cordate, stipules oblong, obtuse, united to petioles; inflorescence dense lateral spike subtended by 4 involucral bracts; spikes 1–3 cm long, cylindric, densely many flowered; peduncles 2–3 cm long. Flowers minute in dense flowered spikes, white; Stamens 3, combined with ovary up to 1/3 of the height.

Flowering and fruiting   April–August

Distribution   Entire North-East Region up to a height of 2000 m

Propagation   By seeds and vegetative method

Parts used   Whole plant

Chemical constituents   The whole plant yields an essential oil consisting of methanonyl ketone, myrcene, α-pinene, *p*-cymene, linalool and geraniol.

Uses   The root extract is reported to possess diuretic action attributable to the presence of quercetin and inorganic salts like potassium chloride and potassium sulphate. The whole plant has beneficial effect in the treatment of haemarrhoids, acute conjunctivitis and ocular infections caused by *Bacillus* sp.

Hoya globulosa Hk.f.

Family   Asclepiadaceae

Description   Stout climber, stem woody, leaves elliptic or oblong, cuspidate or acuminate, coriaceous, hairy, midrib stout, base rounded, petiole peduncle and pedicels villous, 2.5 cm, calyx segments, rounded, corolla 1.5 cm across, cream coloured, almost glabrous within, lobes short, incurved, coronal processes short, broadly elliptic, pink, concave above, inner angle produced into an erect spur, which is shorter than the large broad anther tips; seeds 0.7 cm, slender.

Flowering and fruiting   March–May

Distribution   Himalayas and sub-Himalayas; common in Meghalaya in dense evergreen forests, usually at higher elevations.

Propagation   By seeds and vegetative method

Parts used   Leaves

Chemical constituents   Amyrins, apigenin glucoside, benzoic acid, chlorogenic acid. Chrysogeriol glycoside, cosmosin, lupeol, lipids and nyctanthic acid derivatives.

Uses   Leaf paste is useful for dog bite wound. The leaves are used to make for body rashes. An infusion of the leaves is applied as a lotion and taken internally for skin inflammation. The juice from the plant is applied to body burn.

Hymenodictyon excelsum (Roxb.) Wall.

Family   Rubiaceae

Indian names   *Bhramarchhallika, Bhrangavrksha* (Sanskrit); *Bandaru, Bauranga* (Hindi); *Bataga, Gandele* (Kannada); *Ittiyila, Nishakatampa, Vellakatampa* (Malayalam); *Ilaimergai, Vellaikkadambu* (Tamil); *Bandara-chettu, Bandaru* (Telugu).

Description   Tree deciduous, 8–30 m high; crown ovoid with symmetrically spreading, horizontal branches; bark brownish grey; leaves obovate or ovate, broadly oblong or elliptic, shortly acuminate, base cuneate, greyish, tomentose beneath, panicle decurved or deflexed, ascending at tip; flowers white, fragrant, style much exerted, subpersistent; capsule 1.5–2.5 cm long; elliptic, reddish, brown when ripe.

Flowering and fruiting   June–January

Distribution   Throughout India, frequently in deciduous forests of Meghalaya.

Propagation   By seeds

Parts used   Leaves

Uses   Leaves are useful for dysentery and diarrhoea.

Hyptianthers stricta (Roxb.) Wight and Arn.

Family   Rubiaceae

Description   An aromatic evergreen shrub or small tree; stems brown, reticulatey fissured; branches thin, 4-angled or somewhat compressed; leaves decussate, narrow, elliptic, oblong or linear, entire, somewhat slightly undulate, finely caudate, acuminate, glaborous and shining above, pubescent on nerves beneath when young; lateral nerves 6–7 on either half; stipules caudate, acuminate, persistant; flowers sessile in dense axillary cymes; anther 4–5, sessile, disc epigynous, annual, pulvinate; ovary 2-celled with 4–10 pendulous ovules in each cell; style solitary; stigma bifid, white, brush-like; berries, more or less globose, seeds angled and compressed; testa fibrous, embryo small.

Flowering and fruiting   March–October

Distribution   Throughout North-Eastern India

Propagation   By seeds

Parts used   Leaves

Uses   Infusion is given to expectant mothers.

Hyptis suaveolens Poit.

Family   Lamiaceae

Indian names   Ganga basil (English); *Bhustrna* (Sanskrit); *Vilaiti-tulsi* (Hindi); *Ganga thulasi* (Kannada); *Danthi thulasi, Seema thulasi* (Telugu); *Damana* (Oriya).

Description   Hairy undershrubs, 3 m high; stems obtusely quadrangular, aromatic; leaves ovate, elliptic, acute, obtuse or mucronate, base narrowed, cuneate, hairy; flowers blue, 0.5–1.5 cm long with spiny lobes; spike like inflorescence 30–60 cm long; fruits a nutlet, 0.4 cm long.

Flowering and fruiting   September–February

Distribution   Native of tropical America, fairly naturalised in India, fast establishing in Meghalaya along rivers and roadsides in sandy soils.

Propagation   By seeds and vegetative method.

Parts used   Leaves

Chemical constituents   Miniterpenes, *p*-cymenes, thymol, γ-terpinene, α-thujene, mycrene, hyptolide, essential oils, ursolic acid.

Used   Leaves have wound-healing qualities and are epecially used in treating cuts in the case of diabetes; chest pains and painful breathing.

I

I

Ichnocarpus frutescens (L.) R. Br. [Endangered]

Family   Apocynaceae

Indian names   *Ananta, Bhadra, Chandana* (Sanskrit); *Dudhi, Dudhilata, Kalidudhi* (Hindi); *Gorwiballi, Gouriballi* (Kannada); *Paravalli* (Tamil); *Palvalli, Parvalli* (Malayalam); *Illukkatti, Karampala* (Telugu).

Description   Tree; bark reddish brown, lenticellate, much branched, climbing, rusty villous, evergreen, laticiferous, woody; leaves simple, opposite, 3.5–10 × 1–4 cm, elliptic, lanceolate, oblong, acute, base, narrowed, rusty pubescent; panicles up to 10 cm long; main nerves 5–7 pairs;

flowers greenish white, 0.6–0.9 cm across, fragrant, numerous in axillary or terminal panicles or cymose clusters; corolla lobes spreading, twisted; follicles 6–15 cm long, linear, cylindric, erect or spreading; fruits straight or slightly curved, cylindrical follicles, usually 2, divaricate; seeds black, cosmose, coma white.

Flowering and fruiting   April–June (sometimes in November)

Distribution   Indo-malaya, extending to Australia, nearly throughout India, rare in Meghalaya, found only in southern parts of Garo Hills, in dry scrublands, associated with *Vitis* spp.

Propagation   By seeds and vegetative method

Parts Used   Roots

Uses   The roots are sweet, refrigerant, febrifuge, aphrodisiac, diaphoretic, diuretic, depurative, demulcent and tonic. They are useful in burning sensation, hyperdipsia, fever, seminal weakness, nephrolithiasis, skin diseases, leprosy, vomiting, diabetes and general weakness.

Ilex embeliodes Hk. f. [Endemic/Rare]

Family   Aquifoliaceae

Description   A small tree with bark thin, nearly smooth, distantly warty and with faint horizontal wrinkles; shoots and inflorescence finely puberulous; leaves 2-farious, deep green, elliptic to oblong, lanceolate, acudate with a blunt apiculate tip, with few distant spinous teeth or nearly entire, shining above, finely puberulous along midrib on both surfaces, ultimately glabrous, base acute or cuneate, petiole channelled, up to 0.25 in. long; flowers 4-merous, white, male flowers peduncled umbellules, clustered at the ends of branches or leaf axils, often on very short branches, sometimes solitary on peduncles of umbellules; female flowers in fascicles at the leaf axils often obtuse, petals ovate or oblong, cuneate at the base; stamens nearly as long as the petals; drupe globose with stones.

Flowering and fruiting   April–December

Distribution   In Khasi hills, 4000 to 6000 ft. in Meghalaya.

Propagation   By seeds and vegetative method

Parts used   Bark and root

Uses   Bark and root decoction is given for extreme cough and cold.

Ilex khasiana Purk. [Endemic/Rare]

Family   Aquifoliaceae

Description   Tree 15–20 m high; crown lax with branches horizontally spreading; bark greyish brown, lenticellate; leaves oblong, lanceolate, elliptic, acuminate to caudate, base rounded or obtuse, cuneate; umbels compound, repeatedly forked, cymose; flowers 0.3 cm across, corolla spreading and deflexed; stamens equalling the corolla; fruit purplish red, 0.8 cm across, globose.

Flowering and fruiting May–December

Distribution Confined in Meghalaya to sacred grooves, in Khasi Hills

Propagation By seeds and vegetative method

Parts used Bark and root

Chemical constituents Contains ilicin, xanthin, theobromine and caffeic acid. Theobromine ia a caffieine–type alkaloid, used to treat asthma.

Uses Bark and root decoction is used in cold and cough and in tuberculosis.

Imperata cylindrica Beauv.

Family Poaceae

Indian names *Darbha, Darbhah* (Sanskrit); *Ulu, Dabh, Sauraun* (Hindi); *Sannadabbehullu, Sanna dabbe hullu* (Kannada); *Dharbai pullu, Tharpaipullu* (Tamil); *Darbhappullu, Kodi-pullu* (Malayalam); *Dharba, Darbha gaddi* (Telugu).

Description Perennial herb up to 1.5 m high, rhizome hard, coriaceous, descending deeply into the soil, culms erect, with wiry hairs; leaves narrow and long, prominently nerved, scabrous on the upper surface, margins sharp edged; inflorescence of many spikes, silvery white, densely silky; grain small elliptic to oblong, brown, light, loose.

Flowering and fruiting May–September

Distribution Common weeds everywhere especially at Jhum cultivation areas in Meghalaya.

Propagation   By seeds and vegetative method

Parts used   Rhizome

Chemical constituents   The rhizome contains glucose, fructose and organic acids.

Uses   Rhizome is diuretic and febrifuge and is used in the treatment of urodynia, pollakiuria, haematuria, fever and spleen complications. A decoction of rootstock is given for diarrhoea, dysentery and gonorrhoea.

Indigofera tinctoria L.

Family   Fabaceae

Indian names   *Akika, Anjanakesika* (Sanskrit); *Gouli, Lil, Nil* (Hindi); *Ajara, Ajura, Hennunili* (Kannada); *Asidai, Attipurashadam* (Tamil); *Amari, Amen, Avari* (Malayalam); *Aviri, Neelichettu* (Telugu).

Description   Shrub up to 2 m high; leaves compound, 2.5–5 cm long, leaflets opposite, membranaous, turning blackish when dry, 9–13 cm, obovate. Racemes, nearly sessile, 5–10 cm; flowers many in nearly sessile, lax, spicate racemes which are much shorter than the leaves, red or pink; fruits cylindric pods, pale greenish grey when young and dark brown on ripening with 10–12 seeds.

Distribution   Throughout North-East Region

Propagation   By seeds and vegetative method

Parts used   Whole plant

Chemical constituents   It is a rich source of potash, the ash containing as much as 9.5 per cent of soluble potassium salts.

Uses   The roots, stem and leaves are bitter, thermogenic, laxative, trichogenous, expectorant anthelmintic, tonic and diuretic, and are useful for promoting growth of hair and in gastropathy, splenomegaly, cephalalgia, cardiopathy, chronic bronchitis, asthma, ulcer and skin diseases.

Ipomoea alba L.

Family   Convolvulaceae

Indian names   Moonflower (English); *Dudhiakalmi* (Hindi); *Chandra kaanthi, Chandra pushpa* (Kannada); *Nagamukkori, Chantirakanti* (Tamil); *Munda-valli* (Malayalam); *Naagaramukkatte, Pandithivankaaya* (Telugu).

Description   Climbing shrubs, leaves ovate, ovate-orbicular, acuminate, base cordate or subcordate, glabrous; cymes few flowered, flowers 8–12 cm long, white or often purplish; capsules 1–2.5 cm long, yellow-seeded.

Flowering and fruiting   October–February

Distribution   Native of tropical America, common in Meghalaya, particularly along marshy areas.

Propagation   By vegetative method

Parts used   Tuberous roots

Uses   The roots are sweet, refrigerant, laxative, aphrodisiac, diuretic and tonic; they are useful for burning sensation, hyperdipsia, constipation, strangury, renal vessel calculi, diabetes and general weakness.

Ipomoea nil (Linn) Roth.

Family   Convolvulaceae

Indian name   Morning glory, Pharbitis seeds (English); *Krsnabijah* (Sanskrit); *Kaladana, Jharmaric* (Hindi); *Gowri beeja* (Kannada); *Kakkattan, Sirikki, Kotikkakkattan* (Tamil); *Thaliyari* (Malayalam); *Kollivittulu* (Telugu).

Description   Extensive twiners; branches hirsute; leaves broadly ovate, usually trilobed, often entire, densely hairy beneath, lobes subacute or acuminate, 3–7 nerved at base; cymes umbellate, 5–7 cm across, hairy; flowers 5–7 cm across, bluish-purple or pinkish; calyx lobes much enlarged in fruits; capsules 4–6 seeded.

Flowering and fruiting   July–February

Distribution   Throughout tropics and subtropics; common in Meghalaya particularly at higher elevations.

Propagation   By seeds and vegetative method

Parts used   Seeds

Chemical constituents   Seeds contain palmitic, stearic, arachidic, behenic, linolenic and oleic acids. Deep violet and red violet flowers contain peonidin chloride and 3,5-diglucoside of peonidin chloride.

Uses   The seeds are useful in inflammations, constipation, verminosis, skin dieases, leucoderma, scabies, dyspepsia, flatulence, bronchitis, gout, hepatopathy and fever.

Ipomoea purpurea (L.) Roth.

Family   Convolvulaceae

Indian names   Morning glory (English); *Kaphlagla* (Hindi); *Karukolli* (Telugu).

Description   Climber with retrorse hairs; leaves orbicular, ovate, cordate, entire, membranous; petiole 4–7 cm long; flowers purplish to white, umbelled in apex of a peduncle which is longer than the petiole, each pedicel supported by 3 subulate bracts; calyx segments short, acuminate, or unequal with spreading hairs at base; ovary 3-celled; seeds usually 6, glabrous.

Flowering and fruiting   July–September

Distribution   Largely grown in warmer parts of India, common in Meghalaya; cultivated apparently wild in Khasi hills (Barapani).

Propagation   By seeds and vegetative method.

Parts uses   Whole plant

Uses   The plant is an astringent, acrid, refrigerant, stomachic, laxative, diuretic and tonic. It is useful in skin diseases, boils, swelling, wounds, ulcers, carbuncle, dropsy, amenorrhoea, vomiting and burning sensation.

Ipomoea quamoclit L.

Family   Convolvulaceae

Indian names   Cupid flower (English); *Kamalata, Tarulata* (Sanskrit); *Kamalata* (Hindi); *Kaama lathe, Kaama mallige* (Kannada); *Vishnu-krant, Kasi ratnam* (Tamil); *Suryakanti* (Malayalam); *Ksiratnam, Suryaratnam* (Telugu).

Description   Plant is a slender twiner, leaves pinnate, segments numerous, linear, few flowered, corolla crimson or white, tube narrow, anther exerted.

Flowering and fruiting   May–August

Distribution   Entire North-East Region, cultivated as ornamental plant

Propagation   By seeds and vegetative method

Parts used   Whole plant

Uses   A paste made from the leaves and roots is applied as a poultice to backache and sored muscles. The plant extract is used for breast pain and ulcer. The plant is considered to be a purgative and used as an antidote for snake bite. The leaf paste is used in ulcer and piles and carbuncle.

Ixora acuminata Roxb.

Family   Rubiaceae

Description   Shrub or undershrub, branchlets ribbed or angled; leaves very variable, elliptic or linear, elliptic, acuminate, coriaceous, glabrous, lateral nerves 9–11 on either half, base usually conical or wedge-shaped; petiole 1; 5–2.5 cm long, floral pair of leaves ovate or obovate, sessile, auricled or with rounded base; flowers 5–16 cm across, 4–6 cm long, white; calyx pinkish or reddish; corolla tube slender, lobes oblong, spreading; drupes 1; 5–2 cm long, ovoid, ellipsoid.

Flowering and fruiting   May–December

Distribution   Bangladesh, India, largely in the North-East, common in Meghalaya, particularly under the sal forests.

Propagation   By seeds and vegetative method

Parts used   Root, leaves and flower

Uses   Roots are useful in hiccup, fever, gonorrhoea, anorexia, diarrhoea, dysentry and sores. Leaves are useful in diarrhoea. The flowers are useful in leucoderma, haemoptysis, bronchitis, sores and ulcers.

Ixora nigricans R. Br. Ex. J.E.Sm.

Family   Rubiaceae

Indian names   *Adyaala, Aelegaara* (Kannada); *Mashagani, Udappu* (Tamil); *Katkura, Lokhandi* (Marathi).

Description   Large evergreen shrubs; leaves 6–9 × 2–4 in elliptic, lanceolate or oblanceolate, shortly acuminate, entire, subcoriaceous, glabrous above, glabrous or minutely puberulous beneath, lateral nerves 8–10 on either half, terminating in intermarginal veins; petiole 5–10 cm long, stipules cuspidate; flowers white, about 2.5 cm long, sub-sessile, on short peduncled or sub-sessile cymes. Calyx glabrous, teeth as long as the tube; corolla tube slender, lobes oblong; style exerted; fruit size of a pea; seeds ventrally concave.

Flowering and fruiting   April–November

Distribution   Southern and North-East India; frequent in Meghalaya in evergreen forests

Propagation   By seeds and vegetative method

Parts used   Roots, leaves and flowers

Chemical constituents   The roots contain acrid aromatic oil, tannin, fatty acids and white crystalline substance. Root bark contains II-octadecadienoic acid, mannitol and myristic acid.

Uses   Roots are useful in fever, gonorrhoea, anorexia, diarrhoea, dysentery, sores, chronic ulcer and skin diseases. The leaves are useful in diarrhoea. The flowers are astringent, bitter, carminative, digestive and constipative.

J

Jasminum amplexicaule Buch. Hamm. ex. Don.

Family   Oleaceae

Indian name   Jasmin (English).

Description   Bushy shrubs; leaves ovate, lanceolate, acute or acuminate, base rounded or subcordate, glabrescent except the pubescent nerves; cymes capitate; flowers 2.5–3 cm long, white, fragrant; fruits ellipsoid, red when ripe.

Flowering and fruiting   August–April

Distribution   Indo-Malaya, confined in North-East India, frequent in Meghalaya in evergreen forests and forest margins.

Propagation   By vegetative method

Parts used   Leaves and flowers

Uses   Dried leaves are useful for stomach ulcers. The flowers are sweet, acrid, refrigerant, laxative, and digestive and they are useful in inflammations, rheumatism and cephalalgia.

Jasminum subtriplinerve Blume.

Family   Oleaceae

Description   Small trailing shrub, branches glabrous; leaves opposite, glabrous, ovate, elliptic, lanceolate, shortly acuminate or sharply acute, entire, coriaceous, glabrous above and shining on both surfaces, 3-nerved, often above the base, midrib impressed above, base rounded or subcordate, often oblique, petiole up to 2 cm long, jointed.

Flowering and fruiting   April–December

Distribution   North-East India and adjacent countries; common in Meghalaya.

Propagation   By seeds and vegetative method

Parts used   Leaves

Chemical constituents   The leaves contain alkaloids, resins and flavonoids

Uses   The leaves posses antibacterial and anti-inflammatory properties. They are used in treating post-partum hyperthermic infection, lymphadenitis, metritis, galactophoritis, leucorrhoea, ostealgia, impetigo, and haematometra.

Justicia gendarussa Burm. f.

Family   Acantheceae

Indian names   *Ghutakeshi, Gandharasa* (Sanskrit); *Kalabashimb, Mili-nargandi* (Hindi); *Aduthodagidda, Karalakkigidde* (Kannada); *Karinochil, Vatamkolli* (Malayalam); *Tev, Bakas* (Marathi); *Karunochi, Vadaikkutti* (Tamil); *Aaddasaramu, Gandharasamu* (Telugu).

Description   An undershrub 1 m high, stem green, smooth; leaves lanceolate, bluntly acuminate, undulate or crenulate, subcoriaceous, pubescent when young, glabrous on maturity; lateral nerves slender, 5–7 on either half, base acute or cuneate, petiole 3–12 cm long; flowers white with purple spot inside, clustered in interrupted spikes; bracts linear, calyx segments linear; corolla about 1.5 cm long; fruits glabrous capsules.

Flowering and fruiting   February–September

Distribution   In North-Eastern India, particularly in Khasi and Jaintia hills up to 1500 m, also cultivated as a hedge plant.

Propagation   By vegetative method

Parts used   Roots and leaves

Uses   The leaves and roots are acrid, bitter, thermogenic, anodyne, emetic, expectorant, diaphoretic, emmenagogue, antiperiodic and insecticidal. Leaf is useful for fractures and dislocated bones. Root is useful for chronic indigestion, dysentery, rheumatism and fever. Leaf juice is used for cold, internal haemorrhages and cough.

K

Kaemferia galanga L.

Family Zingiberaceae

Indian names Galanga (English) *Candrani, Chandramulika* (Sanskrit); *Candramula, Chandramula* (Hindi); *Kachchura, Kacora* (Kannada); *Kaccolam, Kacholum* (Tamil); *Kacholam, Kachoram, Kaccuri* (Malayalam); *Candramula, Chandramoola* (Telugu).

Description Perennial, herbaceous; Rhizome includes little ovate tubers; Leaves 2–3, with broad blade, spreading flat on the ground, round, ovate or sub-arbicular, petioles short, channelled, flowers white 6–12 on short scape, lip bilobed with black or purple spots.

Distribution Entire North-East Region

Propagation By vegetative method

Parts used Rhizomes and leaves

Chemical constituents   The rhizomes contain essential oil (2.4–3.9%), consisting of *p*-methoxy transcinnamate, ethyl-methoxy transcinnamic acid, *p*-coumaric acid, *n*-pentadecane, 3-careen and borneol.

Uses   Rhizome is very useful in pectoral and abdominal pains, headache, toothache and cold. Decoction powder or pill form is used for oral administration. The leaves are used in lotions and poultices for sore eyes, swellings, rheumatism and fevers.

Kaemepferia rotunda L.

Family   Zingiberaceae

Indian names   *Bhucampaka, Bhuchampaka* (Sanskrit); *Bhuichampa, Bhuyicampa* (Hindi); *Nelasampige, Kalluloove* (Kannada); *Nerppicin, Konda-kalava* (Tamil); *Cennalinirkilannu, Cennalinirkkilannu* (Malayalam); *Bhucampakamu, Bhuchampakamu* (Telugu).

Description   Aromatic herb with tuberous rhizomes; leaves simple, few, erect, oblong, ovate, lanceolate, acuminate, 30 cm long, 10 cm wide; flowers fragrant white, borne in a crowded spike, opening successively, lip purple or lilac.

Distribution   Throughout North-East Region, particularly in Meghalaya.

Propagation   By vegetative method

Parts used   Rhizome

Chemical constituents   The oil contains cineol and probably methyl chavicol.

Uses   The rhizomes are thermogenic, aromatic, stomachic, anti-inflammatory, emetic and vulnerary. They are useful in gastropathy, wound, ulcers, blood clots and tumours. Juice is used as eye drop for eye disorders. Rhizome powder is applied on wounds and bruises.

Kalanchoe laciniata (Lamk) DC.

Family   Crassulaceae

Indian names   *Astibhaksha, Hemasagara* (Sanskrit); *Hamsagar, Zakhm hyat* (Hindi); *Kalanaru, Gandu kalinga* (Kannada); *Kattukkalli, Malakkalli* (Tamil); *Sima-jamudu* (Telugu).

Description   Erect, stout perennial fleshy herbs; stem succulent; leaves large, very variable, succulent, deeply pinnatifid twice or thrice; flowers erect, protandrous, yellow, orange, magenta, in paniculate cymes; stamens and carpels as many as petals, pistil with a scaly gland at the base.

Distribution   Throughout North-Eastern India, particularly in valleys, in moist and shaded places.

Propagation   By seeds and vegetative method

Plant parts   Leaves

Chemical constituents   The whole plant contains organic acids like citric, isocitric and malic acids.

Uses   Leaves are boiled and taken for treatment of urinary disorder. Five leaves are taken daily for kidney stone. Leaf is also useful for soars, wounds, gastric complications and insect bite.

Kalanchoe pinnata (Lamk.) Pers.

Family   Crassulaceae

Indian names   *Parnabijah* (Sanskrit); *Jakh me hayat* (Hindi); *Gandukalinga* (Kannada); *Runakkalli, Malaikkalli* (Tamil); *Ilamulachi, Illayinmelthai* (Malayalam); *Simajamudu* (Telugu).

Description   Herbaceous perennial succulent plant up to 40 cm high; stems tubular and glabrous, mottled with purple. Leaves opposite variable from unifoliate to pentafoliate, blade thick and fleshy; inflorescence in terminal cyme with long stalk, flowers pendulous, red or orange-red; fruit of 4 follicles.

Distribution   Throughout North-East India in moist-shaded places.

Propagation   By seeds and vegetative method

Parts used   Whole plant

Chemical constituents   The whole plant contains organic acids like citric, isocitric and malic acids.

Uses   The fresh leaves, which possess antibacterial and demulcent properties, are used in headache, boils, soars, wound, gastric complications, kidney stones, insect bites and burns.

Knemna angustifolia (Roxb.) Warb.

Family   Myristicaceae

Description   Large tree up to 20 m high with a dense, ovoid crown; branches horizontal, slender, bark dark grey, vertically fissured or furrowed; leaves 8–30 × 2.5 cm, oblong, lanceolate or elliptic, acute or acuminate, base rounded, cuneate, glaucous beneath; Inflorescence on short peduncles; flowers rusty, tomentose outside, pink or reddish pink; fruit 2–3 cm long.

Flowering and fruiting   November–June

Distribution   Bangladesh, North-East India, frequent in Meghalaya.

Propagation   By seeds and vegetative method

Parts used   Seeds

Chemical constituents   The seeds contain oil, phytosterol and phenolic acid.

Uses   It is used for spleen disorders, breathing disorders and tastelessness.

Kydia calycina Roxb.

Family   Malvaceae

Indian names   *Bharanga, Potari* (Hindi); *Benda, Bellaka* (Kannada); *Vendai, Vattakannu* (Tamil); *Vellachadachi, Benda* (Malayalam); *Kondapathi, Potri chettu* (Telugu); *Dieng misiri* (Khasi); *Boldubak* (Garo).

Description   Tree 10–15 m tall, deciduous; branchlets tomentose; leaves broader than long, orbicular, cordate to truncate at base, palmately 3–5 lobed at apex, 6–16 cm in diameter, palmately lobed 5–7 cm; flowers white, bracteoles 4–6, obovate or spathulate, oblong, parallel nerved, accrescent; calyx 5 lobed, adpressed and incurved in fruit; anther sessile; ovary ovoid or sub-angular capsules, yellowish, villous, seeds curved, black.

Flowering and fruiting   August–March

Distribution   Throughout India, common in Meghalaya at low elevations

Propagation   By seeds

Parts used   Roots

Uses   Root paste mixed with oil is used as a febrifuge and in rheumatism.

L

Lagerstroemia indica L. [Rare]

Family   Lythraceae

Indian names   Crepe flower (English); *Siddhesvara* (Sanskrit); *Telingachina, Har singar* (Hindi); *Pavalakkurinji, Sinappu,* (Tamil); *Chinnagoranta* (Telugu); *Pharash* (Bengali); *Jarol* (Manipuri).

Description   Small tree up to 10 m tall, usually branching from base; bark white or greyish white; leaves obovate, orbicular, rounded, sub-obtuse or often notched, base narrowed or cuneate, entire, glabrous, when mature except along the midrib; corymbs 10–20 cm across; flowers pink or white, 2–3 cm across; petals 1–1.5 cm long; fruits up to 1 cm across, obconic.

Flowering and fruiting   May–August

Distribution   Native in China, cultivated throughout India and Meghalaya.

Propagation   By seeds

Parts used   Leaves

Uses   Leaves are useful in diarrhoea and dysentery.

Lannea coromandelica (Houtt.) Merr.

Family   Anacardiaceae

Indian names   *Ajasringgi, Jhingini* (Sanskrit); *Jhingan, Ghinghan* (Hindi); *Manjistha, Godda* (Kannada); *Oti, Udi, Othiyamaram* (Tamil); *Kalasu, Karasu, Karayam, Otiyamaram* (Malayalam); *Oddimanu, Gumpena* (Telugu).

Description   Large or middle sized deciduous trees with a straight bark, grey or greyish black, rough, exfoliating in round flakes; leaves crowded, up to 45 cm long, leaflets ovate, lanceolate or elliptic, acuminate, rounded or oblique at base, glabrous; panicles tomentose, up to 25 cm long; flowers minute, yellowish green; drupes up to 1.5 cm long, laterally compressed with a pitted stone.

Flowering and fruiting   February–June

Distribution   Throughout India, common in Meghalaya particularly in deciduous secondary and open forests in Garo hills.

Propagation   By seeds and vegetative method

Parts used   Bark and leaves

Uses   The bark is acrid, astringent, sweet, stomachic and anodyne. It is useful in cuts, wounds, bruises, ulcer, ophthalmia, gout, odontalgia, diarrhoea and dysentery. The leaves are useful in elephantiasis, inflammations, sprains and bruises.

Lantana camera L.

Family   Verbenaceae

Indian names   *Jhingini, Caturangi* (Sanskrit); *Caturang, Ghaneri* (Hindi); *Kadugulabi, Lantavanigidda* (Kannada); *Arisimalar, Unnicceti* (Tamil); *Arippu, Puchedi* (Malayalam); *Pulikampa, Akshinte poolu* (Telugu).

Description   Prickly shrubs, often straggling; bark grey or greyish brown, peeling off in long strips; leaves 2.5–9 × 2.6 cm, ovate, acute or acuminate, cuneate, crenate; spikes capitate, 2.5–4 cm across; flowers 0.8–1.2 cm long, usually orange, sometimes white or pinkish; fruits 0.4–0.6 cm across, shining purplish blue when ripe.

Flowering and fryiting   Throughout the year, more during June–December.

Distribution   Native of America; runs wild throughout India, both in the hills and plains.

Propagation   By seeds and vegetative method

Parts used   Whole plant

Chemical constituents   Catalase, amylase, invertase, lipase, tannase and glucosidase. Appreciable amounts of tannins and sugar and crystalline glucoside have been separated from the resin by ether extraction.

Uses   The plant is useful for tetanus, malaria, epilepsy and gastropathy. Root decoction is a good gargle for odontalgia and is used by hill tribes for all types of dysentery. Leaf powder is useful for cuts, wounds, ulcers and swellings.

Leea indica (Burm. F.) Merr.

Family   Leeaceae

Indian names   *Chatri, Kukkurajihva* (Sanskrit); *Kurkurjihva* (Hindi); *Gadhapatri* (Kannada); *Naikki, Ottanali* (Tamil); *Kudanjazhuku, Njallu, Njazhuku* (Malayalam); *Ankadora* (Telugu).

Description   Shrubs or small trees; leaves 2–3 pinnate, leaflets many, oblong, lanceolate, long acuminate, base rounded or truncate or acute, umbel 10–12 cm long; flowers pale green; berries depressed, globose, obscurely angled, red or black, 3–6 seeded.

Flowering and fruiting   June–December

Distribution   Common throughout India; common in Meghalaya in deciduous forests, forest margins and secondary forests.

Propagation   By seeds and vegetative method

Parts used   Roots and leaves

Uses   The roots are useful in diarrhoea, dysentery, hyperdipsis, ulcers and skin diseases. The young leaves are used for digestive purposes.

Leucas aspera Spreng.

Family   Lamiaceae

Indian names   Thumbe (English); *Dronapuspi, Ksavaka* (Sanskrit); *Chota halkkusa, Chota-kalkusha* (Hindi); *Tumbe, Thumbe* (Kannada); *Tumpai, Kaviz thumbai* (Tamil); *Tumba, Tumpa* (Malayalam); *Tammachettu, Tummacettu* (Telugu).

Description   Erect, much branched annual herb, with quadrangular stems and branches; leaves linear or oblong or lanceolate, obtuse, pubescent up to 7.5 cm long and 1.25 cm wide; flowers pure white, small dense terminal or axillary whorls; fruits nutlets, oblong, brown, smooth.

Flowering and fruiting   October–February

Distribution   Throughout India on wastelands and roadsides up to 900 m.

Propagation   By seeds

Parts used   Leaves and flowers

Chemical constituents   Alkaloids and glucoside

Uses   The leaves and flowers are acrid, thermogenic, carminative, digestive, anthelmintic, expectorant, antibacterial and depurative. They are useful in verminosis, arthralgia, chronic skin eruption, cough and catarrh in children, intermittent fevers.

Leucas cephalotes Spreng.

Family   Lamiaceae

Indian names   *Chitrakshupa, Chitrapathrika* (Sanskrit); *Deldona, Goma madhupati* (Hindi); *Dronapushpi* (Kannada); *Thumbai, Peruntumpai* (Tamil); *Tumba* (Malayalam); *Peddathummi, Thummi* (Telugu).

Description   A stout course herb with hairs spreading; leaves 3.7 × 2.5–3 cm, elliptic, lanceolate, membranous, more or less pubescent, petiole 0.2–0.5 cm long; bracts prominently nerved.

Flowering and fruiting   September–January

Distribution   Fairly common in North-East India.

Propagation   By seeds

Parts used   Leaves

Uses   The leaf juice is useful in chronic skin eruption and painful swelling.

Leucus plukereti (Roth.) Spreng.

Family   Lamiaceae

Description   Herb much branched, erect, annual 40–60 cm high; leaves subsessile, linear or narrow, oblong lanceolate; flowers usually white, in axillary, whorls, or terminal; calyx tubular, oblique; corolla tube, bilabiate; stamens 4, ascending, anthers conniving, cells divaricate; style subulate; nutlets lobe, ovoid.

Distribution   Entire North-Eastern India, frequently in Khasi and Jaintia hills, Meghalaya.

Propagation   By seeds and vegetative method

Parts used   Leaves

Uses   Leaves are useful as antipyretic and also as an external application for psoriasis, chronic skin eruption and painful swelling. Flowers are used for cold and cough.

Ligustrum lucidum Ait.

Family   Oleaceae

Description   Small tree; bark grey, croky, inside dull white with coarse strands of dark brown, sometimes dirty chocolate brown colour; branchlets closely lenticallate; leaves elliptic, lanceolate to ovate, lanceolate, acuminate, subcoriaceous, glabrous, lateral nerves 8–10 on either half, irregular, slender, base obtuse, often cuneate; flowers white, faintly scented, sessile in bracteate terminal, glabrous, panicles; bracts linear, lanceolate; calyx tube, almost truncate or obscurely toothed; corolla lobes oblong, rounded; fruits elongated.

Flowering and fruiting   July–November

Distribution   Throughout India, common in Meghalaya, particularly in Khasi hills up to 5000 feet.

Propagation   By seeds and vegetative method

Parts used   Bark and leaves

Uses   Decoction of bark and leaves are useful in rheumatism.

Lindera latifolia H.k.f. [Endemic/Rare]

Family   Lauraceae

Description   Moderate sized tree; bark grey, warty, reddish or yellowish brown; branchlets, undersurface of leaves and inflorescence, grey tomentose; leaves obovate, broadly oblanceolate or elliptic, acuminate to sub-acute, thin, lateral nerves 8–12 on either side, impressed above; petiole 1–2 cm long; umbels, 10–12 flowered on rather stout peduncles solitary, clustered or fascicled on short stalks, pubescent, filaments hairy; fruits globose, turning brown to dark chocolate on ripening.

Flowering and fruiting   February–October

Distribution   In Khasi hills up to 4,500–6,000 ft.

Propagation   By seeds and vegetative method.

Parts used   Aerial parts

Uses   It is useful for gastric or abdominal pains, hernia pain, rheumatic bone and joint pains and pains from external injury.

Lindera pulcherrima (Nees) Benth.

Family   Lauraceae

Indian names   *Dadia* (Hindi); *Dieng jaburit* (Khasi).

Description   Small or middle sized trees up to 15 m high; branchlets slender, bark blackish brown; leaves oblong, lanceolate, ovate, long, caudate, acuminate, base rounded, shining, green above; umbels in sessile clusters; flowers 1–1.5 cm, yellow; fruits ellipsoid.

Flowering and fruiting   March–December

Distribution   Temperate Himalayas, very common in Meghalaya, particularly in Khasi hills.

Propagation   By seeds and vegetative method

Parts used   Root tubers and bark

Uses   It is useful for gastric or abdominal pains, hernia pain, rheumatic bone and joint pain, pain from external injuries and chest pain. Bark used in cold, cough and worm.

Lindernia anagallis (Burm. f.) permell.

Family   Scrophulariaceae

Description   Plant herb; leaves ovate, obscurely serrate, truncate, cordate, shortly petiole; pedicels axillary, petals pink, stamens 4, anther and filament appendiculate; fruiting sepals divided almost to base, less than half the length of the capsule.

Distribution   Whole of Meghalaya and adjoining areas

Propagation   By seeds

Parts used   Leaves

Uses   Paste of leaves is useful for head ache and stomach ache.

Lasia spinosa (L) Thw.

Family   Araceae

Description   Stout herb with thickly creeping rhizomes; leaves long petiole, hastate, pinnatifid; spadix short, cylindrical, claret-coloured, densely packed with pink flowers, fruit berry.

Distribution   Throughout North-East Region in swampy and marshy areas and Songsak areas of Garo hills.

Propagarion   By vegetative method

Parts used   Root and leaves

Uses   Leaves and root juice are anthelmintic. Leaf paste is useful in cuts, diarrhoea, dysentery, intestinal worms, tuberculosis and fistula. Root decoction is useful in throat and piles problems.

Litchi chinensis Sonner.

Family   Sapindaceae

Indian names   Lychee, Nephelium litchi (English) *Lichi* (Hindi); *Litchi hannu* (Kannada); *Lichi* (Marathi); *Lachu* (Mizoram); *Lachu, Lichi* (Oriya).

Description   Middle sized tree, up to 15 m high, branching nearly from base; crown dense, large, ovoid; bark grey, nearly smooth; branchlets lenticellate, rachis 5–15 cm long; leaflets oblong, lanceolate, elliptic, acute to acuminate, base obtuse, cuneate, glabrous and shining above; panicles up to 50 cm long; flowers 0.3–0.5 cm across; petals absent; stamens exerted; drupe fleshy with a brown muricated rind turning red; seeds shining brown.

Flowering and fruiting   January–June

Distribution   Cultivated in India for delicious fruits

Propagation   By vegetative method

Parts used   Fruits

Uses   The fruits are a source of tartaric acid. Young fruits are used for pickling. It has emetic and digestive properties.

Litsea citrata BL.

Family   Lauraceae

Indian names   *Zeng-jil, Zeng jir* (Garo); *Siltemur* (Nepali).

Description   Small tree, deciduous, aromatic; bark green, warty, thin, yellowish, turning brownish, young shoots silky; leaf buds naked, leaves somewhat inequilateral, lanceolate, narrow, ovate, acuminate, membranous, bright green above, lateral nerves 8–13, slender, midrib often purplish below, base somewhat oblique, acute; petiole slender; flowers in capitate umbels, solitary or in corymbs; bracts 4, ovate, membranous, glabrous, ciliate at the edges; peduncles slender; sepals membranous, obvate, sub-equal.

Flowering and fruiting   November–July

Distribution   Common in Meghalaya, particularly in Garo hills

Propagation   By seeds

Parts used   Fruits

Uses   Fruit is useful for medical perfumes, deodorant and as an insect repellent.

Litsea cubeba (Lour.) Pers.

Family   Lauraceae

Indian names   *Kankolam* (Sanskrit); *Dieng sying* (Khasi); *Zengjir* (Garo); *Sernam* (Mizoram).

Description   Small or middle sized tree up to 15 m high; crown lax, ovoid with usually slender, drooping branches; bark greenish brown; leaves lanceolate, tapering to the tip, long, acuminate, base obtuse, glaucous beneath; umbels solitary or in short corymbs; flowers yellow or pale yellow; fruits ovoid-ellipsoid.

Flowering and fruiting   November–July

Distribution   Indo-Malaya, fairly common in Meghalaya usually in lower elevations, Bagmara in Garo hills.

Propagation   By seeds

Parts used   Bark, leaves and fruits

Uses   Fruits medicinal perfume, deodorant and as an insect repellent.

Litsea lancifolia (Roxb.) Wall.ex. Hook. F.

Family   Lauraceae

Description   Shrubs or small trees, up to 8 m high, slender; bark dark grey; leaves lanceolate, base narrowed, cuneate, glabrous; shortly peduncled; flowers yellowish; fruits ellipsoid-oblong, 1–1.5 cm long, bluish.

Flowering and fruiting   April–November

Distribution   Eastern tropical and sub-tropical himalayas, frequent in Meghalaya as undergrowth in dense forests.

Propagation   By seeds

Parts used   Aerial parts of the plant

Uses   Plant is used for cuts, wounds and ulcers.

Litsea monopetala (Roxb.) Pers.

Family   Lauraceae

Indian names   *Meda, Katmarra* (Hindi); *Gajapippali, Hemmudi* (Kannada); *Muchaippeyetti, Picinpattai* (Tamil); *Tumitla* (Manipuri); *Chiru mamidi, Meda* (Telugu).

Description   Middle sized evergreen tree, up to 20 m high; crown narrow, oblong; bark dark grey, often vertically fissured; leaves obovate, oblong or oblanceolate, coriaceous, glabrescent above and rusty pubescent beneath. Flowers pale yellow, in pedunculate umbellate heads. Perianth segments 5, almost free stamens 9–12; short racemes; fruits 0.8–1 cm long, ovoid, blackish when ripe.

Flowering and fruiting   April–November

Distribution   Nearly throughout Northern India, frequent in Meghalaya, often planted; leaves used in rearing muga silkworm.

Propagation   By seeds

Parts used   Leaves, bark and seeds

Uses   Bark and leaves are astringent and they are useful for ulcers. Leaf paste is used in boils and fruits are useful for rheumatism.

Litsea saliciflora Roxb. Ex. Nees.

Family   Lauraceae

Description   Small tree or shrubs; branches silky, pubescent, slender, horizontal; bark greyish or blaze yellowish, turning dark brown; leaves very variable, elliptic, lanceolate, oblong, acuminate, base cuneate, nerves prominent beneath; umbels fascicled, shortly peduncled, yellowish; fruits ellipsoid, seated on the slightly enlarged, sub-cupular perianth.

Flowering and fruiting   February–June

Distribution   Entire North-East Region particularly in Garo and Khasi hills.

Propagation   By seeds

Parts used   Leaves

Uses   Aerial parts are used in ulcers, dysentery, diarrhoea, indigestion and abdominal pains.

Lonicera japonica Thumb.

Family   Caprifoliaceae

Indian name   Japanese honeysuckle (English).

Description   Shrub erect, evergreen; stem brownish red; leaves opposite, tomentose; flowers usually in pairs, in axillary or sub-terminal peduncles or sessile in leaf axils, often connate by their ovaries, subtended by a bract and bracteoles, fragrant, white when blooming then changing to yellow; calyx tube ovoid; corolla tubular or funnel shaped; stamens 5, inserted on the corolla tube; anthers usually exerted; ovary 2–3 celled; style filiform; stigma capitate; ovules many in each cell in double rows; fruit fleshy, berry, globose, black when ripe.

Flowering and fruiting   April–November

Distribution   North-Eastern India; grows wild in hill areas

Propagation   By seeds and vegetative method

Parts used   Whole plant

Chemical constituents   The whole plant contains tannins, saponins, luteolin, inositol and cryptoxanthin.

Uses   The stem is antibacterial and antiallergic. It is recommended in the treatment of boils, impetigo, urticaria, allergic rhinitis, fever, malaria, measles, syphilis and rheumatism. Leaves are astrigent used for gargling; flowers are used for pulmonary diseases.

Luvunga scandens (Roxb.) Buch-Ham. [Extremely/Rare]

Family   Rutaceae

Indian names   *Dhankshika, Dhmanksholi, Gandhakokila* (Sanskrit); *Jeeanthi balli, Lavangalathe* (Kannada); *Lavangalata* (Manipuri); *Lavangalata* (Bengali).

Description   Scandent evergreen shrub, generally tufted from the ground with strong axillary, sharp, straight or slightly recurved spines; bark ash coloured, somewhat rough, about 2.5 cm thick, cream coloured inside; leaves alternate, simple on young shoots; leaflets oblong, lanceolate, narrowed at both ends, entire, firmly coriaceous, dark green, glabrous, minutely punctate both above and beneath; flowers white, fragrant, about 2.5 cm diameter, in axillary branched racemes with cymose branches; calyx cup-shaped, minutely 4–6 toothed; petals linear-oblong, fleshy, recurved; filaments 8–10 glabrous, united below; anthers linear; ovary 3-celled, 1 ovule in each cell; fruit an oblong berry, 3-lobed, 1–3 seeded with an aromatic pulp; seeds ovoid, pointed, radical superior, cotyledons green, fleshy, albumen 0.

Flowering and fruiting   February–September

Distribution   In Meghalaya, North-Eastern India

Propagation   By seeds

Parts used   Leaves

Lycopodium clavatum L. [Vulnerable]

Family Lycopodiaceae

Indian names Vegetable sulphur, Running clubmoss (English); *Bendarli* (Marathi); *Leishang* (Manipuri).

Description Creeping evergreen plant growing to 12 cm; has numerous stragging branches covered with bright green linear leaves, and scaly spikes bearing yellow spores.

Distribution Club moss is found through regions of northern hemisphere. It is common on mountains and in moorland. The plant is gathered in summer.

Parts used Plant and spores

Chemical constituents Club moss contains about 0.1–0.2 per cent alkaloids, polyphenols, flavonoids and triterpenes.

Uses Club moss is diuretic, sedative and antispasmodic, and it is particularly useful for treating chronic urinary complaints. The herb may also be taken for indigestion and gastritis. The spores may be applied to relieve itching.

Lyonia ovalifolia (Wall) Druce.

Family Ericaceae

Indian names *Aiyaar, Airean, Ailan* (Hindi); *Diengla samiang, Jirhap* (Khasi); *Tlangham* (Mizoram).

Description Middle sized crooked tree or large shrubs, 1–10 m high; crown lax, stems twisted; bark reddish-brown, thin; young pairs pubescent or glabrous; leaves ovate to oblong, elliptic, abruptly acuminate, base rounded, cordate; racemes 4–15 cm long; flowers recurved, creamy white or pale pink; calyx deeply lobed; corolla rounded; capsules subglobose, depressed above; 0.3–0.5 cm across.

Flowering and fruiting   May–October

Distribution   Temperate Himalayas and at higher elevations in Meghalaya

Propagation   By seeds

Parts used   Root

Uses   Root used for rheumatism, scabies, eczema, cuts and wounds.

M

Macropanax undulatus Wall ex. D. Don. Saem.

Family   Araliaceae

Description   Tree up to 15 m high; crown dense, much spreading, bark blackish brown, smooth or nearly so; leaves 3–5 foliate, common petiole 6–22 cm long, leaflets 4–16 × 1–4 cm, elliptic, oblong, lanceolate, often broader above the middle, acuminate, base rounded or cuneate, subentire, coriaceous, glabrous, lateral nerves 5–7; flowers greenish yellow; calyx lobes small; petals 1.5–2 mm long, deciduous; fruits ellipsoid, oval, 0.5–0.8 cm across with persistent bilobed stylar column.

Flowering and fruiting   June–February

Distribution   Burma, Bhutan, Bangladesh and India; common in Meghalaya in evergreen forests.

Propagation   By seeds

Parts used   Leaves

Uses   Leaves are useful for dysentery and diarrhoea.

Macrosolen cochinchinensis (Lour.) Tiegh.

Family   Lorantheceae

Description   Branching drooping; leaves elliptic, oblanceolate, lanceolate, acute or acuminate, base narrowed, cuneate; racemes fascicled, up to 3.5 cm long; flowers greenish or greenish yellow, 1–1.5 cm long; stigma purplish; fruits globose, pale-yellow.

Flowering and fruiting   December–May

Distribution   Indo-Malaya; nearly throughout India; parasitic on host-like *Artocarpus*.

Propagation   By seeds

Parts used   Leaves

Maesa chisia D. Don.

Family   Myrsinaceae

Indian names   *Susi porma* (Assamese); *Bilouni* (Nepali).

Description   Shrub about 5 m high, branches lenticellate, spreading; bark dark brown or reddish brown; leaves ovate, lanceolate, crenate or serrulate, usually long acuminate, membranous, glabrous; inflorescences, racemose glabrous, lax, often branched; calyx segments not ciliate; flowers pentamerous, often unisexual, dioecious in axillary or terminal panicled racemes; fruits white, spongy.

Flowering and fruiting   February–November

Distribution   Throughout North-Eastern India.

Propagation   By seeds

Parts used   Leaves, stem and fruit

Uses   Both leaves, bark and stem exhibit insecticidal activity; juice is used as eye drop. Fruits are anthelmintic.

Maesa ramentacea Wall.

Family   Myrsinaceae

Indian names   *Bol jakhandok, Thebeloa* (Garo); *Dieng soheit iar* (Khasi); *Lajachio* (Naga); *Kokkidi* (Telugu).

Description   Large shrubs or small tree; bark dark brownish, warty, thick having vertical lenticels; greenish white, turning brownish; leaves ovate, lanceolate to elliptic, entire, slightly recurved, acuminate rarely acute, thinly coriaceous, glabrous with long transparent bars, lateral nerves more prominent below, 6–9 on either half, base rounded or acute; petiole up to 2 cm long; flowers white, sometime longer, fruit succulent, dull brownish white; seeds angular, dark brown or blakish.

Flowering and fruiting   January–November

Distribution   Entire North-East Region; commonly in Garo, Khasi and Jaintia hills.

Propagation   By seeds

Parts used   Stem, leaves and fruit

Uses   Leaves, bark and stem are useful in insecticidal activity. Fruits are anthelmintic; juice is used as eye drop.

Mahonia acanthifolia G. Don.

Family   Berberidaceae

Description   Erect shrubs, up to 5 m high; bark greyish brown, branches angled; leaves lanceolate, oblong, lanceolate, acute, cuneate, spinous, serrulate shining on both surface; flowers yellow with very small outer sepals; sepals 5 mm long; petals nearly equalling the size of sepals; berries, oblong, elliptic, deep purple when ripe.

Flowering and fruiting   August–February

Distribution   In Meghalaya and Manipur

Propagation   By seeds

Parts used   Bark

Uses   Decoction of bark is useful as eye drop for eye rash.

Mahonia nepaulensis DC.

Family   Berberidaceae

Indian names   *Kasmal* (Hindi); *Pual-eng* (Mizoram); *Mullukkala, Mullukkatampai* (Tamil).

Description   An erect shrub or sparingly branched small tree with soft corky bark; leaves sheathed at the base with a pair of subulate stipules; leaflets 2–12 pairs, besides the terminal coriaceous, glabrous, shining above, pale beneath, with 3–8 large spinous teeth on either side; raceme 10–25 cm long, erect, persistent, covering the ends of the branchlets; bracteoles small, broadly ovate or oblong, concave. Sepals 6, glandular at the base; stamens 6, sensitive; anthers dehiscing by ascending valves. Ovary 1-celled, style short, stigma capitate; large berries, elliptic or globose, purple.

Flowering and fruiting   October–January

Distribution   Khasi hills, 1200–1400 m, on sunny spurs and open hill slopes.

Propagation   By seeds

Parts used   Bark

Uses   Decoction is used as eye drop for eye rash or moles.

Mallotus philippensis (Lam.) Muell. Arg.

Family   Euphorbiaceae

Indian names   Kamala tree (English); *Bahupushpa, Chandra* (Sanskrit); *Kambhal, Kambila* (Hindi); *chandrahittu, ettunalige* (Kannada); *Avam, Kabilam, Kabilappodi* (Tamil); *Chenkolli, Kampipala* (Malayalam); *Adavigubbatuda, Benduruppu* (Telugu).

Description   A small or middle-sized evergreen tree, sometimes a much branched large shrub; young parts densely covered with minute red hairs; leaves alternate borne on long stalks, variable in shape, 7–20 cm long, their lower side dotted with reddish gland, prominently veined; flowers minute, male and female on separate plants; female flowers in erect 5–8 cm long spikes; male flowers yellow, in 8–15 cm long, drooping bunches; fruit 8–15 mm, roundish, 3-lobed, densely covered with a reddish brown, powdery substance and minute hairs, which are easily rubbed out.

Flowering and fruiting   October–January

Distribution   Throughout tropical region in India from an altitude of 1,500 m in the Himalayas, southwards up to Kerala, common in Meghalaya.

Propagation   By seeds and vegetative method

Parts used   Fruits

Chemical constituents   Linoleic, oleic, lauric, myristic, palmitic and stearic acids.

Uses   The red glandular and hairy substance separated from the fruits forms the drug. The hairs are acrid, thermogenic, purgative, digestive, styptic, vermifuge, alexipharmic and depurative. They are useful in verminosis, constipation, flatulence, wounds, ulcer, cough, renal and vesical calculi, polsorious afflictions, scabies, ringworm, herpes and skin afflictions.

Melastoma malabathricum L.

Family   Melastomaceae

Indian names   *Tinisah* (Sanskrit); *Phutki* (Hindi); *Ankerki, Doddanekkare* (Kannada); *Nakkukaruppan, Katalai* (Tamil); *Kadah, Katali, Totukara* (Malayalam); *Pattudu, Pathudu* (Telugu).

Description   Bushy shrub, stem strigose with long, subulate to short echinate scales. Leaves oblong, lanceolate to elliptic, acuminate, and scabrous with hairs adpressed for their entire length, under surface scaly; flowers 4–7 cm across, mauve purple, usually in fascicles of 1–5 at branch tips; calyx densely scaly; fruits 0.6–0.8 cm across, truncate and purple.

Flowering and fruiting   October–December

Distribution   Throughout India; common in Meghalaya particularly in open wastelands and near water courses.

Propagation   By seeds

Parts used   Leaves, bark and flowers

Uses   Flowers are useful for piles, haemarrhages and necrosis. Juice and paste of leaves are useful for cuts and wounds.

Mangifera indica L.

Family   Anacardiaceae

Indian names   Mango (English); *Alipriya, Amra* (Sanskrit); *Amra, Amba* (Hindi); *Amba, Ballimavu* (Kannada); *Mamaram, Mankai,* (Tamil); *Manga, Gomanga* (Malayalam); *Amramu, Elamavi* (Telugu).

Description   A large evergreen tree with rough thick dry grey fibrous bark; leaves crowded at the ends of branches, oblong or obovate, lanceolate, bluntly acuminate, entire but often with wavy margins, coriaceous, glabrous, mature, dark glossy green, pinkish when very young, turning yellow before falling; flowers about 35 cm across, greenish yellow, scented; calyx 4–5 partite, segments, imbricate, ovate, concave, pubescent outside; petals imbricate much longer than the calyx.

Flowering and fruiting   February–July

Distribution   Throughout North-East Region

Propagation   By seeds and vegetative method

Parts used   Bark

Chemical constituents β-carotenes and xanthophylls, neo-β-carotene U and neo-β-carotene B, are present in a few varieties. It also contains 2-octene, α- and β-pinene, α-phellandrene, limonene, dipentene, nerol, geraniol, neryl acetate, citronellal, etc.

Uses   The roots and bark are astringent, acrid, refrigerant, styptic, antisyphilitic, vulnerary, antiemetic, anti-inflammatory and constipating. They are useful in metrorrhagia, colonorrhagia, pneumorrhagia, leucorrhoea, syphilis, wounds, ulcers, vomiting, uteritis, diarrhoea, dysentery, diphtheria and rheumatism. Paste of roots and bark is mixed with water and is taken for stomach ache and fever. Rind of unripe fruit is ground with mentha and a little honey given in bleeding dysentery. Rind of unripe fruit mixed with curd is a remedy against cholera. Tender leaves and seed kernel powder is useful for diabetes. Ashes of leaves are useful for burns and scalds.

Manihot esculenta Crantz.

Family   Euphorbiaceae

Indian names   Tapioca, Manioc (English); *Darukandah, Kalpakandah* (Sanskrit); *Sakkarkand* (Hindi); *Maragenasu* (Kannada); *Maravallikkilangu* (Tamil); *Kappa, Marachini* (Malayalam); *Karrapendalamu* (Telugu).

Description   A shrub 2–5 m in height with stems and branches of varying colour, marked by leaf scars and clusters of tuberous roots; leaves palmate, pale green, 5–9 lobed, lobes lanceolate, acuminate; flowers large, monoecious, in racemes, males above, females below; calyx campanuate, 5 lobed, petals absent; stamens 10 in two whorls.

Distribution   Throughout India and common in Meghalaya.

Propagation   By roots

Parts used   Tuberous roots

Chemical constituents   The tuber contains small quantities of albumin, globulin and glutellin; prolamine is present in negligible amounts. The essential amino acids present in the total proteins are arginine, histidine, isoleucine, leucine, lysine, methionine, threonine and valine.

Uses   The juice of tubers is useful for constipation and indigestion, boiled tubers for diarrhoea and fresh tubers for poultice on sores and boils. The leaves are infused in the bath water to treat fever.

Melia azedarach L.

Family   Meliaceae

Indian names   *Akshadru, Arista* (Sanskrit); *Bakain, Bakarja* (Hindi); *Arebevu, Bettada-bevina* (Kannada); *Malaivempu, Malaiveppam* (Tamil); *Aryaveppu, Malaveppu* (Malayalam); *Koda-vepa, Konda-vepa* (Telugu).

Description   Middle sized tree; crown lax; bark grey, greyish brown, smooth at first, rectangular scaly in old trees; leaves bi or tri-pinnate, 30–90 cm long; leaflets 2,5–5 × 1–2.5 cm, ovate, lanceolate, oblanceolate or elliptic, acuminate, base cuneate, oblique, serrate to entire, stellate, tomentose when young; flowers 1–1.5 cm across, purple; drupe ovate or ellipsoid, globose with 4 seeds, yellow when ripe.

Flowering and fruiting   June–December

Distribution   Indo-Malaya; cultivated everywhere for wood, usually cultivated in Meghalaya.

Propagation   By seeds and vegetative method

Parts used   Bark

Chemical constituents   The stem and root barks contain the alkaloid azaridine, sterols and tannins. The leaf contains paraisine and flavonoid. The seeds contain stearic, palmitic, oleic and linoleic acids.

Uses   Root bark is used for ascariasis and oxyuriasis.

Melodinus monogynus Roxb.

Family   Apocynaceae

Indian names   *Haladi kanagalu* (Kannada); *Tylli siertub* (Jaintia); *Soh brab* (Khasi).

Description   A large climber with milky juice, glabrous; branches smooth brownish, old leaves yellow; leaves elliptic, oblong or lanceolate, acuminate, margins obscurely recurved, lateral nerves 15–20 on either half, slender, intermediate nerves numerous, often forked, reticulation transverse, base cuneate; flowers white, terminal, trichotomously branched, puberulous, paniculate cymes; calyx segments ciliate, ovate, oblong; corolla tube widening upwards, villous within, lobes oblong, obtuse, oblique, coronal scales villous, bifid at the apex; berry globose, smooth, orange coloured 10 cm across.

Flowering and fruiting   April–October

Distribution   In North-East India, particularly in Khasi and Jaintia hills.

Propagation   By seeds

Parts used   Fruits

Uses   Cultivated for the fruits which are largely pickled or eaten raw; they are reputed to have antiscorbutic properties.

Mentha arvensis Linn.

Family   Lamiaceae

Indian names   Japanese mint, Field mint (English); *Pudina, Putiha* (Sanskrit); *Podina, Ban pudina* (Hindi); *Chetamargugu, Chetni-marugu* (Kannada); *Pudina, Puthina* (Tamil); *Puthina* (Malayalam); *Igaenglikura, Pudina* (Telugu).

Description   Perennial herb, 30–50 cm high, stems quadrangular, erect or prostrate, rooting at the nodes; leaves shortly petioled or sessile, oblong, opposite, ovate or lanceolate serrate, whorls, axillary capitate, calyx teeth triangular or lanceolate, softly tomentose on both sides, margins serrate; inflorescence in axillary capitate whorl; flowers small, white or lilac.

Distribution   Grows wild in the hills and is cultivated every where; also runs wild in Khasi hills.

Propagation   By vegetative method

Parts used   Whole plant

Chemical constituents   Plant yields an essential oil consisting of L-menthol 65–85 per cent menthyl acetate, L-menthone, L-α-pinene, limonene.

Uses   The dried plant is a refrigerant, stomachic, diuretic and stimulant. Leaf juice is used for stomach ache. It is anthelmintic and is also used as an ear drop.

Meriandra benghalensis Benth.

Family   Lamiaceae

Indian names   *Kafurkapat* (Hindi); *Karpoora vaasane aele mara* (Kannada); *Cayailai, Ceci* (Tamil); *Kanghuman* (Manipuri); *Simakarpuram, Seema karpooram* (Telugu).

Description   A shrub strongly smelling of camphor; leaves 5–8 cm long, obtuse, base rounded; petiole 0.5–0.8 cm long. Flowers white in globose whorls, in terminal spikes, possesses the properties of *Salvia officinalis*.

Distribution   Throughout North-East Region

Propagation   By seeds

Parts used   Stem and leaves

Uses   This is a tonic, carminative, astringent and antiseptic. Decoction is given for urinary infections.

Merremia umbellata (L.) Hall. f.

Family   Convolvulaceae

Indian names   *Sithri bodu* (Garo); *Kolavarvalli* (Tamil); *Kolavara* (Malayalam); *Catukattutivva, Kappativva* (Telugu).

Description   Climbers or twiners; stems angular or terete; leaves 3–7 × 1–3.5 cm, ovate, oblong, elliptic or lanceolate, acute, often apiculate, base cordate, hairy or glabrous; flowers white or dull white; capsules ovoid.

Flowering and fruiting   February–October

Distribution   Tropics, Africa to Australia; throughout India except North-West, common in Meghalaya at lower elevations, Balphakram and Garo hills.

Propagation   By seeds and vegetative method

Parts used   Whole plant

Uses   The plant is bitter, acrid, refrigerant, diuretic, alterant, anthelmintic, carminative and digestive. It is useful in cardiac diseases, gastropathy, metropathy, fever, anaemia, strangury, otalgia and rat bite.

Merremia vitifolia (Burm. F.) Hall. f.

Family   Convolvulaceae

Indian names   *Dukhumi bidu, Dukhumi bider* (Garo); *Navalicha vel, Navli* (Marathi);

Description   Hairy twiners or trailers; leaves 2–8 cm across, broadly ovate, lobes acute or acuminate, base deeply cordate, more hairy above; cymes 1–7 flowered, erect; flowers bright yellow, 5–7 cm across, spreading; calyx spareely hairy; capsules 1.5 cm in diameter; 4-seeded.

Flowering and fruiting   February–July

Distribution   Indo-Malaya; throughout India; common in Meghalaya, Bagmara, Garo Hills.

Propagation   By seeds and vegetative method

Parts used   Whole plant

Uses   The plant is useful for haemarrhoids, uropathy, inflammations and general debility.

Mesua ferrea L. [Near endemic]

Family   Clusiaceae

Indian names   Ironwood (English); *Ahikesara, Ahipuspa* (Sanskrit); *Gajapushpam, Nagakesara* (Hindi); *Kanchana, Nagakesara* (Kannada); *Naganchembagam* (Malayalam); *Naganchambagam* (Tamil); *Gajapushpamu, Geja-pushpam* (Telugu).

Description   A medium sized, large handsome, glabrous, evergreen tree, up to 18–30 m in height and reddish brown bark which peels off in thin flakes; leaves simple, opposite, thick, lanceolate, coriaceous, covered with waxy bloom underneath, red when young, acute or acuminate nerves.

Distribution   Throughout India, in evergreen forests up to 1,500 m.

Propagation   By seeds and vegetative method

Parts used   Flowers

Chemical constituents   Palmitic, stearic, linoleic acid, stearo diolein, palmito-diolein, triolein and linoleodiolein.

Uses   The flowers are bitter, astringent, acrid, anodyne, digestive, anthelmintic, diuretic, expectorant, stomachic, febrifuge and cardiotonic. Flowers are useful in asthma, cough, scabies, vomiting, dysentery, ulcers and burning sensation of the feet, dipsia, and cardiac debility.

Meyna laxiflora Robyna.

Family   Rubiaceae

Indian names   *Nagakesarah, Nagapuspah* (Sanskrit); *Acchoora mullu, Chegu gadde* (Kannada); *Manakkarai* (Tamil); *Segagadda, Veliki* (Telugu); *Soh mon* (Khasi); *Thitchkeng* (Garo).

Description   Middle sized trees, 10 to 15 m in height; crown compact, oval; bark brown or deep grey, often with short spines; leaves elliptic, ovate, acuminate, base shortly cuneate, glabrous or nearly so, tufted hairy at nerve. Stipules long, cuspidate; cymes axillary, very shortly peduncled or fascicled; flowers greenish white; corolla villous within; lobes spreading or reflexed; stigma 5-lobed; drupes 2.5–4 cm long, yellowish when ripe.

Flowering and fruiting   April–November

Distribution   Entire North-East India, in lower elevations

Propagation   By seeds

Parts used   Root, bark and leaves

Uses   The root and bark are useful for gastropathy and fever. Leaves are useful in haemorrhages, wounds, dyspepsia, skin diseases, leprosy and fever.

Michelia champaca L. [Endangered]

Family   Magnoliaceae

Indian names   Champak, Sapu (English); *Anjana, Atigandhaka* (Sanskrit); *Campa, Campaka* (Hindi); *Champaka, Kendasampige* (Kannada); *Sambagam, Shampangi* (Tamil); *Champakam, Chempakap-pu* (Malayalam); *Campangi, Campangipuvvu* (Telugu).

Description   A tall tree, 20–25 m in height and 50–80 cm in diameter with straight stem and smooth bark, grey, greyish brown, rough; crown lax, oval; branches tapering to a long point;

leaves simple, alternate, oblong, 10–25 × 4–7 cm, ovate, lanceolate, acuminate, base cuneate, glabrous and dark green and shiny above, pale and glabrescent beneath, petiole 2.5–3 cm; flowers 3–4 cm long, axillary, pale or orange yellow, segments of perianth 15–20, ovaries pubescent; flowers 2.5 cm in diameter, very fragrant; peduncles short; buds silky; sepals oblong, acute; petals linear; fruit 7–15 cm long, cone-like, drooping; ripe carpels ovoid, ellipsoid.

Flowering and fruiting   May–April

Distribution   Indo-Malaya; usually cultivated for their fragrant flowers; cultivated as well as wild in foot hills.

Propagation   By seeds and vegetative method

Parts used   Whole plant

Chemical constituents   Oleic and palmitic acids.

Uses   The root and root barks are purgative and emmenagogue and are useful in the treatment of inflammation, constipation, amenorrhoea and dysmenorrhoea. The stem bark is astringent, febrifuge, diuretic, stimulant and expectorant and is useful in chronic gastritis, fever, strangury, cough, and bronchitis and cardiac debility. Flower buds and fruits are bitter, acrid, digestive, carminative, anthelmintic, diaphoretic, expectorant, stimulant, stomachic and antipyretic. They are useful in dyspepsia, nausea, burning sensation, skin diseases and ulcers. The seeds and fruits are useful in the treatment of psoriasis.

Micromelum minutum (Forest. f.) Wt. and Arn.

Family   Rutaceae

Indian name   *Nauterimnam* (Mizoram).

Description   Shrub or small slender tree up to 15 m in height; leaves alternate, pinnately compound with 7–12 unequal sides, leaflets each with a short petiole, the entire leaf up to 50 cm long; Flowers white, fragrant, borne in many flowered terminal or axillary panicles; fruit red, drupe with small puncate dots on the surface, up to 1 cm long.

Distribution   Sikkim, Assam and Meghalaya in North-East Region.

Propagation   By seeds and vegetative method

Parts used   Bark and leaves

Chemical constituents   Butyl 7-methoxyfindersine, imperatorin, limettin, micromelin, microminutinin, minumicrolin, murralongin.

Uses   Bark is useful to reduce headache. In Garo hills, leaf juice is used for preventing vomiting.

Micromelum pubescens non. BL.

Family   Rutaceae

Indian names   *Gobar-hura, Sagladi* (Assamese); *Marsusepel, Silkhol* (Garo); *Kakkaipalai* (Tamil).

Description   Small evergreen tree, 0.5–1 m high; young parts pubescent; bark yellowish grey, somewhat rough outside, thin, yellowish inside; leaves 23–50 cm long, rachis pubescent; leaflets generally alternate, occasionally sub-opposite, obliquely ovate, lanceolate, wavy or obscurely crenulated at the margins, thinly coriaceous, glabrous above, tomentose or pubescent beneath along the nerves, closely gland dotted; flowers dull white, strongly scented, in spreading terminal corymbose or decomposed panicles or cymes; calyx truncate or with triangular lobes; petals 5, valvate, narrow, oblong, pubescent; stamens 10, alternate, shortly capitate; ovules 2, superposed in each cell; fruit a dry berry.

Flowering and fruiting   January–April

Distribution   Fairly common in Meghalaya, ascending to 1200 m in Khasi hills.

Propagation   By seeds and vegetative method

Parts used   Root and bark

Chemical constituents   Butyl-7-methoxyfindersine, limettin, micromelin, micromelium microminutin, murralongin and murrangain derivatives.

Uses   Root is used with betel leaf in cough. Bark is chewed but not swallowed to reduce headache.

Mikania micrantha Kunth. ex. HBK.

Family   Asteraceae

Indian names   Climbing hempine (English); *Jpanhlo* (Mizoram).

Description   Shrubs or erect or twisting herbs, climbing, glabrous, leaves opposite, petiolate, the blade up to 19 cm long, cordate to triangular with a broad cordate base; flowers minute, white or cream coloured borne in small densely packed heads, achenes blackish-brown, pappus more than twice longer than the achenes.

Flowering and fruiting   October–February

Distribution   Tropical America, introduced in various parts of India, common in nearly all forest cleared areas in Garo hills.

Propagation   By seeds and vegetative method

Parts used   Whole plant

Chemical constituents   27 terpenoid constituents, taraxasterol, coumarin and stigmasterol.

Uses   The plants are used to stop bleeding, for gastritis, insect bites and various skin irritations.

Miliusa roxburghiana Hk. f. and Th.

Family   Annonaceae

Indian names   *Chag-ladoi* (Assamese); *Dieng jwat* (Jaintia).

Description   Small deciduous tree, scarcely more than 50 cm in height with spreading branches and pubescent young shoots, bark grey dark-brown, turning deeper brown after exposure; leaves somewhat aromatic, elliptic, oblong or lanceolate, acuminate, thinly coriaceous, upper surface glabrous except the puberulous midrib, lower surface pubescent, midrib, arched and looped to form an intramarginal nerve away from the edge, base more or less rounded; petiole pubescent; flowers dioecious, axillary, solitary or 2–3 together on pedicle; sepals and outer-petals alike, small lanceolate, reflexed, tomentose outside, inner petals red, ovate, subacute, more or less serrate at the base.

Flowering and fruiting   March–December

Distribution   Occurs in most districts ascending to 4000 feet in Khasi Hills.

Propagation   By seeds

Parts used   Leaves

Uses   The bruised leaves are used by the daffias as smelling salt to reduce headache.

Millettia caudata Baker.

Family   Fabaceae

Description   Shrubs straggling up to 5 m height; branches thin and sparcely lenticelled; leaves up to 35 cm long; leaflets oblong lanceolate long acuminate, base rounded; racemes up to 25 cm long; flowers reddish; petals densely silky; pods oblong, straight, obliquely jointed, 3–5-seeded, adpressed, hairy.

Flowering and fruiting   May–January

Distribution   In North-East India and Bangladesh, rare in Meghalaya; occurs along the southern slopes of Khasi and Jaintia hills in evergreen forests.

Propagation   By seeds and vegetative method

Parts used   Root and veins

Uses   Root constituents have antitussive properties. They are useful in debility, anorexia, phlegm cough, headache, dry cough and dysuria.

Mimosa pudica L.

Family   Mimosaceae

Indian names   Sensitive plant, Humble plant (English); *Lajjalu, Samanga* (Sanskrit); *Lajjavanti, Lajvanti* (Hindi); *Nacikegida* (Kannada); *Tottalcurunki* (Tamil); *Tottalvadi, Tindarmani* (Malayalam); *Manugumaramu* (Telugu).

Description   Undershrubs, deciduous, straggling and spreading, densely prickly and bristly all over; leaves sensitive, pinnae 4, digitate, petiole, bristly, stipules linear, lanceolate; bristly leaflets, 12–20 pairs, obliquely narrow, oblong, and undersurface bristly; flowers pink with several stamens up to 8 mm long, borne on globose heads; fruits a flat, hairy legume (pod) breaking into 2–4 one-seeded segments.

Flowering and fruiting   July–January

Distribution   Afganistan and nearly throughout India, common in secondary forests in Garo hills, Meghalaya.

Propagation   By seeds and vegetative method

Parts used   Whole plant

Chemical constituents   Amino acid, norepinephrine, gentisic acid, jasmonic acid and D-panitol.

Uses   Leaves are useful for haemarrhoids and urinary infections. A decoction of leaf is used as a cure for diarrhoea, skin diseases, hydrocele and for dissolving kidney and gall bladder stones. Decoction of roots is useful for urinary infections, dysentery, fever, syphilis, leprosy, stomach worms, venereal diseases, insect bites, insomnia, nervousness and piles. The leaf paste is applied on boils.

Mimusops elengi L. Baker.

Family   Sapotaceae

Indian names   *Bakulah, Chirapushpa* (Sanskrit); *Bakul, Bolsari* (Hindi); *Bakula, Buckhul* (Kannada); *Ilanci, Magilam* (Tamil); *Elengi, Elangi* (Malayalam); *Kesara, Nemmi* (Telugu).

Description   A large tree generally smaller and handsome in cultivation; young parts rusty, pubescent; Bark grey, fissured; leaves elliptic, acuminate, shining, glabrous, lateral nerves numerous, not very conspicuous, perpendicular to the midrib, sub-parallel, base acute or rounded; petiole 1.5–2.5 cm long; flowers 8-merous (rarely 6), creamy white, fragrant, star-like, solitary or in fascicles; calyx rarely in two rows; corolla caducous, lobes usually 24 in two rows, all lanceolate and almost similar; Stamens 8, staminodes rather petaloid, membranous, fimbriate; anther lanceolate; ovary hirsute, 6–8 celled; fruit usually glabrous, ovoid or ellipsoid, about 2.5 cm long, yellow orange; seeds usually solitary, ovoid, compressed, brown and shining.

Flowering and fruiting   May–July

Distribution   Throughout North-East India, common in Meghalaya.

Propagation   By seeds and vegetative method

Parts used   Bark, flowers, fruits and seeds

Chemical constituents   Saponin, rhamnose, arabinose, tannin, palmitic, linoleic and oleic acids.

Uses   The bark, flowers and fruits are useful as gargle for odontopathy, and ulemorrhagia; tender stem is used as toothbrushes. Powder of dried flowers is a brain tonic, and flowers are used for preparing a lotion for wounds and ulcers.

Mitragyna rotundifolia (Roxb.) O. Kuntze.

Family   Rubiaceae

Indian names   *Bhumikadamba* (Sanskrit); *Thinglung, Lungkhup* (Mizoram).

Description   Middle sized trees up to 15 m high with spreading, lax crown; bark brown or greyish-brown; branchlets quadrangular; leaves 9–20 × 4.5–15 cm, orbicular, broadly obovate, elliptic, obtuse to shortly acuminate, base rounded or subcordate, entire, glabrous, pale beneath; flowers pink or greenish-white; ovary ovate-lanceolate, syncarpous 1–1.5 cm in diameter.

Flowering and fruiting   October–March

Distribution   Confined to North-East India, frequent in Meghalaya along the southern slopes of Garo hills.

Propagation   By seeds and vegetative method

Parts used   Roots and bark

Chemical constituents   Pyroligneous acid, methyl acetate, ketones and aldehydes, methanol, tar and pitch.

Uses   Roots and bark are acrid, bitter, stomachic and febrifuge. They are useful in gastropathy and fever.

Momordica charantia L.

Family   Cucurbitaceae

Indian names   Bitter gourd, Carilla fruit (English); *Karavellam* (Sanskrit); *Karela, Kareli* (Hindi); *Karate, Hagalakayi* (Kannada); *Pavakkai, Paval* (Tamil); *Kaipaka, Pavaka* (Malayalam); *Kakara* (Telugu).

Description   Plant short climber; leaves 2.5–10 cm, deeply 5–7 lobed; lobe lobulate or sinuate, dentate; flowers omnoecious, yellow, fruit 2.5–16 cm long, tapering at both ends, longitudinally ribbed with rows of triangular tubercles, fleshy but dehiscent; seeds immersed in bright red pulp.

Flowering and fruiting   May–August

Distribution   Entire North-East India, cultivated and sometimes apparently wild.

Propagation   By seeds and vegetative method

Parts used   Fruits

Chemical constituents   Mycose, vicine, steroidal glucoside, momorcharaside A and B, cucurbitane triterpenoids like momordicines I and II, spinasterol, taraxerol, locophenol and diosgenin.

Uses   Fruits are useful for leprosy, malignant ulcers, stomach worms, fever and phlegm, dysentery and diabetes.

Morinda angustifolia Roxb.

Family   Rubiaceae

Indian names   *Chhengrong* (Garo); *Dieng siroi* (Khasi); *Thingaieng, Lum* (Mizoram).

Description   Shrubs up to 10 m high; crown lax; bark reddish brown or greyish, peels off in thin papery flakes; leaves variable in size 6–35 × 2–10 cm, oblanceolate, oblong; elliptic, acuminate, base narrowed to the petiole, nerves much prominent beneath; heads pedunculate; flowers 2–3 cm long; white drupes turbinate, 0.2–0.3 diameter.

Flowering and fruiting   February–October

Distribution   Burma, Bangladesh, Himalayas and sub-Himalayas; very common in Meghalaya.

Propagation   By seeds and vegetative method

Parts used   Roots

Chemical constituents   The roots contain resins, vitamin C, an essential oil, and phytosterols.

Uses   Leaves are useful in treating impotency, spermatorrhoea, menstrual disorders, hypertension and rheumatism.

Moringa oleifera Lam.

Family   Moringaceae

Indian names   Horse-radish tree, Drumsticks (English); *Madhusravah, Sigruh* (Sanskrit); *Sahinjan, Sainjna* (Hindi); *Murunga, Nuggi* (Kannada); *Murungai* (Tamil); *Muringa* (Malayalam); *Mulaga, Munaga* (Telugu).

Description   Middle sized trees; bark greyish with black patches, warty or more or less nearly smooth and brownish when young; leaves up to 1 m long; leaflets variable in size, ovate or orbicular obvate, oblong, rounded or emarginated at tip, often oblique at base, glabrous, glaucous beneath; panicles up to 30 cm long; flowers 2.5–3 cm long, white, fragrant; sepals petaloid, petals white, spathulate, oblong; capsules linear-oblong; seeds 2.5–4 cm long, trigonous, oblong, winged along the angles.

Flowering and fruiting   January–June

Distribution   Throughout India, cultivated and also run wild in Meghalaya at lower elevations.

Propagation   By seeds and vegetative method

Parts used   Root, bark, leaves and seeds

Chemical constituents   The essential amino acids present in the total proteins are arginine, histidine, lysine, trytophan, phenylalanine, methionine, threonine, leucine, isoleucine and valine.

Uses   Roots are bitter, acrid, thermogenic, digestive, carminative, anthelmintic, constipating, anodyne, anti-inflammatory, sudorific, diuretic, ophthalmic, rubefacient, expectorant, haematinic, antilithic, and vesicant. They are useful in dysentery, anorexia, verminosis, diarrhoea, colic flatulence, otalgia, paralysis, amenorrhoea and fever. The bark is useful in ascites and ringworm. The leaves are useful in scurvy, wounds, tumours, helminthiasis and seeds are useful in neuralgia, intermittent fevers and ophthalmopathy.

Mucuna bracteata DC.

Family   Fabaceae

Indian name   *Wakmi* (Garo).

Description   Climbers; young parts grey pubescent; leaves up to 25 cm long; leaflets lateral obliquely ovate or deltoid, terminal ovate-rhomboid, acute or subacute, base rounded, truncate, glabrous or glabrescent above, adpressed, pubescent beneath; racemes up to

10 cm long; flowers blackish purple, calyx and corolla scattered bristly hairy; pods greyish brown, tomentose, hairs itching, 4–5 seeded.

M

Flowering and fruiting   January–December

Distribution   North-East India; very common in Meghalaya.

Propagation   By seeds and vegetative method

Parts used   Root, leaves, seeds and hairs

Chemical constituents   The seeds contain dihydroxyphenylalanine of DOPA glutathione, lecithin, gallic acid and a glucoside and a number of alkaloids including pruriemmine, prurienidine and five others.

Uses   Roots are useful in constipation, nephropathy, strangury, dysmernorrhoea, amenorrhoea, elephantiasis, dropsy, neuropathy and ulcers. Leaves are useful in ulcers, inflammation, helminthiasis and general debility. Seeds are useful in gonorrhoea, sterility and general debility.

Mucuna pruriens (L.) DC.

Family   Fabaceae

Indian names   Common cowitch, Cowhage (English); *Atmagupta, Kapikacchuh* (Sanskrit); *Gonca, Kaunc* (Hindi); *Nasuganni* (Kannada); *Punaikkali* (Tamil); *Naykkurana, Chorivalli* (Malayalam); *Pilliadugu* (Telugu).

Description   Climber; young parts pubescent; leaves up to 40 cm long; leaflets lateral obliquely ovate or deltoid, terminal, ovate, rhomboid, acute, truncate, rounded or acute; racemes 10–20 cm long; flowers dark-purple, calyx bilipped, grey, tomentose, sparsely bristly hairs; pods oblong, rounded at tip, ribbed, 5–6 seeded.

Flowering and fruiting   January–December

Distribution   Tropics nearly throughout India, common in Meghalaya.

Propagation   By seeds and vegetative method

Parts used   Roots and seeds

Uses   Roots are used as a nerve tonic and diuretic; seeds are astringent, anthelmintic, nervine tonic and aphrodisiac.

Murraya koenigii (L.) Spreng.

Family   Rutaceae

Indian names   Curry leaf (English); *Kaidaryah, Kalasakah* (Sanskrit); *Katnim, Mithanim* (Hindi); *Karibevu* (Kannada); *Kariapala, Kariveppu* (Malayalam); *Kariveppilai, Karuvembu* (Tamil); *Karivepaku* (Telugu).

Description   An aromatic shrub or small tree; leaves pinnate, leaflets mostly ovate, crenate, dentate; cymes terminal; ovary 2-celled, style cylindric, stigma capitate; fruit ovoid or subglobose, seeds embedded in mucilage.

Flowering and fruiting   February–October

Distribution   Throughout India, common in Meghalaya, particularly in Khasi hills as undergrowth in deciduous forests.

Propagation   By seeds and vegetative method.

Parts used   Leaves, bark and roots

Chemical constituents   Major aroma constituents in the oil are β-caryophyllene, β-gurjunene, and β-phellandrene.

Uses   Leaf infusion is used in diarrhoea and dysentery. Root and bark paste is useful for skin eruptions.

Murraya paniculata (L.) Jack.

Family   Rutaceae

Indian names   Chinese box (English); *Ekangi, Muramamsi* (Sanskrit); *Marchula, Bilgar* (Hindi); *Angaarakana mara, Kaadu karibevu mara* (Kannada); *Maramulla* (Malayalam); *Kadarkongi, Kattu-karuveppilai* (Tamil); *Naga golugu, Gana renu* (Telugu).

Description   Small tree or shrub, bark smooth, whitish grey leaves; acuminate, base acute, entire; flowers 1–1.5 cm long, pure white, sepals minute, ovate, acute, petals 1 cm long, oblong, lanceolate; fruits ovate, ellipsoid, red or orange red, 1–2 seeded.

Flowering and fruiting   February–May

Distribution   Throughout India, common in Balphakram sanctuary, South Garo hills, Meghalaya, where it forms dense forest undergrowth along the river side.

Propagation   By seeds and vegetative method

Parts used   Leaves

Uses   The leaves and roots are given for beneficial results in the treatment of influenza, headache and abdominal pains.

Mussaenda belilla Buch. Ham.

Family   Rubiaceae

Indian names   *Belilla* (Malayalam); *Bili yele hoo, Bili yele gida* (Kannada); *Bhoorthkashi, Bhuthkes, Churthkasi* (Marathi).

Description   Shrubs; branchlets and leaves glabrous, stipules long, hairy; leaves elliptic, orbicular; cymes densely hirsute, one of the calyx lobes transformed into a showy petiolate bract red yellow, tube 3 cm long; berry ovoid, 1 cm across.

Flowering and fruiting   May–November

Distribution   Confined to North-East India

Propagation   By seeds and vegetative method

Parts used   Leaves and fruits

Uses   Juice of leaves and fruit is applied in cases of weak eyesight. Infusion of the dried shoots is given to children to relieve cough and cold, asthma, flatulence and jaundice.

Mussaenda roxburghii Hook. F.

Family   Rubiaceae

Indian names   *Gardek* (Garo); *Jalai, Dieng jalongtham* (Khasi); *Hanurei* (Manipuri); *Vokep* (Mizoram); *Naolungkamcha* (Nepali).

Description   Erect or sub-scandent, hirsute or villous shrubs; leaves elliptic or oblong, lanceolate, acuminate, base cuneate, pubescent or tomentose beneath, stipules ovate, lanceolate; cymes trichotomous, compact; flowers orange red, calyx lobes foliaceous, petaloid, lobes ovate, corolla tube hirsute, throat villous; berry globose, hairy.

Flowering and fruiting   March–December

Distribution   Burma, Bangladesh, Bhutan and Nepal; confined in North-East India, common in Meghalaya.

Propagation   By seeds and vegetative method

Parts used   Root, bark and leaves

Uses   Powder of root and bark is useful for ulcer of the mouth. Leaf juice is useful for cuts and wounds.

Myrica esculenta Buch. Ham.

Family   Myricaceae

Indian names   Box Myrtle (English); *Katphala, Mahavalkala* (Sanskrit); *Kaiphal, Kaphal* (Hindi); *Kaitariyam, Cavviyaci* (Tamil); *Sohphi* (Khasi).

Description   Tree 5–15 m high, crown dense; bark blackish brown, minutely pustuled; leaves oblong, oblanceolate, elliptic, acute, base narrowed, cuneate, glabrous, entire or remotely serrate; panicles 6–12 cm long, green, turning yellow; fruits 1–4 cm long, ovoid-oblong, tubercled, turning red, soury.

Flowering and fruiting   October–July

Distribution   Indo-Malaya and subtropical Himalayas; largely in North-East Region, very common and often gregarious in higher elevation in Khasi hills.

Propagation   By seeds and vegetative method

Parts used   Fruit and bark

Uses   The bark is astringent, carminative and antiseptic. A decoction of the bark is useful in asthma, diarrhoea, fevers, lung afflictions, chronic bronchitis, dysentery and diuresis. The bark is also used as fish poison. Fruits are sedative, stomachic and carminative.

Myrioneuron nutans Wall.

Family   Rubiaceae

Description   Small shrub, sometimes climbing; branches stout, old stem white with soft corky bark; leaves ovate, elliptic, acuminate, subcoriaceous, glabrous above, stipules erect, oblong, lanceolate, about 2.5 cm × 0.5 cm; flowers white, in dense pedunculate, bracteate, corymbose cymes from the axis of the uppermost leaf; bracts rigid, lanceolate; calyx teeth subulate, exceeding the corolla, persistent; corolla tubular; stamens 5, adnate to the corolla tube; filaments short, subulate, anthers linear; ovary 2-celled; style short, stigma 2, linear, oblong; berries white, globose, covered by the scarious calyx-teeth, many seeded; seeds black, minute, angular.

Flowering and fruiting   October–February

Distribution   North-East Region, common in Khasi hills.

Propagation   By seeds

Parts used   Stem

Uses   Stems are useful for various eye diseases.

Naravelia zeylanica (L.) DC.

Family   Rununculaceae

Indian names   *Dhanavalli* (Sanskrit); *Neendamalli, Balluli hambu* (Kannada); *Karuppakkoti, Poitalacci* (Malayalam); *Nintavalli, Vatamkolli* (Tamil); *Pulla bachala, Mukkupeenasa teega* (Telugu).

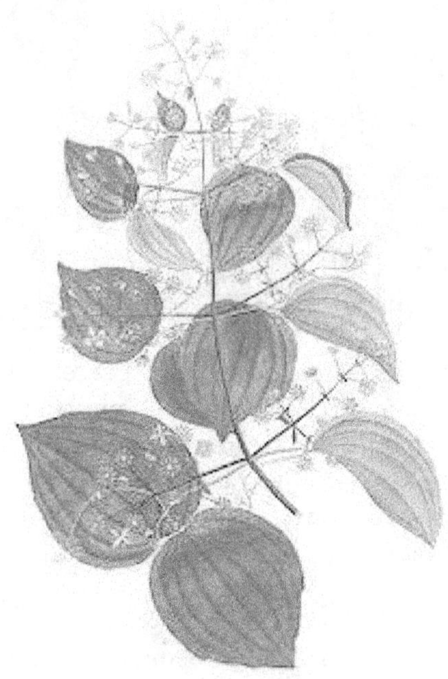

Description   Woody climber, generally spreading on bushes in scrub jungles; branches sulcate; leaves pinnate leaflets become reduced to tendrils; common petiole 3–8 cm long, leaflets usually two, often in unequal pairs, 5–15 × 5–10 cm, ovate, rotundate, acuminate, rounded or cordate and often abruptly cuneate at the base, entire or distantly pubescent; flowers greenish; sepals densely tomentose outside, caducous; petals usually 12 about 0.8 cm long; stamens numerous; fruits, a head of tailed achenes, pointed at both ends, shortly stipulate, more or less twisted brown when mature, tail very slender, finely hairy.

Flowering and fruiting   October–May

Distribution   Both the slopes of Garo and Khasi hills, up to 2,500 ft at Meghalaya.

Propagation   By seeds and vegetative method

Parts used   Root and stem

Uses   The roots and stems have a strong smell, and are used by tribes for cephalalgia.

Nepenthes khasiana Hook. f. [Endemic]

Family   Nepenthaceae

Indian name   *Tiew rakot* (Khasi).

Description   Usually scandent shrubs, climber, branched. Leaves simple, alternate, 5–50 × 1.5–8 cm, oblong, lanceolate or acuminate, base attenuate, amplexicaulous, glabrous, often reddish, leaf tips modified into pitcher; pitchers subcylindric, 10–20 × 3–7 cm, constricted towards mouth; inflorescence terminal or lateral racemes up to 75 cm long; flowers 0.8–1.5 cm across, greenish red or greenish brown; capsules 2–3 cm long, ellipsoid, oblong, supported by the perianth; seeds 0.5–0.7 cm long, spindle-shaped.

Flowering and fruiting   June–December

Distribution   Endemic to Meghalaya; one of the threatened species of India; the habit of this species varies considerably. Confined at Bagmara and Nangalbibra of South Garo Hills, Meghalaya.

Propagation   By vegetative method

Parts used   Pitcher

Uses   Stored water in the pitcher is useful in maternity pain (according to the natives of Garo Hills). Juice of unopened pitcher mixed with rice and beer is consumed to relieve urinary troubles and stomach diseases. Juice is also used as eye drop to remove cataract and cure night blindness. It is also used to cure skin diseases including leprosy.

Nephrolepis cordifolia (L.) Prest.

Family   Nephrolepidaceae

Description   It is an erect, tufted wiry fern with tuberous rhizomes and long pinnate fronds.

Distribution   North-Eastern region, up to 1500 m

Propagation   By vegetative method

Patrs used   Tubers and fronds

Uses   A decoction of fresh fronds is useful for cough. Fronds are chewed for running nose.

Nerium indicum Mill.

Family   Apocynaceae

Indian names   *Asvaghna, Asvahana* (Sanskrit); *Chandni, Kanir* (Hindi); *Chandaatha, Kanagalu* (Kannada); *Alari, Arali* (Tamil); *Arali* (Malayalam); *Ganneru, Erra ganneru* (Telugu).

Description   Evergreen shrub; leaves usually ternate, linear, lanceolate, tapering at both ends, coriaceous, glossy, green above, paler beneath, nerves obscure, numerous, parallel; flowers usually rosy, in terminal cymes; calyx divided nearly to the base, segments transverse, about 0.6 cm long; corolla tube funnel-shaped, hairy within, lobes laciniate; stamens at the top of the corolla tube; anther adhering to the stigma; ovary of 2 distinct carpels, style dilated upwards, stigma sub-globose; follicles connate, seeds linear, villous with a terminal coma of brownish hairs.

Flowering and fruiting   May–June

Distribution   Entire North-East India; cultivated in Khasi hills

Propagation   By cutting

Parts used   Roots and leaves

Chemical constituents   All parts of the plant are poisonous. Roots, bark and seeds contain cardio-active glycosides, formerly designated as neriodorein, and karabin. They are reported to have a paralysing action on the heart, like digitalin and a stimulating action on the spinal cord, like strychnine.

Uses   The roots are useful for cardiac asthma, renal and vesical calculi, stomachalgia, arthralgia, leprosy, pruritus and ulcers. The leaves are useful for repellent and are used for scabies, haemorrhoids.

Nicotiana tabacum L.

Family   Solanaceae

Indian names   Tobacco (English); *Dhumapatram, Dhumrapatra* (Sanskrit); *Bujjerbhang, Tamaku* (Hindi); *Hogesoppu, Tambaku* (Kannada); *Pugaiyilai, Biramam* (Tamil); *Pogala, Puga-yila* (Malayalam); *Dhumra-patramu, Dhumrapatramu* (Telugu).

Description   Erect, viscidly, pubescent, herb; leaves large oblong or elliptic, base cuneate; corymb compound; ultimate branches short; flowers light red, white or pink in many flowers panicled racemes; fruits narrowly elliptic, ovoid capsule; seeds many, brown, with fluted ridges.

Flowering and fruiting   December–March

Distribution   It is cultivated all over North-East Region

Propagation   By seeds and vegetative method

Parts used   Leaves

Chemical constituents   Nicotine, nornicotine, anabasine, nicotyrine, nicotimine, nicotelline and myosamine.

Uses   The leaves are useful for odontalgia, dental caries, inflammations, helminthiasis, dyspepsia, flatulence, bronchitis, asthma, scabies, skin diseases, ulcers and painful tumour. Leaf juice is used as anthelmintic and paste is applied on boils.

Nyctanthes arbor-tristis Linn.

Family   Oleaceae

Indian names   Coral Jasmine, Queen of the night (English); *Harashingarpushpaka, Parijatah* (Sanskrit); *Harsinghar, Karasli* (Hindi); *Goli, Harisringi* (Kannada); *Parisadam, Pavalamalligai* (Tamil); *Parijatam, Pavizhamalli* (Malayalam); *Kapilanagadustu, Karuchiya* (Telugu).

Description   Hardy large shrub or small tree up to 10 m in height with grey or greenish white rough bark and sharply quadrangular strigose young branches; leaves simple, opposite, ovate, acute or acuminate, distantly toothed or entire, coriaceous, scabrid, lateral nerve 4–6 on either half, with short bulbous hairs, densely pubescent beneath with adpressed hairs, main nerves few; flowers small, sweet scented, sessile, 3–7, together on hairy quadrangular peduncles of various length which are arranged in cymose panicles, white with bright orange corolla tubes 3–7 in each head, in trichotomous cymes; fruit capsules, compressed; seed exalbuminous.

Distribution   Throughout India

Propagation   By seeds and vegetative method

Parts used   Leaves and flower

Chemical constituents   The flower contains D-mannitol, tannin and glucose. The oil consists of glycerides of linoleic, oleic, lignoceric, stearic, palmitic and myristic acid. The leaves of the plant contain tannic acid, methyl salicylate, an amorphous glycoside and mannitol.

Uses   The leaves are thermogenic, antibacterial, chelagogue, anthelmintic and laxative. Leaves are useful in inflammations, dyspepsia, helminthiasis, pruritus, dermatopathy, chronic fever, asthma and cough. The flowers are useful in inflammations, ophthalmopathy, flatulence, dyspepsia and greyness of hair. Seed powder is useful for scurvy and affliction of scalp.

Ocimum basilicum L.

Family   Lamiaceae

Indian names   Sweet basil, Common basil (English); *Barvari* (Sanskrit); *Babul Babuyiulsi* (Hindi); *Kamakasturi* (Kannada); *Tirunittru, Tiruniruppachai* (Tamil); *Ramathulasi* (Malayalam); *Bhutulasi* (Telugu).

Description   Slender, much scented and much branched herb, generally purple coloured; stems glabrous, more or less pubescent, hairy at the nodes; leaves ovate, acute, entire or more or less lobed or toothed, glandular; petiole 0.8 cm long; flowers pale purple, nearly single racemes; pedicels shorter than the calyx, much deflexed in fruit, upper lip of calyx rounded, shorter than the teeth of the lower lip; nutlets ellipsoid, black, slightly pitted and become very gelatinous in water.

Flowering and fruiting   September–March

Distribution   Throughout India; cultivated at Khasi hills.

Propagation   By seeds and vegetative method.

Parts used   Leaves

Chemical constituents   Essential oil is ether methyl cinnamate, linalool or methyl chaviocol in different samples; other major components include ocimene. Linalyl acetate, eugenol and anethole.

Uses   The leaves are used in the treatment of bronchitis. The juice is said to cure ringworm, and an application of bruised leaves relieves pain from scorpion stings. They are used in catarrh, chronic diarrhoea, dysentery, nephritis and in several other ailments.

Ocimum sanctum L.

Family   Lamiaceae

Indian names   Holy basil (English); *Amrita, Apetarakshasi* (Sanskrit); *Baranda, Kala-tulasi* (Hindi); *Kalatulasi, Sritulasi* (Kannada); *Kirusnatulaci* (Tamil); *Krishnatulasi, Nallatrittavu* (Malayalam); *Brynda, Krishnatulasi* (Telugu).

Description   A perennial herb with woody root-stock; branchlets purplish, softly hirsute or pubescent; leaves 1–2 × 0.8–2.5 cm, ovate or oblong, distantly serrate, crenate or entire, acute, membranous, pubescent, base acute; flowers purplish, raceme, often panicle, up to 20 cm long; pedicels usually longer than the calyx; bracts reflexed; calyx glabrous within glandular and pubescent without, enlarged in fruit; upper lip obovate and acute at the base in fruit, shorter than the lower lip. Corolla upper lip hairy on the back; filaments of the upper stamens bearded at the base; nutlets reddish with black markings, ellipsoid.

Flowering and fruiting   May–January

Distribution   In North-East Region, cultivated in Meghalaya

Propagation   By seeds and vegetative method

Parts used   Whole plant

Chemical constituents   Eugenol, eugenol methanol ether and carvacrol, methyl chavicol, cineole and linalool.

Uses   It is the sacred plant of the Hindus. The leaves are medicinally efficacious and are much used for catarrh and other pulmonary afflictions. The stems are made into rosaries and worn largely by Vaishnavas. The leaves and seeds of the plant are medicinal; The oil obtained from leaves has the properties of destroying bacteria and insects. The juice or infusion of the leaves is useful in bronchitis, digestive complaints. It is applied locally on ringworm and other skin diseases. It is dropped into ears to relieve earache. A decoction of leaves is used in Indian homes to cure common colds. Seeds are useful in complaints of urinary system. Decoction of root is given for malarial fever to bring about sweating.

Olax acuminata Wall. ex. Benth.

Family   Olacaceae

Indian names   *Bol-narang, Moen* (Garo); *Duritodo* (Oriya); *Yegadedi* (Telugu).

Description   Large shrubs or small tree up to 40 cm high; branches dark green, smooth and with shollow but long longitudinal fissures; bark thin, greenish white inside; branchlets angular; leaves elliptic or oblong, lanceolate, acute or acuminate, membranous, glabrous, shining above, pale beneath; inflorescence axillary, scarcely exceeding; flowers solitary or fascicled; bracteoles, ovate, caducous; buds elliptic, ovoid; calyx very small, salver-shaped, almost imperfectly 3-toothed, acccrescent in fruit; petals oblong, valvate, each bearing at the base one short fertile stamen along the middle and two long staminodes, one on either side of the stamen; anthers 2-celled adnate, oblong; style simple; drupe ellipsoid or ovoid, oblong, seated inside the cup-shaped accrescent calyx, orange red when fully ripe; stone 1-celled, 1-seeded.

Flowering and fruiting   January–July

Distribution   Entire North-East Region; common in Garo, Khasi and Jaintia hills of Meghalaya.

Propagation   By seeds

Parts used   Leaves

Uses   Leaves are used as cathartic.

Olea dioica Roxb.

Family   Oleaceae

Indian names   *Akkasalle, Edale, Bilisarali* (Kannada); *Kolipayar, Idalai koli* (Tamil); *Kari-vetti, Edana* (Malayalam); *Karambu, Paarajaamba* (Marathi); *Bonbholuka* (Assamese).

Description   Slender tree or a shrub; bark greyish brown, warty, having vertical fissures, sometimes peeling off in flakes; leaves very variable, usually elliptic, lanceolate, entire or serrate, acute or acuminate, lateral nerves 8–12 on either half, impressed above, prominent beneath, sub-parallel, arcuate, base cuneate; petiole 0.5–3 cm long; flowers small, white, dioecious in axillary lax panicles with very slender branching; female flowers–calyx 4 toothed; teeth triangular or acute, corolla absent; male flowers-corolla deeply lobed; lobes elliptic to obtuse; drupe blue when ripe, ovoid.

Flowering and fruiting   March–November

Distribution   North Cachar, Khasi and Jaintia hills

Propagation   By seeds and vegetative method

Parts used   Leaves and oil

Chemical constituents   Leaves contain oleoropine, oleoasterol and leine. Oline oil contains about 75 per cent oleic acid, a mono-unsaturated fatty acid.

Uses   Leaves are useful for control of blood pressure and help in the function of the circulatory system.

Onosma emodi Wall.

Family   Boraginaceae

Description   Herb with hispid hair; leaves sessile, lanceolate, usually alternate; flowers in capitate cymes, pedicels long; calyx ovate, acuminate, hispid, spreading in fruit; corolla purplish, ventricose, narrowed at the mouth, hairy; anthers inserted, style exerted; nutlets 4, tuberculate.

Distribution   Throughout North-Eastern India.

Propagation   By seeds

Parts used   Fruit

Uses   Flower is a good refrigerant, demulcent, relieves excessive thirst and restlessness.

Ophiopogon intermedius D. Don.

Family   Liliaceae

Indian names   *Piyajimurba* (Bengali); *Ching-charot* (Manipuri); *Murunga, Nuggi* (Kannada); *Keeri poondu* (Tamil).

Description   Rootstock short, leaf margins minutely serrate, scape slender as long as the leaves, flowers solitary, white, anther linear, oblong, filaments very short.

Flowering and fruiting   March–August

Distribution   Khasi hills, Meghalaya

Propagation   By vegetative method

Parts used   Root tubers

Chemical constituents   Root tubers possess mucilage and sugars, β-sitosterol, ophiopogenins A,B,C and D and ruscogenin.

Uses   Root tubers possess expectorant and antitussive properties. They are useful in sore throat, bronchitis, tuberculosis, haematemesis, and evening fever with thirst, rheumatism, and hypogalactia. Root paste is applied in cuts and wounds.

Ophiorrhiza ochroleuca Hk.f.

Family   Rubiaceae

Description   Small shrub, glabrous; leaves elliptic or lanceolate, acuminate, lateral nerves 10–15 on either half; petiole 0.8–2.5 cm long, stipules 2–fid, lanceolate; flowers second on the branches of axillary or terminal dichotomous pubescent cymes; peduncles and branches stout; bracts 0, or obscure and caducous. Calyx teeth very short; corolla tubular, pubescent within but not winged at the back; stamens 5; ovary 2-celled, style filiform, ovules many on basal ascending placenta; capsule coriaceous, compressed, glabrous, pedicelled, 10 cm across; seeds many, minute, angled, embryo clavate, albumen fleshy.

Flowering and fruiting   March–September

Distribution   North-East India, common in Khasi hills

Propagation   By seeds and vegetative method

Parts used   Leaves

Uses   Leaf juice is used for relieving headache.

Oreocnide integrifolia (Gaud.) Miq.

Family   Urticaceae

Indian names   *Banrhea, Chho-oi-paroli* (Assamese); *Sejugbu, Gingsining* (Garo); *Dieng teingbah* (Khasi).

Description   Small evergreen tree; young parts pubescent or tomentose; bark greyish, warty with lenticels, thick; leaves elliptic, oblong, ovate, acuminate, usually entire, sometimes obscurely crenate; flower heads dichotomously branched; cymes, hispid; male flowers 3–4-merous, stigmas papillose, ciliate, the bracts become succulent and glossy. When the fruit ripens it is seated on the cavity of the cup-shaped bract.

Flowering and fruiting   January–November

Distribution   Occurs throughout Meghalaya up to 4,500 ft.

Propagation   By seeds

Parts used   Root

Uses   Root is mashed with ginger and consumed to cure rashes and skin dieases.

Oroxylum indicum (L.) Benth. [Endangered]

Family Bignoniaceae

Indian names Indian trumpet tree (English); *Shyonaka, Tunukah* (Sanskrit); *Sonapatha, Syona* (Hindi); *Alangi, Tattuna* (Kannada); *Paiyaralantai, Palaiyudaichi* (Tamil); *Palakapayani, Payyalanta* (Malayalam); *Dundillum, Pampana* (Telugu).

Description Small or middle sized trees, 5–15 m height, laxly branched or unbranched, branches erect or suberect; bark greyish, corky. Rachis very soft, cylindric, swollen at the junction of the branches; leaves 1–2 m long, leaflets 5.5–15 cm long, broadly ovate, orbicular, obtuse acuminate, base usually oblique, often cordate or subcordate, pale beneath, racemes terminal, stout, up to 2 m long, erect; flowers 8–10 cm across, greenish yellow, purple tinged, fleshy; pods black, drooping, 40–80 cm long (flat capsules, up to 1 m long tapering to both ends); seeds white, winged all around, 3–8 cm long, Fruiting occurs when the tree is totally leafless.

Flowering and fruiting June–March

Distribution Indo-Malaya; throughout the greater part of India; it is found in deciduous forest belts of Meghalaya at lower elevations.

Propagation By seeds and vegetative methods

Parts used Roots, leaves, fruits and seeds

Chemical constituents The stem and root barks contain three flavone colouring matters, viz. oroxylin-A, baicalein and chrysin. The seeds contain a yellow crystalline principle and baicalein and its glucoside named tetuin.

Uses The roots are sweet, astringent, anti-inflammatory, anodyne, appetizing, carminative, digestive, anthelmintic, febrifuge and tonic. They are used in inflammations, dropsy, strains, cough, asthma, bronchitis, anorexia, diarrhoea, dysentery, vomiting, wounds and fever. The leaves are stomachic and anodyne. The tender fruits are used as expectorant, carminative and stomachic and are useful in cough and bronchitis. The seeds are purgative.

Osbeckia nepalensis Hook. f.

Family   Melastomataceae

Indian names   *Bagaphatkala* (Assamese); *Builukhampa* (Mizoram).

Description   Shrubs up to 2 m high; stems obscurely quadrangular; leaves 6–14 × 1.5–4 cm, broadly ovate, lanceolate, pilose, adpressed, hairy along nerves; panicles up to 25 cm long; flowers white, densely adpressed, scaly; bracts equalling the calyx lobes, lanceolate; capsules campanulate truncate, densely hairy.

Flowering and fruiting   August–February

Distribution   Himalayas from Nepal to Burma; common in shady locations of Meghalaya.

Propagation   By seeds

Parts used   Root and bark

Uses   Bark juice is useful for stomach ache and root paste can cure dysentery.

Osbeckia stellata Buch-Ham.ex. D.Don.

Family   Melastomaceae

Indian names   *Dieng sohkthem* (Khasi); *Builukham* (Mizoram).

Description   Shrubs up to 2.5 m high; stems reddish, covered with spreading bulbs; leaves ovate, lanceolate, elliptic, acute, base rounded, scabrous, hairy on both surfaces, usually 5-nerved from base; corymbs, flowers pink, purple or white; capsules ovoid, 1–1.5 cm long.

Flowering and fruiting   August–March

Distribution   Common in Meghalaya, usually in secondary forests.

Propagation   By seeds

Parts used   Root and bark

Uses   Bark juice is useful for stomach ache. Roots are useful in dysentery.

Oxalis corniculata L.

Family   Oxalidaceae

Indian names   Indian sorrel (English); *Ahangeri, Amlapatrika* (Sanskrit); *Cukatripati, Amrulsak, Tinpatiya* (Hindi); *Hulihunice, Teltuppi* (Kannada); *Puliyarai* (Tamil); *Puliyaral, Puliyarala* (Malayalam); *Ambotikura, Pulichintaku* (Telugu).

Description   Herb with procumbent; stem, leaflets obcordate; stipules adnate to the petiole; peduncles 2 or more flowered; Bulbs are formed at the top of swollen roots; flowers yellow throughout the year.

Flowering and fruiting   April–September

Propagation   By bulbs

Parts used   Leaves

Chemical constituents   Glyoxylic acid, oxalic acid, vitexin and isovitexin, vitexin, D-glucopyranoside and beta-tocopherols.

Uses   The plant is used as a remedy for convulsions in children and for healing fractured bones. Leaf is cooling, astringent, refrigerant, antiscorbutic and appetising. It is useful for fever. The plant is used to treat burns, wounds and body sores. Leaf paste is used as antidote for snake bite.

P

Paederia scandens (Lour.) Merr.

Family   Rubiaceae

Indian names   *Padurilata* (Assamese); *Pashum* (Garo); *Mei iwtung* (Khasi); *Vawihuihhrui* (Mizoram).

Description   Slender twining shrubs; leaves simple, opposite, long petioled, ovate, oblanceolate; flowers in axillary and terminal, 2–3 dichotomously branched panicled cymes; Ovary 2-celled, style slender, stigma 2, ovules solitary in each cell; fruits compressed or globose, seeds much compressed dorsally.

Flowering and fruiting   February–October

Distribution   North-East Region; common in Garo and Khasi hills, Meghalaya

Propagation   By seeds and vegetative method

Parts used   Whole plant, root and leaves

Chemical constituents   The leaves contain an essential oil and alkaloid α-paederine and β-paederine.

Uses   The leaves have got antidysenteric properties, which will help, in bacillary dysentery. The entire plant is useful for rheumatic afflictions.

Pandanus fascicularis Lamk.

Family   Pandanaceae

Indian names   Screw pine, Umbrella tree (English); *Ketaki* (Sanskrit); *Kewra* (Hindi); *Kaida, Katthaale* (Kannada); *Thazhai, Tazhai* (Tamil); *Kaida, Pukkaita, Thazhampu* (Malayalam); *Gaajangi, Gedaji* (Telugu).

Description   Bushy shrub with thick terete stilt roots; leaves glaucous green, long, caudate, acuminate, coriaceous with spines on the margins and on the midrib; flowers unisexual in spadix; spadix of male flowers, 25–50 cm long with numerous sub-sessile cylindric spikes, 5–10 cm long enclosed in long, white, caudate acuminated spathes; spadix of female flowers solitary 5 cm in diameter.

Distribution   Common in North-East Region; abundant in Cherapunjee area, Meghalaya.

Propagation   By vegetative method

Parts used   Stem, leaves and essential oil

Chemical constituents   The essential oil contains methyl beta-phenylethyl ether, dipentene, linalool, phenylethyl acetate, citral phenylethyl alcohol, and ester of tyalic acid.

Uses   Leaves are stimulant, diaphoretic and antispasmodic. They are useful in leprosy, smallpox, scabies and diseases of blood. In animals stem paste is useful for ulcers.

Parabatium micranthum (DC.) Pierre.

Family   Apocynaceae

Description   Woody climbers; branchlets pendulous; bark rough, white; leaves ovate, oblong, lanceolate, acuminate, often caudate, base obtuse or cuneate; cymes panicled, up to 25 cm long; flowers yellow, follicles spreading, divaricate.

Flowering and fruiting   April–November

Propagation   By seeds and vegetative method

Parts used   Leaves

Uses   Leaves are applied in the form of paste on fractured bones to hasten the joining of fractured portion and also in headaches and fevers.

Parkia roxburghii G. Don. [Rare]

Family   Mimosaceae

Indian names   *Maniouri urohi* (Assamese); *Aoelgap* (Garo); *Zawngtah* (Mizoram).

Description   Large tree, up to 25 m high; crown lax, branches spreading; bark greyish brown, young parts lenticellate; leaves up to 60 cm long; leaflets oblique-oblong, acute, base obliquely truncate, glabrous, tomentose along rachis; heads constricted at base; flowers narrow, tubular, yellow, dull-white; corolla lobes spreading; stamens exerted; pods flat, dark brown, twisted, drooping from the peduncle.

Flowering and fruiting   November–March

Distribution   North-East Regions; rare in Meghalaya

Propagation   By seeds

Parts used   Seeds

Uses   The seeds are useful for the curing of piles.

Pratia begonifolia Lingl.

Family   Lobeliaceae

Description   Small, erect, trailing herb, roots come out from the nodes; leaves small, alternate, petioled, cordate, ovate, dentate; peduncles axillary flowered; flowers green, marked with pink.

Distribution   In North-Eastern states up to 600–2000 m altitudes, common in Meghalaya.

Propagation   By seeds

Parts used   Whole plant

Uses   Decoction of plant is diaphoretic. It is useful for urinary trouble to dissolve stones to cure dysentery and asthma. Plant juice is used to coagulate blood and stop excessive bleeding.

Pavetta indica L.

Family   Rubiaceae

Indian names   *Carnicara, Kakachedi, Papata* (Sanskrit); *Kankra, Karnikara, Kathachampa* (Hindi); *Nitile, Pappadi, Patta* (Kannada); *Pavattai, Peramalli* (Tamil); *Mallikamuti, Pavatta* (Malayalam); *Duyipapata, Kondapapata* (Telugu).

Description   Large deciduous shrub or small tree up to 7 m high; bark greyish, slightly rough; leaves variable, elliptic, ovate or oblanceolate, obtuse, acute or acuminate, subcoriaceous, pubescent or glabrous, lateral nerves 10–15, base tapering, petiole 1.5–4 cm long; stipules with acute tip; flowers white in terminal or lateral corymbose panicles; bracts broad, membranous; calyx unusually pubescent, corolla salver-shaped, tube long, very slender, lobes controlled in bud; stamens 4 on the mouth of the corolla, filaments short, anthers exerted; style filiform; fruit globose, glossy, blackish-green.

Flowering and fruiting   June–December

Distribution   Entire North-East Region; common in Garo hills, Meghalaya.

Propagation   By seeds

Parts used   Root, leaves and fruits

Uses   The roots are purgative and diuretic and are used in visceral obstruction, jaundice, headache, urinary dieases and dropsical afflictions. Leaves are used as a lotion for ulcerated nose.

Pedilanthus tithymaloides L. Poit.

Family   Euphorbiaceae

Indian names   Bird cactus (English); *Kannatikkalli* (Tamil); *Kancipala, Kantapala* (Telugu); *Vilaayathi shera* (Marathi).

Description   Laticiferous shrub, grown on hedges; leaves ovate-lanceolate, succulent; cyathia in terminal crowded cymes, scarlet bright red or orange, slipper-shaped.

Distribution   Entire North-East Region

Propagation   By seeds

Parts used   Roots and leaves

Uses   Leaves are useful for headache. The root is a powerful emetic. The latex of leaves is useful for venereal diseases and leucoderma.

Pegia nitida Colebr.

Family   Anacardiaceae

Description   Large scandent hairy shrubs; bark blackish or brownish, peeling off in strips; leaves 20–40 cm long, leaflets oblong, lanceolate or elliptic, acuminate, subcordate or rounded at base, crenate, villous, hairy on both surface; panicles with dense, minute flowers; flowers yellow or white; petals spreading; drupe blackish when ripe, obliquely ovoid.

Flowering and fruiting   January–June

Distribution   Himalayas, Burma, Bangladesh; common in Meghalaya at lower elevations.

Propagation   By seeds

Parts used   Whole plant

Uses   The plant juice is applied on cuts and wounds.

Peliosanthes teta Andr.

Family   Liliaceae

Description   Herb with root stock horizontal; leaves long petioled, nerves 10–20; scape stout, shorter than the leaves; raceme many-fid; flowers purplish or bluish green. Seeds olive blue.

Distribution   In North-East India, particularly in Khasi hills

Propagation   By seeds and vegetative method

Parts used   Tuber

Uses   Tuber is used to reduce fever.

Persea gamblei (King. Ex. HK. f.)

Family   Lauraceae

Indian names   *Bhadrol, Badro* (Hindi); *Mojli* (Assamese); *Omgthat* (Garo).

Description   Large trees up to 20 m high, with a spreading dense crown; bark dark grey or brown, warty, straight, buttressed at base; leaves oblanceolate, oblong, elliptic, acute, acuminate, base narrowed, attenuate; panicle, axis reddish; flowers greenish, yellow; fruits globose, black.

Flowering and fruiting   February–August

Distribution   North-East India, common in Meghalaya at lower elevations.

Propagation   By seeds and vegetative method

Parts used   Whole plant

Chemical constituents   Leaves and bark contain volatile oil, tannins, flavonoids and the fruit pulp contains sesquiterpenes, vitamins A, B_1 and B_2.

Uses   To alleviate muscular pain, plant juice is useful.

Phaseolus lunatus L.

Family   Fabaceae

Indian names   Rangoon bean, Burma bean, Duffin bean, Butter bean, Lima bean (English); *Aksipidaka* (Sanskrit); *Lobia* (Hindi); *Dabbale beans* (Kannada); *Kaci-k-kollu* (Tamil).

Description   Slender-stemmed, biannual twiner; leaves trifoliate, leaflets oval, triagular with narrowed or acuminate apex; raceme with small, greenish yellow flowers and a beanpod containing kidney-shaped seeds.

Distribution   Cultivated throughout North-East Region

Propagation   By seeds and vegetative method

Parts used   Leaves

Uses   Leaf juice is anthelmintic and it is useful to reduce fever.

Phoebe attennuata (Nees) Syst.

Family   Lauraceae

Description   Large tree; bark dark, grey, exfoliating in papery flanks, inside greyish, brown; leaves crowded at the end of the branchlets; oblong or oblanceolate, entire, coriaceous, base cuneate or narrowed into a short petiole; flowers tomentose in pedunculate spreading panicles; perianth rigid, segments coriaceous, stamens in series, somewhat shorter than the perianth, filaments linear, narrow, longer than the anthers; ovary globose, depressed, style slightly longer, filiform, stigma oblique, narrowly ellipsoid.

Flowering and fruiting   March–October

Distribution   Khasi hills up to 1000 m.

Propagation   By seeds

Parts used   Wood and leaves

Uses   The wood is one of the most valuable local timbers commercially. The leaves are used as medicine.

Phoebe lanceolata Nees.

Family   Lauraceae

Indian name   *Uningthou* (Manipuri).

Description   Small or middle-sized trees, up to 10 m high; branches spreading, bark whitish, smooth or nearly so, leaves linear, lanceolate, elliptic, acuminate, base narrowed to the petiole; panicles 4–18 cm long; flowers greenish white or yellow, glabrous; fruits ellipsoid, blackish purple.

Flowering and fruiting   March–December

Distribution   Indo-Burma, subtropical Himalayas and southern India, common in Meghalaya usually along river bank, in Khasi and Jaintia hills.

Propagation   By seeds

Parts used   Leaves and fruits

Uses   Leaves are used as cattle and buffalo fodder. Ash of the berries is applied on sores.

Pholidota imbricata (Roxb) Lindl.

Family   Orchidaceae

Indian names   *Valiyatekkamaravazha* (Malayalam); *Tolasi* (Marathi).

Description   A very common epiphyte; the tufted conical pseudobulbs are 2.5–8 cm long and narrow, or short and broad; leaves solitary, 15–30 cm, elliptic, lanceolate, petioled, 3-nerved;

raceme long-peduncled, elongate, drooping, 8–20 cm long; bracts 1.5 cm broad; flower in compact chains, white or pink; sepals 1.5 cm long, lateral, connate at the base; lip-4-lobed, 5-nerved, 3-median nerves thickened at the base; column circular when spread out; rostellum truncate; capsule ellipsoid.

Distribution   In North-East Region, particularly in Khasi hills, Meghalaya.

Propagation   By vegetative method

Parts used   Bulb

Uses   Pseudobulb is useful for cuts.

Pholidota pallida Lindl.

Family   Orchidaceae

Indian name   *Welliathekamaravara* (Malayalam).

Description Perennial herb, bracts semi circular, dorsal sepal orbicular, 3-nerved, lateral cymbiform, petals linear oblong, falcate 1-nerved, lip with large broad rounded sides and 2 smaller terminal lobes.

Distribution Parts of North-East India, particularly in Khasi hills

Propagation By vegetative method

Parts used Bulb

Uses Juice of bulb is used for cuts as a haemostat.

Phyllanthus emblica L.

Family Euphorbiaceae

Indian names Emblic myrobalan, Indian gooseberry (English); *Amalaka, Dhatri* (Sanskrit); *Amalak, Amlika* (Hindi); *Nellka, Amalaka* (Kannada); *Nellikai* (Tamil); *Nellikka, Nellimaram, Nellikkaya* (Malayalam); *Amalakamu, Usarika* (Telugu).

Description Small to medium sized deciduous tree, 8–18 m in height with thin light grey bark, exfoliating in small thin irregular flakes; leaves simple, light green, having appearance of pinnate leaves; flowers greenish yellow, in axillary fascicles, unisexual, male numerous on short slender pedicels, female few, subsessile, ovary 3-celled; fruits globose, fleshy, pale yellow with six obscure vertical furrows enclosing 6 trigonous seeds in 2-seeded 3 crustaceous cocci.

Flowering and fruiting April–August

Distribution Indo-China, frequent in Meghalaya.

Propagation By seeds and vegetative method.

Parts used Root, bark, leaves and fruits

Chemical constituents   The major amino acids present are alanine, aspartic acid, glutamic acid, lysine, and proline and vitamin. The fruit ash contains, chromium and copper.

Phyllanthus niruri L.

Family   Euphorbiaceae

Indian names   Niruri (English); *Adhyanda, Ajada* (Sanskrit); *Bhuinanvalah, Sadahazurmani* (Hindi); *Kiranelligida, Kirunelli* (Kannada); *Kikkaynelli, Kilanelli* (Tamil); *Kizhakkayinelli, Kizhanelli* (Malayalam); *Nelausirika, Nelavusari* (Telugu).

Description   Shrubs, usually deciduous, sometimes with deciduous branchlets; leaves entire, alternate, distichous; flowers very small, monoecious, in axillary cluster or sub-solitary; female flowers larger; calyx, 5–6 segments, imbricate in 2 series, accrescent in fruit; petals 0; disc in male of small glands, in female of large glands; stamens 3, filaments connate or free, anthers confluent; ovary 3-celled, styles 3, free, connate at base, 2-fid; fruit capsular with three thin or crustaceous 2-valved cocci; seed trigonous, 2 in each cell, rounded at the back, albumen fleshy.

Flowering and fruiting   June–August

Distribution   Throughout India; frequent in Meghalaya along the streams and river banks.

Propagation   By seeds and vegetative method

Parts used   Whole plant

Chemical constituents   In the aerial parts, three crystalline lignins including phyllanthine and hypophyllanthine have been found. Five flavonoids have been identified, quercetin, astragin, quercitrin, isoquercitrin and rutin.

Uses   The plant is an astringent, sweet, cooling, diuretic, alterant, stomachic, constipating and attennuant. It is useful in burning sensation, strangury, gastropathy, sores, burns, diarrhoea, skin eruptions and obesity. It is also used in bleeding gums, smallpox, asthma and syphilis.

Phyllanthus parvifolius Ham.

Family   Euphorbiaceae

Description   Shrub 1.5–2.5 m, with slender upright branches, almost glabrous; ribbed twigs purplish; bark pale, brownish; leaves bifarious, sessile, obvate or elliptic, rounded, entire, membranous, glaucous beneath; lateral nerves 4–5 on either half; stipules hastate; flowers minute, axillary, pedicelled, solitary or few; male flowers have calyx segments rounded, filaments short, spreading, anthers didynamous, disc glandular; female flowers larger, sepals oblong, reflexed in fruit, ovary globose, style arms capillary, disc annular, capsule depressed, globose.

Flowering and fruiting   June–August

Distribution   Khasi and Jaintia hills

Propagation   By seeds and vegetative methods

Parts used   Whole plant

Uses   Plant astringent, diuretic, stomachic and used in jaundice, diarrhoea, dysentery and fever. The plant is useful for burning sensation, sore throat, boils, tongue thrust and arthralgia.

Picrasma javanica Bl. Facing extinction.

Family   Simaraubaceae

Indian names   *Nimtita, Putichhal* (Assamese); *Borjagreng* (Garo); *Thingdamdawi, Khwawsik-damdawi* (Mizoram).

Description   Large tree up to 20 m high; crown dense; bark dark brown to blackish; leaves 15–30 cm long; leaflets usually 7, (sometimes 5 or 9) 4.5–12 × 1.5–4 cm, oblong, lanceolate, elliptic, caudate, acuminate, base cuneate, coriaceous, glabrous, dark green above; inflorescence up to 6 cm across, sparsely hairly; flowers 0.6–1 cm across, white or greenish white; sepals 4, rounded, very small; petals 4, 0.4 cm long, acute; stamens 4, filaments dilated at base, hairy, disc yellowish; drupelets 1–4, globose, obovate, seated on the cushion-like disc.

Flowering and fruiting   April–November

Distribution   Tropical South-East Asia extending to Philippines, North-East India and Andamans. Throughout in tropical evergreen forests in shady localities, principally as a second storey tree, associated with *Garcinia paniculata, Castanopsis indica,* etc.

Propagation   By seeds and vegetative method

Parts used   Bark called Quassia

Chemical constituents   Quassia Bark contains quassinoid alkaloids, α-coumarin and vitamin B.

Uses   The strongly bitter quassia bark supports and strengthens weak digestive systems. It increases bile flow, the secretion of salivary juices and stomach acid, and improves the digestive process as a whole. Quassia is commonly used to stimulate the appetite.

Piper betle L.

Family   Piperaceae

Indian names   Betel pepper (English); *Bhakshyapatra, Bhujangalata* (Sanskrit); *Pan tamboli, Tambuli* (Hindi); *Ambadiyele, Chigurele* (Kannada); *Tambulam, Vettilai* (Tamil); *Vettila, Vettilakkodi* (Malayalam); *Akumadupa, Kammeraku* (Telugu).

Description   Perennial dioecious creeper, stems semi-woody, climber by short adventitious roots; leaves ovate, cordate, elliptic, acuminate, acute, entire, with undulate margin, glabrous, yellowish or bright green, shining on both sides, petiole stout, male spikes dense, female spikes long pendulous.

Flowering and fruiting   April–October

Distribution   Common in Meghalaya usually in tropical evergreen forests

Propagation   By seeds

Parts used   Leaves and roots

Chemical constituents   The whole plant yields essential oil containing eugenol, carvacrol, chavicol, allyl catechol, chavibetol, cineol, estragol, methyl eugenol, p-cymene, cadinene, tannins, carotene, thiamine, riboflavin, nicotinic acid, etc.

Uses   Leaf juice is useful for cuts, wounds and as eye drop.

Piper cubeba L. f.

Family   Piperaceae

Indian names   Java pepper, Cubeb (English); *Cinorana, Kakkola* (Sanskrit); *Kabacini, Sitalachini* (Hindi); *Balmenasu, Gandhamenasu* (Kannada); *Valmilagu* (Tamil); *Valmulaku, Chinimulaku* (Malayalam); *Calavamiriyalu, Tokamiriyalu* (Telugu).

Description   A perennial woody climbing with ash grey climber stems and branches, rooted at the joints; leaves simple, ovate, oblong, smooth, base, cordate or rounded, tip pointed, lower

surface dense provided with minute sunken glands; flowers small, unisexual, female spikes often curved; fruits subglobose berries, somewhat apiculate.

Flowering and fruiting   December–February

Distribution   Khasi hills

Propagation   By vegetative method

Parts used   Unripe berries

Chemical constituents   Cubebin having a bitter taste, cubebol and cubebic acid.

Uses   The berries are useful in odontalgia, cephalalgia, inflammation, wounds and ulcers, catarrh, haemarrhoids, cough, asthma, bronchitis and hay fever.

Piper griffithii Cas DC.

Family   Piperaceae

Indian name   *Ching-marich* (Manipuri).

Description   Climbering on branches of trees; leaves ovate, lanceolate, acuminate, base rounded, cordate or oblique, 5–7 nerved, pale beneath; spikes 1.5–6 cm long, yellowish; fruiting spikes up to 25 cm long, drooping fruits globose.

Flowering and fruiting   April–October

Distribution   North-East India; common in Khasi hills, Meghalaya

Propagation   By vegetative method

Parts used   Whole plant

Uses   An infusion of the leaves is used to treat watery vaginal discharges; plant is also used in bronchitis, gonorrhoea and rheumatism.

Piper longum Linn.

Family   Piperaceae

Indian names   India long pepper, Long pepper (English); *Maghadi, Pippali* (Sanskrit); *Pimpli, Pipal* (Hindi); *Hipli* (Kannada); *Tippali* (Malayalam); *Pippili, Tippili* (Tamil); *Pippallu* (Telugu); *Pimpli, Piplee* (Marathi).

Description   Aromatic herb, trailing on ground, also climbing on tree, glabrous, root-stock jointed; leaves orbicular, ovate or oblong, acuminate, membranous, glabrous, 5–7 nerve, base cordate; petiole long, male spikes slender, yellow; flowers dioecious, bracts stalked, peltate; fruiting spike, fleshy; ovary sunk, more or less confluent; fruit 0.25 cm diameter.

Flowering and fruiting   June–December

Distribution   Moist deciduous and evergreen forests, Khasi and Jaintia hills, Meghalaya.

Propagation   By seeds and vegetative method.

Parts used   Root and fruits

Chemical constituents n-hexadecane, n-heptadecane, n-octadecane, n-nanodecane, n-eicosane, α-thujene, terpinolene, zingiberene, p-cymene, phenethyl alcohol, etc.

Uses   Roots and fruits are useful to treat diarrhoea, indigestion, jaundice, abdominal disorders, asthma, cough, piles, malarial fever, vomiting, thirst, chest congestion and throat infections.

Piper mullesus Buch. Ham.

Family   Piperaceae

Description   Ascending, woody climbers; nodes much swollen and rooting; leaves elliptic, lanceolate, long acuminate, base narrowed, cuneate, lateral nerves much oblique extending to the tip; fruiting spikes 0.5–1 cm long, ovoid, purplish.

Flowering and fruiting   June–November

Distribution   Chiefly North-East India, Khasi hills and Cherapunjee, Meghalaya.

Propagation   By vegetative method

Parts used   Roots and fruit

Uses   Similar to *Piper longum*; the root and fruits are used for mainly cold and cough.

Pittosporum nepaulense (DC.) Rehder and Wilson.

Family   Pittosporaceae

Indian names   *Raini, Tumri* (Hindi); *Ekkadi* (Kannada); *Chettu kasind, Rakamuki* (Telugu); *Dieng duma, Dieng mulosshiing* (Khasi); *Vehkali, Vehyenti* (Marathi).

Description   Middle sized trees, up to 15 m high; crown spreading; bark greyish brown, warty; leaves crowded at the ends of branchlets, oblong, lanceolate, narrowly elliptic, acuminate, base cuneate, glabrous, flowers yellow, sepals ovate, oblong; petals narrowly oblong, ovate, stamens 3–5 mm long; ovary pubescent, style glabrous; capsules 5–8 mm in diameter, crowded by the style base, valves horizontally striate inside, aril scarlet.

Flowering and fruiting   February–December

Distribution   Throughout southern parts of Meghalaya, Garo hills

Propagation   By seeds

Parts used   Leaves

Uses   For cough, throat pain and paste is applied on bruises.

Plantago major L.

Family   Plantaginaceae

Indian names   Common plantain (English); *Asvagola* (Sanskrit); *Lahuriya* (Hindi); *Ishappukol vitai* (Tamil); *Singa gach* (Assamese).

Description   A perennial herb with an erect stout root stock; leaves alternate, radial 12.5 cm long, ovate or oblong, obtuse or sub-acute, entire or toothed, nearly glabrous, base tapering and decurrent into the petiole, commonly 7 nerved; petioles usually longer than blade, broad sheathing at base; flowers scattered or crowded in long slender rather lax spikes, 5–15 cm long; spikes seldom less than 3; bracts 1.5–2 mm long, shorter than the calyx, broadly ovate, oblong, obtuse, with scarious margins; calyx 3 mm long, glabrous, oblong, obtuse or subacute, keeled on the back and with broad scabrous margins, lobes lanceolate, acute; capsule ovoid, dull black, 0.85 mm long usually 8–16 seeds in a capsule.

Distribution   All over temperate Himalayan regions

Propagation   By seeds and vegetative method

Parts used   Whole plants, seeds

Chemical constituents   Alkaloids such as boschinakine and methyl esters of boschniakinic acid, choline, steroids, thioglucosides, acubin, plantaenaloside, acuboside, melitoside, caffeic acid, cinnamic acid, ferulic acid, fumaric acid, salicylic acid.

Uses   It is used for toothache, ear ache and eneuresis, depression and insomnia due to chronic tobacco addiction. Traditionally used for bee sting incised wounds and bleeding piles.

Plumbago zeylanica L.

Family   Plumbaginaceae

Indian names   *Agnika, Agnimata* (Sanskrit); *Lechkuro, Chitarak* (Hindi); *Chitramulike* (Kannada); *Angodiveli, Kanilam* (Tamil); *Koduveli, Tumpakoduveli, Vellakoduveli* (Malayalam); *Agnimatha, Chitra-mulam* (Telugu).

Description   Perennial sub-scandent shrub, glabrous; leaves ovate glabrous; flowers white, in elongated spikes; capsules oblong.

Distribution   Native to southern India and Malaysia and entire North-East Region

Propagation   By seeds and vegetative method

Parts used   Seeds and roots

Chemical constituents   It contains plumbagin, which stimulates sweating.

Uses   In Indian herbal medicine, the leaves and root are used to treat infections and digestive problems such as dysentery. Externally a paste and root is applied to painful rheumatic areas or to chronic and itchy skin diseases. The decoction of the seed is used to reduce muscular pain. The roots possess abortifacient and vesicant properties.

Podophyllum hexandrum Royle.

Family   Berberidaceae

Indian names   *Bankarkatee, Vankakhharu* (Hindi); *Vanatrapusi* (Sanskrit); *Paadbal* (Marathi).

Description   An erect succulent herb, up to 40–60 cm high, with creeping perennial rhizome, bearing numerous roots; leaves 2 or 3, orbicular, reniform, and palmate, peltate, with lobed segments; flowers solitary, white or pink, cupshaped; fruit ovoid, and edible.

Distribution   Entire North-East Region up to 900–1200 m.

Propagation   By seeds and vegetative method

Parts used   Rhizome

Uses   Rhizome and roots are considered to be purgative, emetic and a bitter tonic. They are useful for tumours and skin diseases.

Pogostemon purpurascens Dalz.

Family   Lamiaceae

Indian names   *Shangbrei* (Manipuri).

Description   An erect branched herb; leaves opposite, long petioled, large, membranous, ovate, or lanceolate, toothed base crenate, acute, doubly serrate, softly hairy on both sides; whorls dense flowered, globose; calyx teeth as long as the tube, upper lip of corolla purplish; nutlets ovoid, compressed, inner face angled.

Distribution   Mostly cultivated in warm and moist regions in China, India, Malaysia and Philippines.

Propagation   By stem cutting

Part used   Leaves

Uses   Decoction of leaves is useful to reduce headache and giddiness.

Polyalthia longifolia Benth. and Hk. f.

Family   Annonnaceae

Indian names   Indian willow, Mast tree (English); *Asoka, Devadaru* (Sanskrit); *Asoka, Devadar* (Hindi); *Assoti, Puttrajivi* (Kannada); *Asogu, Nettilingam* (Tamil); *Aranamaram, Ashokam* (Malayalam); *Asokamu, Devadaru* (Telugu).

Description   Handsome evergreen tree with a conical crown and dark greyish brown bark; leaves narrowly lanceolate, pointed, rather membranous quite glabrous, shining above with wavy edge, somewhat aromatic, lateral nerves up to about 30 on either half, vary oblique, base cuneate; petiole long; flowers yellowish-green, in fascicles or very short umbels; sepals about 5 cm long, ovate, lanceolate, densely pubescent; petals tapering from a slightly expanded base, puberulous; ripe capsules numerous, black, on glabrous stalks; seed smooth, shining.

Flowering and fruiting   March–September

Distribution   Indo-Malaya; North-East India, common in Meghalaya

Propagation   By seeds and vegetative method

Parts used   Bark

Uses   The bark is bitter, acrid, cooling, febrifuge and anthelmintic and used in skin diseases, hypertension, diabetes and helminthiasis.

Polygonum chinense L. [Endangered]

Family   Polygonaceae

Indian names   *Jangli palak* (Hindi); *Bilee kanagilu* (Kannada); *Jaryndem* (Khasi); *Piripu* (Malayalam); *Kelnap, Modhusoleng* (Assamese).

Description   A rambling or erect shrub, reaching 12 cm; stems and branches many from the roots and shoots; shoots flowers angled and grooved, glabrous or sparsely glandular, pubescent; leaves 7–12 cm linear, oblong deltoid, ovate or entire truncate, rounded, acute or subcordate. Heads very variable in number 2–3 cm in diameter; divaricated; flowers white, pink or purplish; perianth 5 cleft, stamens 8, styles 3, connate below, nuts variable in size enclosed in the dry or fleshy perianth. Petioles hardly winged usually 2-auricled at the base, stipules long, oblique; heads, panicles or corymbose in peduncles usually glandular, hairy; involucral, bracts glabrous, usually acute; fruiting perianths dry or fleshy, nut trigonous. Gland-dotted or membranous, flat or undulate.

Distribution   Grows wild throughout India

Propagation   By seeds and vegetative methods

Part used   Leaves

Chemical constituents   The whole plant contains rubin, rheumemodin; oxy-methyl-anthraquinone, glucosides, myrcyl alcohol.

Uses   The leaves have antibacterial properties and are effective in the treatment of furuculosis, impetigo, ulceration of the helix, scabies, cold sore, and streptococcal skin infections. Snake bite is treated by oral administration of the juice of the fresh leaves, the residue being applied to the bite as a plaster.

Polygonum hydropiper L.

Family   Polygonaceae

Indian names   Red knees, Water pepper (English); *Kari agrada gida, Kari sanni* (Kannada); *Patharua bihalagani* (Assamese); *Lilhar* (Manipuri).

Description   A glabrous, often glandular, reddish, annual or perennial herb up to 80 cm high; roots tufted or shortly creeping; stems and branches rather stout, leafy, glabrous, nodes often swollen; leaves linear, lanceolate or oblong with resinous cavities, rarely more than 10 cm long, very variable in width and in the length of the apex, usually covered with impressed glands; stipules glabrous; racemes flexuous, leafy at the base; flowers pink or red in slender racemes; nuts granulated and finely dotted.

Distribution   Entire North-Eastern India

Propagation   By vegetative method

Parts used   Whole plant

Uses   It is a diuretic, emmenagogue, stimulant; extract of plant is used as an oral contraceptive.

Polygonum orientale L.

Family   Polygonaceae

Indian names   *Bara-panimarich* (Bengali); *Chakhong* (Manipuri).

Description   Gregarious undershrubs in marshy areas; leaves broad, rarely lobed; stipule tubular; flowers bisexual, minute, clustered, axillary or terminal; perianth 4–5 rarely 3 cleft, 2 outer segments; stamens 5–8; ovary compressed or trigonous, style 2–3, erect greenish.

Flowering and fruiting   March–November

Propagation   By seeds and vegetative method

Parts used   Whole plant

Uses   It is used for urinary tract infection and gonorrhoea.

Polygonum perfoliatum L.

Family   Polygonaceae

Indian names   *Ma sein thli* (Khasi); *Lilhar* (Manipuri).

Description   Straggling or rambling shrubs, stems retrosely spiny; leaves 3–4 cm across, ovate, acute or obtuse, peltate, glaucous, membranous; foliaceous stipules, deltoid leaves and prickly petioles; racemes 2–3 cm long, usually terminal; flowers 0.2 cm long, nuts globose, 0.5 cm long, purplish, shining.

Flowering and fruiting   May–October

Distribution   Entire North-East India

Propagation   By seeds and vegetative method

Parts used   Leaves and fruits

Uses   It has anti-inflammatory, antipyretic activities and used for liver protection.

Polygonum posumbi Bch. Ham.

Family   Polygonaceae

Description   Small creeping herb, stem extensively creeping below and peduncles quite glabrous; leaves 2.5–10 cm long, petioled, elliptic, lanceolate, caudate, acuminate, glabrous or sparsely hairy, stipules sparingly strigose, cilia stiff, longer than the tube; flowers in racemes. Peduncles and erect racemes filiform sometimes very long, bracts minute close or distant, very shortly ciliate; perianth very small eglandular. Nut perfectly smooth and polished.

Distribution   Entire North-Eastern Region

Propagation   By vegetative method

Parts used   Roots

Uses   Root is an astringent and used in bowel complaints, diarrhoea, dysentery and sores.

Pongamia glabra Vent.

Family   Fabaceae

Indian names   *Angaravalli* (Sanskrit); *Karanjaka, Kiramal* (Hindi); *Huligili, Karanja* (Kannada); *Punga maram* (Tamil); *Minnari, Pungammaram* (Malayalam); *Kanuga* (Telugu).

Description   Medium sized semi-evergreen glabrous tree with a short bole and spreading crown up to 18 m high; bark greyish green or brown; leaves compound, leaflets 5–7, ovate-acuminate or elliptic; flowers lilac or pinkish white, fragrant in axillary racemes; fruits thick, smooth, compressed, with a short beak; seeds 1 or 2 per pod.

Distribution   Throughout India, common in Khasi hills, Meghalaya.

Propagation   By seeds

Parts used   Leaves and roots

Uses   Juice of leaves and root is useful for cough and leprosy.

Portulaca oleracea L.

Family   Portulacaceae

Indian names   Common purslane (English); *Brihalloni, Ghotika* (Sanskrit); *Badi noni, Bara laniya* (Hindi); *Budagora* (Kannada); *Karikkirai, Paruppuukkirai* (Tamil); *Kozhuppa, Kozhuppachira* (Malayalam); *Peddapuvilakura* (Telugu).

Description   Prostrate annual herb; stem succulent, green or reddish; leaves alternate, fleshy, shining glabrous, oblong or truncate, base attenuate, apex truncated, nerves inconspicuous; flowers bright yellow in terminal sometimes axillary cluster, without stalks; capsule globose or ovoid, opening transversely; seeds numerous, shining black.

Distribution   In entire North-East Region up to 1500 m altitude, commonly found in wild in wet places.

Propagation   By seeds and vegetative method

Parts used   Whole plant

Chemical constituents   The plant contains vitamins C and B and mineral salts like P, Ca, Mg, Na, and K, organic acids like nicotinic and oxalic, noradrenalin and the biflavonoid liquirtin.

Uses   The plant is used as an antibacterial, anti-inflammatory and anthelmintic. Decoction gives cure to jaundice. Prickly heat and burning sensation. It is also used in cardio vascular diseases, dysuria, haematuria, gonorrhoea, dysentery and ulcers of the mouth, scurvy and liver dieases.

Potentilla integrifolia Lindl.

Family   Rosaceae

Description   Large shrubs or small trees, with yellowish lenticellate bark, spinous large branches and yellowish brown woody aromatic root; leaves simple, opposite, sometimes, whorled, elliptic, ovate, membranous when young, coriaceous when mature; flowers small, greenish yellow or greenish white; fruit globose drupes, black when ripe.

Distribution   Meghalaya, Assam at an altitude of 1200–4000 m.

Propagation   By seeds

Parts used   Root

Uses   Root extract is used as tonic.

Pouzolzia frondosa Don.

Family   Urticaceae

Description   A large shrub; leaves usually alternate, 3-nerved, usually smaller above; stipules free; pistillode, clavate or oblong, stigma villous on one side; achenes enclosed by the marcescent calyx.

Flowering and fruiting   Summer–Winter

Distribution   Entire North-East India

Propagation   By seeds

Parts used   Roots

Uses   Herb is crushed and applied to body as a refrigerant and used for bone fracture.

Pouzalzia sanguinea (BL.) Marr.

Family   Urticaceae

Description   Small trees; bark dark brown, branchlets pubescent; leaves lanceolate, elliptic, lanceolate, acuminate, base rounded or obscurely cordate, snow white beneath; flowers in clusters 0.5–1 cm across, often on leaflet interrupted spikes, greenish white or yellow.

Flowering and fruiting   August–February

Distribution   Frequent in Meghalaya in high rainfall areas, Khasi hills and Mawsinrum.

Propagation   By seeds

Parts used   Twigs

Uses   The paste of the twigs is useful for various healing purposes.

Premna latifolia Roxb.

Family   Verbenaceae

Indian names   *Bakar, Gin* (Hindi); *Agnimantha* (Kannada); *Munnai, Chedi munnai* (Tamil); *Kondamanga, Takli* (Telugu); *Dieng lamarawai* (Khasi); *Dukhemi, Tuthekmi* (Garo).

Description   A middle sized tree with spreading crown, young shoots pubescent, young stem spinous, bark grey, somewhat rough outside; blazes slightly greenish white, soft and crisp; leaves unpleasant smelling, ovate or elliptic, usually entire, undulate, acuminate, thinly coriaceous, flowers greenish, usually in terminal compound corymbose villous cymes. Calyx obscurely 2-lipped, 4–5 toothed, accrescent in fruit, up to 4 cm long; corolla about 2-fid; drupe globose, black on ripening, hardly verrucose.

Flowering and fruiting   April–June

Distribution   Common throughout the Meghalaya

Propagation   By root cutting

Parts used   Roots and leaves

Chemical constituents   There are three chief alkaloids, premine, ganiarine and ganikarine.

Uses   The roots are useful in neuralgia, inflammations, cardiac disorders, hepatopathy, cough, asthma, bronchitis, leprosy and skin diseases. The leaves are useful in dyspepsis, flatulence, cough and catarrh, fever, rheumatalgia, and tumours.

Prismatomeris tetrandra K. Schum.

Family   Rubiaceae

Indian name   *Dieng sohsri* (Khasi).

Description   Small tree or shrubs up to 5 m high; crown spreading; bark pale or greyish brown, rough or nearly so; leaves oblong, elliptic or lanceolate, acuminate, base rounded, cuneate, glabrous, petiole cuspidate; flowers on slender 2–3 cm long pedicels, pure white; fruits globose, purplish black, shining.

Flowering and fruiting   April–December

Distribution   Indo-Malaya; common in Meghalaya as an evergreen forest undergrowth

Propagation   By seeds

Parts used   Leaves

Uses   This plant has got horticultural value. Leaves are used for stomach troubles.

Prunus cerasoides D. Don.

Family   Rosaceae

Indian names   *Padmagandhi, Padmaka* (Sanskrit); *Patmakath, Patmakh* (Hindi); *Padmaka* (Kannada); *Patumugam* (Tamil); *Pathimukham, Pathumukham* (Malayalam); *Padmakla* (Telugu).

Description   Large tree, up to 20 m high; crown oval, branches spreading, bark brown or greyish brown, horizontally lenticellate, peeling off in horizontal flakes and strips; leaves ovate, lanceolate to elliptic, caudate, acuminate, base rounded or truncate, sharply glandular, serrate, petiole with 1–2 glands at tip; flowers pink or white; calyx tube narrow, glabrous, base persistent in and supporting the fruit, lobes ovate, acute; ovary glabrous, drupe, ellipsoid or oblong, yellow, sometimes tinged with red, stone bony, unevenly furrowed on the ventral face.

Flowering and fruiting   October–June

Distribution   Khasi hills up to 1800 m

Propagation   By seeds and vegetative method

Parts used   Bark

Uses   It is useful in burning sensation, sprains, neuralgia, wounds, ulcers, leprosy, skin decolorations, diarrhoea, asthma, vomiting and intermittent fevers. Decoction is taken to cure venereal diseases.

Prunus ceylanica (Wight) Miq.

Family   Rosaceae

Indian names   *Gandhi gach* (Assamese); *Bol mangsam* (Garo); *Dieng sohkhyrnem* (Khasi); *Attainari, Katilai* (Tamil).

Description   Large trees, up to 25 m high; crown lax, spreading; bark grey or dark grey, fissured and warty; leaves ovate, lanceolate, elliptic, oblong, caudate, acuminate, base narrowed, obtuse, midrib channelled above, pedicels up to 0.4 cm long; flowers white, calyx truncate, lobes minute, petals up to 1 mm long, styles much exerted; drupes, purple, turning black.

Flowering and fruiting   April–October

Distribution   North-East India; frequent in Meghalaya

Propagation   By seeds and vegetative method

Parts used   Seeds

Chemical constituents   Amygdalins, saponins, flavonoids, prunuside.

Uses   It is used in constipation, laryngitis and acute gastritis.

Prunus nepalensis (Ser.) Steud.

Family   Rosaceae

Description   Middle sized trees, 10–15 m high; crown oval, branchlets spreading, bark greyish or dark brown, smooth or shortly fissured; branchlets lenticallate; leaves ovate, lanceolate, acuminate, base rounded or slightly oblique, minutely crenate, serrate, stiplues linear-lanceolate, caducous, racemes 8–15 cm long; flowers white, petals obovate; drupes fleshy, blackish purple when ripe; pyrenes smooth.

Flowering and fruiting   November–August

Distribution   Himalayas, common in Meghalaya at higher altitudes of Khasi and Jaintia hills, often cultivated.

Propagation   By seeds and vegetative method

Parts used   Leaves and fruits

Uses   Fruit is an astringent. Leaf diuretic and used in dropsy, hardwood astringent, acrid, refrigerant.

Prunus persica (L.) Stokes.

Family   Rosaceae

Indian names   Peach (English); *Alukah, Aruka* (Sanskrit); *Aruka* (Hindi); *Pichchisuhanne* (Kannada); *Pichesu* (Telugu); *Narabogori* (Assamese).

Description   Tree up to 8 m high; crown lax, spreading; bark dark grey or blackish; leaves oblong, lanceolate, acute, glabrous at length; flowers axillary or fascicled, pink or rose; calyx lobes densely silky tomentose; petals oblong, rounded at tip; drupes 3–5 cm long, elliptic, ovate with a rugose, red, tinged, ellipsoid pyrene.

Flowering and fruiting   January–August

Distribution   Cultivated in Meghalaya for their edible fruits

Propagation   By seeds and vegetative method

Parts used   Leaves

Chemical constituents   Thiamine, ascorbic acid and vitamin A.

Uses   Leaves yield a volatile oil. Infusion of leaves or bark is given for whooping cough. Flowers anthelmintic.

Pseuderanthemum palatiferum (Wall.) Radlk. Ex.

Family   Acanthaceae

Description   Undershrub, glabrous; leaves elliptic or lanceolate, acuminate, crenulate, membranous, lateral nerves 5–8 on either half, base cuneate or acute; petiole 6–12 cm long; flowers purplish in terminal lax simple or compound racemes, which is often curved and up to 15 cm long; bracts linear, calyx puberulous, segments linear, lanceolate, corolla oblong, 2-lipped, tube linear, cylindric.

Flowering and fruiting   February–April

Distribution   Khasi hills, Meghalaya

Propagation   By seeds

Parts used   Leaves

Uses   Leaves are useful for dysentery, diarrhoea and for stomach ache.

Psidium guajava L.

Family   Myrtaceae

Indian names   Common Guava (English); *Mansala, Perukah* (Sanskrit); *Amrud, Safed safari* (Hindi); *Balehannu* (Kannada); *Peraka* (Malayalam); *Koyya* (Tamil); *Goyya, Tellajama* (Telugu); *Goaachi, Peyara, Piyara* (Bengali); *Madhuriam* (Assamese).

Description   Shrubs or small tree up to 10 m high, stem erect, branched twigs glabrous; leaves opposite, short-petiolate, the blade oval with prominent pinnate veins, 5–15 cm long; flowers somewhat showy, 4–5 merous, petals whitish and up to 2 cm long; stamens numerous; fruits a fleshy yellow globose to ovoid berry about 5 cm in diameter with an edible pink mesocarp containing numerous small hard white seeds.

Distribution   Largely grown in gardens throughout India, sometimes found as an escape in forests. Plantations and other similar habitats.

Propagation   By seeds and vegetative method.

Part Used   Bark, leaves, fruits and root.

Chemical constituents   Tannins, vitamin B and C, 18 sesquiterpenes, 11 monoterpense, eugenol, quercetin, quaverin, gallic acid, asiatic acid, brahmic acid, lupeol, maslinic acids, oleanolic acid, ellagic acid, benzaldehyde, butyl acetate, acetylfuran, and alkaloids zeatin and zeatin nucleotide.

Uses   Bark decoction is taken for stomach ache, also used in fever, headache, gonorrhoea, menstruational disturbances, and sores. The young leaves are used as tonic in digestive disorders. A decoction of young leaves and shoots is prescribed as febrifuge and for antispasmodic baths. The pounded leaves are locally applied in rheumatism. A decoction of the leaves is used as gargle to relieve toothache and gum boils. The flowers are said to cool the body and are used in bronchitis. They are also applied to eyesores. The leaf juice is also used for treating diarrhoea, cough, stomach ache and dysentery.

Psoralea corylifolia L.

Family   Fabaceae

Indian names   Malay tea, Bawchan seed (English); *Bakuchi, Sugandhakantak, Bakuchi* (Sanskrit); *Babchi, Bavanchi, Bukchi* (Hindi); *Karkokil, Karkokilari* (Malayalam); *Karpogam, Karpokarishi* (Tamil); *Bavanchalu, Bhavanji* (Telugu); *Bavachi, Kukuch, Karbekhiga* (Bengali).

Description   An erect annual herb, up to 90 cm high, with blackish glands; leaves distinctly petioled, roundish, oval, 6–9 cm long by 5–7 cm, wide; apex slightly acute, base truncate, dentate at margin, both faces conspicuously dotted with black dots; flowers 10–30 in abundant dense long-peduncled heads; calyx cupuliform, teeth 5, lanceolate, the lowest longest; corolla yellow, little exerted; stamens 10, ovary sessile; fruit an oval pod, short, dry, black, glabrous, surrounded by the persistent calyx.

Distribution   Found on waste places in many parts of India. Mostly cultivated in China.

Propagation   By seeds and vegetative method

Part Used   Seeds and fruits

Chemical constitutuents   Seeds yields raffinose, psoralen, isopsoralen, corylifean, psoralidin, isobavachin, bavachin, neobavaisoflavone, β-sitosterol-D-glucoside, corylidin, stigmasterol, limonene, β-caryophylenoxide. 4 Terpenols like linalool, angelicin, psoralen, and bakuchiol are reported from the seed oil.

Uses   Fruits commonly called seeds are laxative, diuretic, diaphoretic, and aphrodisiac, specially recommended for leucoderma, leprosy, psoriasis, and inflammatory diseases of skin; used both internally and externally.

Psychotria montana Bl.

Family   Rubiaceae

Description   Small evergreen glabrous shrub; leaves elliptic or oblong, acuminate, membranous, glabrous, lateral nerves slender, base tapering; flowers whitish, in terminal or axillary cymes; calyx obscurely toothed, corolla tube elongate, throat bearbed; fruit ovoid, oblong, red, pyrenes plano-convex, scarcely ridge on the back, albumen uniform.

Flowering and fruiting   May–February

Distribution   Bangladesh, Assam and Meghalaya

Propagation   By seeds and vegetative method

Parts used   Leaves

Uses   It is used as remedies for piles stomach cancer, for appendicitis, heart attack, infertility, high blood pressure, shortness of breath in the lower chest and back pain.

Pueraria lobata (Willd.) Ohwl.

Family   Fabaceae

Indian names   *Kudzu vine, Japanese arrowroot* (English); *Kakamudga, Marjaragandhika, Mudgaparni* (Sanskrit); *Sisali* (Hindi); *Suting rit* (Khasi); *Bepui* (Mizoram).

Description   Large climber, brown hairy; leaves up to 25 cm long, leaflets ovate, obliquely ovate, acuminate, base rounded or truncate, entire or up to 3-lobed, pale or glaucous beneath, 3-nerved from base; racemes up to 40 cm long; flowers 2.5 cm long; calyx purplish green, deeply cleft; pods flat, oblong, densely brown hairy up to 16 seeded.

Flowering and fruiting   September–December

Distribution   Common in Meghalaya in partially shaded areas and forest margins, associated with *Mucana* sp., *Rubus* sp., etc.

Propagation   By roots

Parts used   Roots

Chemical composition   Proteins, carbohydrate, glucoside, daidzin.

Uses   Root is used for common cold, fever, thirst, acute gastro-enteritis, diarrhoea, palpitation.

Punica granatum L.

Family   Punicaceae

Indian names   Pomegranate, Balustine flowers (English); *Bijapura, Dadima* (Sanskrit); *Anar, Dhalim* (Hindi); *Dalimbe* (Kannada); *Matalam, Matalanarakam* (Malayalam); *Madulai, Madulam* (Tamil); *Dadima* (Telugu).

Description   Shrub or small tree up to 6 m high; leaves mostly opposite, short petiolate, blades oblong, elliptical up to 8 cm long; flowers showy and up to 6 cm broad, bisexual, 5–8 petals, reddish and up to 2.5 cm long, numerous stamens surrounding a conspicuous hypanthial tube, the flowers usually occurring terminally or in axils; fruit a red spherical berry, flowers available during summer.

Flowering and fruiting   May–September

Distribution   In North-East Region, cultivated for fruits

Propagation   By seeds and vegetative method

Parts used   Leaf, bark and fruits

Chemical constituents Apigenin glucoside, betulinic acid, callistephin, tannins, conine, cyanidin, cyanin, diazein, ellagic acid and its derivative, estrone, gallic acid, genistein, genistin, lipids, piperidine, polyphenols, sedrine, estradiol.

Uses Root and stem bark have astringent and anthelmintic properties. Fruit juice is a very good tonic. Leaf decoction is useful for stomach ache, cholera and dysentery. Seed is used in syphilis. Juice of fruit is used in jaundice and diarrhoea. The juice of flower is useful to stop nose bleeding.

Pyrus pashia D. Don.

Family Rosaceae

Indian names *Mehal, Kainath* (Hindi); *Soh jhur* (Khasi).

Description Medium sized deciduous tree, dark grey to almost black; leaves ovate, elliptic, lanceolate or acuminate or sometimes caudate, finely serrate to creanate, specially when young, glabrous, petiole slender, stipules caducous; flowers white in short corymbs, umbels; petals woolly when young; bracts resembling the stipules, calyx tube, woolly or glabrous; petals ovate; stamens 25–30; carpels 3–5, styles free, more or less woolly at the base; fruit globose, depressed at the top, somewhat rough with raised white specks, dark brown when ripe.

Flowering and fruiting March–September

Distribution Khasi and Jaintia hills, Meghalaya

Propagation By grafting

Parts used Leaves

Uses Leaves are used to stop traces of blood in the urine.

Q

Quercus acutissima Carruth

Family   Fabaceae

Description   Tree, deciduous of moderate size up to 15 m high; crown lax; bark ashy grey, fissured, warty and rough; leaves oblanceolate, elliptic, acuminate, base narrowed, sharply distantly serrate, glabrous above, hairy along the midrib, beneath; male spikes longer than female spikes. drooping, many, together, yellow 5–12 cm long; female clustered at tip, acrons solitary or in pairs, scales numerous.

Flowering and fruiting   June–February

Distribution   Tropical Himalayas, confined to North-East India; very common in Meghalaya.

Propagation   By seeds and vegetative method

Parts used   Stem

Uses   Juice is used in the preparation of medicines used for some minor eye irritations.

R

Randia longiflora Lamk.

Family Rubiaceae

Indian names *Boroki amkora, Pulikaint* (Assamese); *Yengokjaching* (Garo); *Jyrmishiah iewkrot* (Khasi).

Description Large glabrous shrub; bark grey, smooth, light red with brown streaks inside, thorns recurved; leaves ovate, elliptic or oblong, entire, abruptly acute, coriaceous, glabrous, lateral nerves 5–6 on either half, stipules triangular with a broad base; flowers puberulous, white, fragrant in axillary or leaf opposed shortly peduncled trichotomous cymes; calyx tube dilated;

corolla tube much longer than the lobes; stigma bifid; berry obscurely ribbed, black when ripe, about 8 seeded.

Flowering and fruiting   October–January

Distribution   Common in Garo hills, Meghalaya

Propagation   By seeds and vegetative method

Parts used   Bark and fruits

Chemical constituents   Saponin, valerianic, vax, resin, d-mannitol and essential oil.

Uses   Fruit extract is insecticidal and insect repellant and used in insecticide preparation.

Rauvolfia densiflora (Wall.) Benth. ex. Hook. f. [Critically endangered]

Family   Apocynaceae

Indian name   *Dieng larkei, Dieng sohbu blang* (Khasi).

Description   Plant shrubs, often bushy, branchlets densely lenticellate; Leaves 10–30 × 2–6 cm, oblanceolate, elliptic, acuminate, base narrowed, acuminate, smooth, glaucous beneath; petiole winged; cymes 7–10 cm long; flowers white, 0.6–1 cm long; drupelets 1 cm long, oblong, purplish black, shining.

Flowering and fruiting   June–December

Distribution   India and adjacent countries; largely in North-East and Southern India. In Meghalaya, associated with *Luculia poinciana*.

Propagation   By seeds and vegetative method

Parts used   Root and bark

Chemical constituents   The root bark contains 2.64 per cent alkaloids, mostly reserpine, rauvomitine and ajmaline.

Uses   The root bark is widely used for the treatment of hypertension and pshychoses.

Rauvolfia serpentina (L.) Benth. Ex. Kurz. [Endangered]

Family   Apocynaceae

Indian names   Rauvolfia root, Serpentina root (English); *Chandrika, Sarpagandha* (Sanskrit); *Chhotachand, Chota cand* (Hindi); *Garudapatala, Patalagaruda* (Kannada); *Amalpori, Chuvannavilpuri* (Malayalam); *Sarppaganti, Sivan amalpodi* (Tamil); *Patalagandhi* (Telugu).

Description   Herbs to under shrubs, 0.5–2 m tall, bark white, rarely lenticellate; stems woody at base; root stock woody and thick; bark ashy white, thin, latex water; branchlets hairless; leaves 3–4 in a whorl or sometimes opposite, elliptical, lanceolate or obvate, acute or acuminate 8–16 × 3–5 cm, base gradually tapering, apex acuminate, margin entire, papery, hairless, with distinct stalk, lateral nerves 8–12 pairs; cymes long, peduncle rounded rarely recemose; cymes 1–2 in., diameter, many flowered; flowers bisexual, in terminal or axillary umbellate cymes, white with pale purple shade, about 2 × 1 cm, peduncle 2–5 in., stout, branched and pedicles red; bracts obsolete; pedicles 0.25–0.75 in., flowers white or pinkish, nearly 1 in., long; calyx small; corolla tube often curved; tubes not one quarter the length of the tube; throat hairy; fruiting shortly, globosely inflated above the middle; drupes 0.25 in., broadly oblique ovoid; drupes black, united in their lower half, stakless, fleshy, ovoid, about 7 mm long, hairless, shiny, purplish black when ripes; seeds 1, ovoid.

Flowering and fruiting   June–December

Distribution   Moist deciduous to evergreen forests in India and adjacent countries, largely in North-East and Southern India. Common in Meghalaya, associated with *Agapetes* sp. and *Wendlandia* sp., etc.; mostly under cultivation for their medicinal value.

Propagation   By seeds and vegetative methods

Parts used   Roots and leaves

Chemical constituents   Alkaloids, rauvanine, reserpine, reserpinine, serpentine, serpentinine, rauvoxine, rauvoxinine, ajmalinine, alkaloid A, alkaloid D, alkaloid RP1, alkaloid RP2, alkaloid RP3. Chandrine, neoreserpiline, obscuridine, obscurine, pelirine, raucaffricine, raucaffridine, rauwolfine, samatine, raunamine and semperflorine.

Uses   The root is chiefly by used in drug preparations. It is a sedative and hypnotic and used for reducing blood pressure. The drug is now largely used in insanity and high blood pressure. It is more suitable for cases of mild anxiety or patients of chronic mental illness. The drug has tranquilizing effect. Root is also considered as anthelminic and an antidote to snake venom. Its decoction is given during labour pains to increase uterine contraction. Juice of the leaves is given for the cure of corneal opacity of the eyes.

Rhododendron arboreum Sm.

Family   Ericaceae

Indian names   *Pullasa* (Sanskrit); *Burans, Baras, Parag* (Hindi); *Bilipu, Billee* (Kannada); *Pumaram, Billimaram* (Tamil); *Kattupuvarasu* (Malayalam); *Khorom-leishak-angangba* (Manipuri).

Description   Middle sized tree, 6–15 m high; bark, reddish-brown with rectangular flakes and fissures; leaves oblong, oblanceolate, acute, base narrowed or rounded, dark green, nerves impressed above, scanty, tomentose beneath; corymbs terminal, dense, subglobose, 6–15 cm across, flower 2.5–4 cm long; corolla with 5 black patches at base within; anther black; capsule oblong, cylindric, sometimes curved; seeds numerous.

Flowering and fruiting   March–October

Distribution   Burma, common in Meghalaya

Propagation   By seeds, stem cutting and rootsuckers

Parts used   Leaves and bark

Uses   Bark is used in the preparation of snuff; tender leaf for relief from headache, in diarrhoea, dysentery; dried flowers are fried with ghee to check blood dysentery.

Rhus javanica L.

Family   Anacardiaceae

Indian names   Chinese sumac, Japanese galls (English); *Damphela, Deshmeel, Arkhar* (Hindi); *Dieng sohmia* (Khasi); *Khawhma* (Mizoram).

Description   Middle sized tree up to 12 m high; crown lax, spreading, young tomentose; bark grey, ovate, lanceolate, acuminate, rounded; panicles pyramidal up to 45 cm long; flowers yellow, greenish white; petals spreading or deflexed, ovoid, orbicular, pink.

Flowering and fruiting   August–March

Distribution   Confined to North-East Region; very common in Meghalaya, particularly in open lands and forest margins.

Propagation   By seeds and vegetative method

Parts used   Fruits

Chemical constituents   Gallotannin, gallotannic acid, gallic acid, *m*-digallic acid.

Uses   The sour fruits are used for catching fishes by local people and as a medicine for stomach pain, skin trouble, treatment of papilloma, root is used in rheumatism.

Rhus succedanea (non L.) Gamble.

Family   Anacardiaceae

Indian names   Japanese tallow, Wax tree (English); *Kalinga, Karkatahvaya* (Sanskrit); *Kakrasingi, Arkhol* (Hindi); *Karkatakashringi* (Kannada); *Karkkadagachingi, Kadukaipoo* (Tamil); *Kakarashingi* (Telugu).

Description   Middle sized deciduous tree up to 50 m height; bark thick dark grey and rough outside, split with vertical reticulate fissures and exfoliating in irregular flakes, white to brown light brown inside, faintly mottled with horizontal lines or lenticels; branchlets usually glabrous, leaves imparipinnate, crowded at the ends of branches; common petiole swollen at the base; leaflets 3–6 pairs, opposite, ovate, lanceolate or oblong, acuminate, entire, thinly coriaceous, usually quite glabrous, pinkish brown when very young; flowers greenish yellow, sepals ovate, obtuse, petals much larger, oblong; drupes orbicular, compressed.

Flowering and fruiting   May–October

Distribution   In Khasi Hills

Propagation   By seeds and vegetative method

Parts used   Fruit

Uses   Fruit branches are used as astringent, tonic, expectorant, stimulant and used in diarrhoea and dysentery.

Ricinus communis L.

Family   Euphorbiaceae

Indian names   Wild castor, Castor oil plant (English); *Eranda, Pancangulah* (Sanskrit); *Erand, Erandi* (Hindi); *Haralu, Manda, Oudla* (Kannada); *Amanakkam ceti, Amanakku* (Tamil); *Avanakku* (Malayalam); *Erandamu, Amudamu* (Telugu).

Description   An evergreen soft wooded shrub; young parts covered with a glaucous bloom; leaves alternate, 5–23 cm across, palmately lobed, peltate, serrate, membranous; petiole stout, hollow as long as the leaves; flowers monoecious; stamens numerous; ovary 3-celled, style 3, plumose, 2-fid; capsule globose, echinate, about 2–2.5 cm long, splitting into three, 2-valved dehiscent cocci; seeds with large caruncle, testa crustaceous.

Flowering and fruiting   Almost throughout the year

Distribution   All over North-East Region

Propagation   By seeds

Parts used   Root, leaves and seed oil

Chemical constituents   β-amyrin, ricins, 5-dehydro-avenasterol, beta carotene, tannins, lupeol, quercetin, casbene, ellagic acid, quinic acid, ricinine, ricinus agglutinins, vitamins B and B_1. Seed saponins, kaempferol glycoside.

Uses   Seed oil is used as a purgative. Leaves are useful for swellings, boils and joints affected with rheumatism. Dried root is used as febrifuge.

Rosa macrophylla Lindl.

Family   Rosaceae

Indian name   *Ban-gulab, Dand kunja, Karer* (Hindi).

Description   Branchlets weak, prickles infrastipular; leaflets small, 9–15, flowers pale pink, double in corymbose clusters, sweet scented; calyx densely clothed, with long bristly prickles, lobes orbicular; carpel developing into follicle; fruit of one, rarely 2, 1-seeded 2-valved follicular capsules; seed 1, often arillate, exalbuminous, radicle superior.

Flowering and fruiting   Throughout the year

Distribution   Throughout India; common in Meghalaya. Grown often in the hills.

Propagation   By cutting method

Parts used   Fruits

Uses   Fruit is useful for stomach ache.

Rourea minor (Gaertn.) Alston.

Family   Connaraceae

Indian names   *Hullechala balli, Phirangi chakke* (Kannada); *Curigi-tali* (Malayalam); *Bardhara, Saambarbael* (Marathi); *Chandrapudi* (Telugu).

Description   Scandent shrubs; bark blackish brown or dark brown; leaves up to 20 cm long; leaflets oblong, lanceolate, rounded or obtuse at base, glabrous and shining above, glaucous beneath; midrib channelled above; panicles up to 10 cm long; flowers 0.5–1 cm long, white; calyx lobes ovate, orbicular; petals oblong; style exerted; follicles up to 2.5 cm long, oblique, tapering; seed with red arils.

Flowering and fruiting   July–February

Distribution   North-East India, in Meghalaya in wet evergreen forests, Khasi and Jaintia hills.

Propagation   By seeds

Parts used   Leaves

Uses   Leaves are useful for dysentery, diarrhoea and vomiting.

Rubus ulmifolicus Schott.

Family   Rosaceae

Description   Shrubs with curved or deflexed prickles; leaves ovate or suborbicular, 3–7 lobed or angled, lobes ovate, triangular, obtuse, minutely serrate, glabrous or glabrescent above, hairy beneath particularly along nerves, palmately 5-nerved from base; stipules and bracts deeply lobed; racemes short up to 7 cm long; flowers white; calyx lobes ovate, lanceolate, tip 2–3 fid or entire, patently villous; petals equalling the sepals, 1–1.3 cm long; fruits globose, 1–1.5 cm across, scarlet.

Flowering and fruiting   May–November

Distribution   North-East India; frequent in Meghalaya.

Propagation   By vegetative method

Parts used   Leaves

Uses   The leaves are useful for external remedy as eyewash for conjunctivitis, a mouthwash for mouth problems or a lotion for ulcers and wounds.

Rubus ellipticus Sm.

Family   Rosaceae

Indian names   *Hinsalu, Anchhu* (Hindi); *Jotelupoka, Jotelu poka* (Assamese); *Dieng shiahsohprew, Sia sohpru* (Khasi); *Heijampet* (Manipuri); *Hmutau* (Mizoram).

Description   Bushy shrub; leaves up to 15 m high; leaflets ovate orbicular, obovate or broadly elliptic, acute to truncate, base rounded or truncate, sharply double serrate, greyish tomentose beneath; panicles 3–10 cm long; calyx lobes, broadly ovate, mucronate, puberulous without; petals exerted, obovate; fruits orange yellow, succulent.

Flowering and fruiting   January–June

Distribution   Entire North-East India, common in Meghalaya in undergrowth of pine forests.

Propagation   By vegetative method

Parts used   Leaves and fruits

Uses   Decoction as well as powder is taken for stomach ailments.

Rubus niveus Thunb.

Family   Rosaceae

Indian names   Mysore raspberry (English); *Kala hinsalu, Kalianchhi, Anchu* (Hindi); *Gomulli, Mulli* (Kannada); *Dieng soh khawiong* (Khasi); *Heijampet* (Manipuri); *Gowriphal* (Marathi).

Description   Bushy shrubs; branchlets reddish green, glabrous, densely prickly, prickles recurved; leaves up to 15 cm long, rachis acute, base acute, cuneate, sessile, sharply dentate, grey, tomentose beneath; panicles corymbose, terminal and upper axillary; flowers red or pinkish 0.6–0.8 cm across; calyx lobes ovate, lanceolate, up to to 0.5 cm long; petals shorter or equalling sepals; fruits ovoid, globose, 1–1.3 cm long, blackish when ripe.

Flowering and fruiting   March–October

Distribution   Entire North-East Region, up to 500 m height

Propagation   By vegetative method.

Parts used   Shoot and roots.

Uses   The paste is useful to cure head ache.

Rumex maritimus

Family   Polygonaceae

Indian names   *Bon chuka* (Assamese); *Ban-palang* (Bengali); *Jub-palum, Jungli palak* (Hindi); *Torongkhongchak* (Manipuri).

Description   Robust herb with stout perennial rootstock; leaves upper smaller, oblong, elliptic, lanceolate, cordate; flowers small, green, in whorls, arranged on long raceme; perianth 6-partite, margins fringed with usualy hooked setae; stamens 6; ovary trigonous, styles 3, with fringed stigma; nut brown.

Flowering and fruiting   May–October

Distribution   North-East India including Assam

Propagation   By vegetative method

Parts used   Leaves

Uses   Leaf paste is used for leucoderma; roots contain chrysophanic acid, and is used as purgative.

Ranunculus diffusus DC.

Family   Ranunculaceae

Indian name   *Angasia-jhar* (Hindi).

Description   A perennial diffuse or prostrate herb with spreading hairs and fibrous roots, often shooting from the nodes; leaves 3-partite, softly hairy, segments cuneate; flowers solitary or terminal or leaf opposite; sepals hirsute, smaller than the petals; petals white or yellow; receptacles small; achenes compressed, cuneately sub-orbicular.

Distribution   Khasi hills, Meghalaya

Propagation   By seeds

Parts used   Whole plant

Uses   The plant is useful to relieve giddiness.

Rhynchoglossum obliquum Bl.

Family   Gesneriaceae

Indian name   *Kalu-tali* (Malayalam).

Description   Tall, succulent herb; leaves ovate, acuminate, membranous, almost glabrous; lateral nerves conspicuous, numerous, oblique; flowers 5-merous in lax terminal racemes up to 10 in., long; bracteoles filiform; calyx campanulate; corolla tube cylindric, limb bilabiate, upper lip short, 2-lobed; stamens 2, perfect, anthers connivent; ovary ovoid, stigma dilated; capsule included in the enlarged calyx, ellipsoid, membranous, 2-valved; seeds smooth.

Flowering and fruiting   September–October

Distribution   All over North-East Region, Khasi hills, Meghalaya

Propagation   By seeds

Parts used   Leaves

Uses   It is useful for skin diseases.

S

Sabia lanceolata Colebr.

Family   Sabiaceae

Indian names   *Miri, Mandri, Madri* (Garo); *Samtameh* (Khasi); *Khai* (Manipuri).

Description   Bark blackish or dark brown nearly smooth; leaves ovate lanceolate, elliptic acute or acuminate, base rounded or obtuse, glabrous, lateral nerves looping; panicles terminal and axillary; flowers greenish yellow or greenish white; calyx minute, lobes ovate; petals ovate lanceolate; stamens included; fruit of 2 drupelets, 1–1.8 cm long, obovoid, blue when ripe.

Flowering and fruiting   October–March

Distribution   Bangladesh, North-Eastern India; up to a height of 2000 m

Propagation   By seeds

Parts used   Leaves

Uses   Used for fomenting swollen ankles and wrists.

Saccharum officinarum L.

Family   Poaceae

Indian names   Sugar cane (English); *Pundrakah* (Sanskrit); *Paunda, Pundiya* (Hindi); *Ikshu, Ikshudanda, Kabbu* (Kannada); *Karumbu, Pundaram* (Tamil); *Karimpu* (Malayalam); *Ceruku* (Telugu).

Description   Perennial tall grass, with stout and solid, green to purplish stems, 3–5 cm in diameter, and up to 3 m height; leaves sheathing and overlapping, lance-shaped, up to 2 m long and 6 cm broad; flowers in cluster.

Flowering and fruiting   Throughout the year

Distribution   In North-East Region, frequently in Meghalaya

Propagation   By vegetative method

Parts used   Whole plant

Chemical constituents   Abscisic acid, apigenin and its glycoside, arunodin, galactose, gibberellins, luteolin, phytosterol, taraxerol methyl ether, vicenin and xylose.

Uses   The leaf ash is useful in sore eyes and stem juice is used to treat sore throats.

Saccharum spontaneum L.

Family   Poaceae

Indian names   Thatch grass, Kansgrass (English); *Ikshugandha, Kasah* (Sanskrit); *Kas, Kus* (Hindi); *Dharbe kabbu, Gorasu hullu* (Kannada); *Pekkarimpu* (Tamil); *Kusa, Njangana* (Malayalam); *Kaki ceruku, Kaki gaddi* (Telugu).

Description   Perennial herb; stem cylindrical, generally solid at the nodes and hollow in the internodes; flowers usually bisexual; spikelets arranged in spikes, each spikelets is enclosed at the base by two glumes, known as empty glumes, one placed a little above the other and each sessile flowers arises in the axil of a flowering glume which, encloses it at the base and a 2-nerves glume called palea placed opposite the flowering glume. Perianth represented by 2 or 3 minute scales called lodicules, placed within the palea; stamens in one whorl of 3, anthers versatile; carpel-solitary, opposite the palea; ovary superior, 1-celled with a single, erect anatropous ovule, stigma usually 2, lateral and feathery. Fruit caryopsis, rarely an utricle, embryo placed at the base of the seed and outside the endosperm.

Distribution   Entire North-East India, common in East Garo Hills, Meghalaya.

Propagation   By seeds and roots

Parts used   Whole plant and roots

Uses   The root is sweet, astringent, emollient, refrigerant, diuretic, lithotriptic, laxative, aphrodisiac and tonic. Roots are useful in burning sensation, renal and vesical calculi, dyspepsia, haemarrhoids, dysentery, and general debility.

Salix tetrasperma Roxb.

Family   Salicaceae

Indian names   *Burum, Jalavetas* (Sanskrit); *Baings, Baishi* (Hindi); Bayike, *Bayise, Niruvanji* (Kannada); *Attuppalai, Niruvanji* (Tamil); *Arali, Atrupala, Nirunji* (Malayalam); *Etipala, Etipisinika* (Telugu).

Description Middle sized trees, 10–20 m high; crown oval; bark dark grey or blackish-brown; leaves lanceolate, elliptic, acuminate, crenate, silky when young; spikes 6–12 cm long, grey or greyish yellow; capsules stalked.

Flowering and fruiting September–February

Distribution Nearly throughout India, very common in Meghalaya, particularly in Khasi hills.

Propagation By cuttings

Parts used Bark

Chemical constituents Phenolic glycosides-salicylic acid, flavonoids, tannins.

Uses Bark febrifuge. Dried and powdered leaves with sugar are used in rheumatism, epilepsy, piles, swelling, stones in gall bladder.

Sapindus rarak DC. [Rare]

Family Sapindaceae

Description Trees up to 20 m high; bark grey, smooth; leaves up to 50 cm long; leaflets oblong, lanceolate, acute or acuminate, base oblique, pubescent when young; flowers pale white or yellowish; fruits fleshy, cocci 3, usually 1 or 2 aborted, 2–2.5 cm cross.

Flowering and fruiting June–December

Distribution Throughout India, in forests at low elevations and also cultivated. Confined to North-East; rather rare in Meghalaya.

Propagation By seeds

Parts used Fruits

Uses Paste of fruit is useful to reduce acne.

Saprosma ternatum Hook. f.

Family   Rubiaceae

Indian name   *Bhedeli* (Assamese).

Description   An evergreen shrub or small tree, all parts glabrous, branchlets angled; bark greyish brown having minute vertical fissures, brown below the cuticle; leaves usually ternately whorled, coriaceous, glabrous on both surfaces, sparingly scaberulous beneath on the nerves and occasionally on the midrib; lateral nerves 7–8 on either half, tertiaries subparallel, transverse; base acute; petiole, stipules very long, lanceolate with needle-like, points, deciduous; flowers white, in trichotomous or corymbosely fasciculate cymes; calyx truncate or 4–6 toothed; corolla funnel-shaped, lobes 4, valvate in bud; stamens as many as corolla lobes, on the throat of the corolla, anthers sub-sessile; ovary 2-celled, style filiform, stigmatic arms 2, linear; fruit black when ripe, succulent, crowned by the conical disc, 1-seeded, vary rarely two.

Flowering and fruiting   May–July

Distribution   Assam, Khasi and Jaintia Hills, Meghalaya.

Propagation   By seeds

Parts used   Bark and leaves

Uses   Leaf relieves flatulence and stomach ache, bark juice is used in indigestion.

Sarcandra glabra (Thunb.) Nakai.

Family   Chloranthaceae

Description   Perennial herb or shrubs; up to 2.5 m high; leaves shortly petioled, elliptic, lanceolate, acuminate, base cuneate, glabrous, coarsely serrate, shining green above; spikes erect, anthers solitary 4-celled; berries red.

Flowering and fruiting August–January

Distribution Khasi and Jaintia hills

Propagation By seeds

Parts used Whole plant

Uses Decoction is useful for dysentery and used as bath for paralytic patients.

Saraca asoca (Roxb.) de Wild. [Endangered]

Family Caesalpiniaceae

Indian names Ashoka (English); *Asokah, Gatasokah* (Sanskrit); *Ashok, Ashoka* (Hindi); *Aksunkara, Asokada* (Kannada); *Asogam, Asoka pattai* (Tamil); *Ashokam* (Malayalam); *Asokamu, Vanjalamu* (Telugu).

Description Middle-sized evergreen trees up to 15 m high, with numerous spreading and drooping glabrous branches; crown oval, umbrella shaped; dark brown or nearly black, rough lenticellate; leaflets 8–25 ×1.5–6 cm, lanceolate or oblong, long accuminate, glabrous; corymbs 4–15 cm across; flowers bright yellow, orange or red, 2.5–4 cm long, fragrant; stamens much exerted, anthers purple; pods 10–25 × 4–5 cm, oblong, compressed, tapering at both ends, 4–8-seeded.

Flowering and fruiting   February–October

Distribution   Indo-Malaya; nearly throughout India, Central and Western Himalayas, in evergreen forests up to an elevation of about 750 m; rather rare in Maghalaya; occurs in tropical, evergreen forests, usually along river banks.

Propagation   By seeds

Parts used   Bark, leaves, flowers and seeds

Chemical constituents   The bark contains tannin, catechol, sterol and organic calcium compounds. Its methanol fraction contains haematoxylene, tannin, and water-soluble glycoside.

Uses   The bark is bitter, astringent, sweet, refrigerant, styptic, stomachic, and febrifuge. It is useful in dyspepsia, fever, dipsis, burning sensation, ulcers, pimples, etc. The leaves are depurative and their juice mixed with cumin seeds is used for treating stomachalgia. The flowers are considered to be a uterine tonic and are used in vitiated condition of pitta, syphilis, cervical adenitis, hyperdipsia, dysentery, and inflammation. The dried flowers are used in diabetes and haemorrhagic dysentery and seeds are used for treating bone fractures, strangury and vesical calculi. Seeds are reported to be useful in urinary discharges.

Schefflera venulosa (Wright and Arn.) Harms.

Family   Araliaceae

Indian names   *Suinl* (Hindi); *Bilibhuthala* (Kannada); *Modakama* (Tamil); *Unjala* (Malayalam); *Cippari, Gaalana* (Telugu); *Dhobai-lata, Dhovalata* (Assamese).

Description Large scandent or climbing shrub; bark grey, greyish brown; leaves oblanceolate, elliptic, obtuse or acuminate, base obtuse or rounded, cuneate, obscurely 3-nerved, coriaceous, glabrous; inflorescence deciduously tomentose; flowers yellowish green to white; calyx truncate, obscurely lobed; petals free or connate; fruits fleshy, 4–5 mm across, yellow.

Flowering and fruiting March–November

Distribution Throughout Meghalaya, common in Garo Hills, Tura

Propagation By seeds

Parts used Bark

Chemical constituents Root and stem barks contain triterpenoid, saponins that yield oleanolic acid on hydrolysis.

Uses Root and stem bark is useful in rheumatism, lumbago, ostedynia, amnesia, dyspepsia, infantile rickets, oedema, and impetigo.

Schima khasiana [Endemic/Rare]

Family Theaceae

Description Middle-sized tree up to 20 m high; bark reddish brown to dark brown, nearly smooth, branch tips silky tomentose; branchlets lenticellate, glabrate; leaves broadly lanceolate, elliptic, oblong, lanceolate, acuminate, base cuneate, acute, glabrous, coriaceous, sharply serrate; flowers axillary, solitary, white, 5–6 cm across; sepals rounded, silky, tomentose outside; stamens numerous, yellow, filaments yellow; ovary hairy at base; capsules depressed globose.

Flowering and fruiting July–March

Distribution Endemic to Meghalaya, confined to sacred forests at Khasi and Jaintia Hills.

Propagation By seeds

Parts used Bark

Uses Bark irritates skin. It is an anthelmintic and rubefacient.

Schima wallichii L.

Family   Theaceae

Indian names   *Chilauni, Kanak, Makusal* (Hindi); *Noga bhe, Makorisal* (Assamese); *Makrisal* (Bengali); *Dieng nganbuit* (Khasi); *Usoi* (Manipuri); *Khiang* (Mizoram).

Description   Large trees, up to 50 m high; bark reddish brown; young tips silky tomentose; leaves elliptic, lanceolate, oblong, oblanceolate, acute, base cuneate, margin entire, bulbous, hairy beneath; flowers axillary, solitary, white; sepals rounded, ciliate; petals obovate; stamens yellow, nearly free; ovary hairy at base, style cylindric; fruits grey pilose, up to 2 cm across.

Flowering and fruiting   February–January

Distribution   Throughout Garo hills, Meghalaya

Propagation   By seeds

Parts used   Bark and leaf medicinal; leaf paste on cuts and wounds; leaf decoction is used to cure flatulence; powdered bark is given to cattle to kill intestinal worms.

Scoparia dulcis Linn.

Family   Scrophulariaceae

Indian names   *Mrugandhi gida* (Kannada); *Perhpawngchaw* (Mizoram); *Bonogajari* (Oriya); *Seni bon* (Assamese); *Mithapata, Ban dhane* (Bengali); *Sarakkotthini* (Tamil); *Potti boli* (Telugu).

Description   Erect branching annual herb, 0.8 m in height, semi-woody; leaves opposite or ternately whorled, rhomboid or elliptic, serrate, punctate, sessile; flowers small, axillary; calyx segments imbricate in bud, 3-nerved, shorter than the capsule; corolla white, throat densely hairy; filaments woolly at base; capsule small, globose, septicidal; seeds many, obovoid, angled, scrobiculate.

Flowering and fruiting   June–January

Distribution   Khasi and Jaintia hills

Propagation   By seeds and vegetative method

Parts used   Whole plant

Chemical constituents   The roots contain alkaloids and a bitter substance, amellin.

Uses   The plant is used in eye sores and reduces headache and giddiness. The fresh plant is especially active against dry cough.

Scutellaria discolor Wall. ex. Benth.

Family   Lamiaceae

Indian name   *Yenakha* (Manipuri).

Description   A herb with creeping root stock, pubescent; leaves chiefly radical, resolute, orbicular or oblong or elliptic, obtuse, crenate, base rounded or cordate, petiole up to 5 cm long; flowers trumpet shaped, generally forming one cluster, bluish purple on erect racemes; corolla tip whitish above; upper lip hooded; nutlets 4, disclosed after the fall of upper lip, granulate, turbinate.

Flowering and fruiting   September–February

Distribution   Khasi Hills

Propagation   By seeds

Parts used   Whole plant

Uses   Leaf juice is useful in vomiting and nausea, indigestion and constipation. Leaf paste is useful for skin infection. Decoction of plant is useful in fever, bronchitis and stomach trouble.

Securinega virosa (Roxb.) Baillon.

Family   Euphorbiaceae

Indian names   *Dalme, Dhani, Bakarcha* (Hindi); *Gudaphala, Kare hoola* (Kannada); *Veppulathi, Irubulai* (Tamil); *Mekarayi, Balli-chettu* (Telugu); *Dumikron* (Garo); *Dieng krong watlam* (Khasi).

Description   Shrubs, bark brown or reddish brown, papery; leaves ovate or elliptic, obtuse, glabrous, base cuneate or rounded, glaucous beneath; flowers white, minute, dioecious, in axillary clusters from a crown of minute bracts; male flowers many on filiform pedicel, female flowers few; fruits 0.3 cm across.

Flowering and fruiting   April–November

Distribution   Throughout India, very common in Garo hills, Meghalaya.

Propagation   By seeds and vegetative method.

Parts used   Fruits

Uses   Fruit is useful for treatment of stomach ache.

Sedum multicaule Wall.

Family   Crassulaceae

Indian name   *Miragha* (Hindi).

Description   Herb succulent, glabrous, stem much branched from the base; leaves oblong, acute; flowers subsessile; petals yellow; cyme branched, fruit usually elongate, subcorpioid; seeds obovoid covered with minute tubercles.

Distribution   North-East India

Propagation   By seeds

Parts used   Bark and leaves.

Siegesbeckia orientalis L.

Family   Asteraceae

Indian names   *Gobariya, Liskura, Lichakura* (Hindi); *Karuntumpai* (Tamil).

Description   A large erect annual herb, almost shrubby up to 1.5 m high, branched, glandular, pubescent; leaves upper generally smaller, opposite, ovate, triangular, coarsely crenate, deeply and irregularly toothed, acute or acuminate; heads small, yellow peduncled in leafy, panicles, involucral bracts dissimilar, 3-seriate; receptacle concave, with many chaffy scales enclosing bisexual flowers.

Flowering and fruiting   October–December

Distribution   North-East Frontier tract and Khasi hills

Propagation   By seeds

Parts used   Whole plant

Uses   Plant paste is useful for the treatment of cuts and wounds.

Semecarpus anacardium L. f.

Family   Anacardiaceae

Indian names   Marking nut (English); *Agnika, Agnimukha* (Sanskrit); *Belatak, Bhela* (Hindi); *Aginimukhi, Bhallika* (Kannada); *Erimugi, Kalagam* (Tamil); *Alakkuceru, Cheru* (Malayalam); *Bhallatamu, Jidi* (Telugu).

Description   Middle-sized deciduous tree up to 15 m high; crown oval; bark blackish or dark brown; leaves broadly obovate, elliptic oblong, oblanceolate, obtuse, cuneate at base; flowers yellow or greenish yellow; drupes ovoid, oblong, black when ripe, seated on a fleshy pseudocarp.

Flowering and fruiting   May–March

Distribution   Throughout India, in semi-evergreen and moist deciduous forests; Khasi hills, Meghalaya

Propagation   By seeds

Parts used   Fruits

Chemical constituents   Anacardic acid, cardol, catechol, anacardol and fixed oil, semicarpol.

Uses   The fruits are acrid, bitter, astringent, thermogenic, digestive, carminative, purgative, liver tonic, expectorant, urinary astringent, etc. They are useful for beriberi, cancer, sciatica, neuritis, cough, asthma, dyspepsia, helminthiasis, leucoderma, scaly skin and cardiac diseases.

Shorea robusta Gaertn. F. [Near threatened]

Family   Dipterocarpaceae

Indian names   Sal (English); *Agnivallabha, Ashvakarna* (Sanskrit); *Sal, Damar, Dhonah* (Hindi); *Ashvakarna, Asina* (Kannada); *Attam, Venkungiliyam* (Tamil); *Karimaruthu, Kungiliyam* (Malayalam); *Gugal, Guggilamu* (Telugu).

Description   Tall deciduous trees; crown oval, bark brown with prominent longitudinal fissures; leaves ovate oblong, acuminate or acute, base cordate or truncate, pinkish when young, orange yellow at senescence, glabrous above, puberulous along nerves beneath, lateral nerves 11–16; flowers 1–2 cm long; sepals 5, 2–3 mm, ovate, densely tomentoes outside; petals 0–8 to 1–3 cm long, creamy white, ovate, lanceolate, silky without; stamens broaded; ovary globose; fruit 1–5 to 2 cm in diameter, beaked, enclosed by the persistent wing-like calyx, with 3 long (6–8 × 0.5–1.5 cm) and 2 small wings.

Flowering and fruiting   March–July

Distribution   India and Nepal, in Meghalaya and Assam. This species occupies the major part of the deciduous tract-extending whole of Garo Hills to the Northern slopes of Khasi hills.

Propagation   By seeds

Parts used   Bark, leaves, fruits and resins

Chemical constituents   Presence of 2-(2-iminoacetic acid)-benzo-furanone, glucoside of 4 hydroxychalcone, hopeaphenol, triterpenoids and a terpene alcohol, furfural, dimethyl ether of homocatechol, alkylbenzene derivatives, lignan, tannin, amino acids and fatty acids.

Uses   The bark and leaves are astringent, acrid, cooling, anthelmintic, anodyne, constipating, urinary astringent, depurative and tonic. They are useful in diarrhoea, dysentery, gonorrhoea, leprosy, cough, haemarrhoids and anaemia. The fruits are sweet, astringent, cooling, aphrodisiac, cholegogue and tonic, and are useful in dipsia, burning sensation, tubercular ulcers, seminal weakness and dermatopathy. The resin is astringent, sweet, acrid, cooling, anodyne, vulnerary, antibacterial, deodorant, constipating, detergent and tonic. It is useful in hyperhydroses, wounds, ulcers, burns, fever, burning of eyes and ophthalmodynia.

Sida acuta Burm. f.

Family   Malvaceae

Indian names   Blue okra (English); *Brihannagabala, Pata* (Sanskrit); *Bariara, Kareta, Kharenti* (Hindi); *Bhimankaddi, Vishakaddi* (Kannada); *Arivalmanaippundu, Arivalmukkan* (Tamil); *Cheruparuva, Kurunthotti* (Malayalam); *Chittimu, Gayapaku* (Telugu).

Description   Undershrubs; young parts puberulous; leaves linear, lanceolate, elliptic, acute, base truncate, serrate; flowers solitary; ovary of 5–9 carpels, awned; seeds rugose.

Flowering and fruiting   July–November

Distribution   Throughout the hotter parts of India, in Garo and Khasi hills, Meghalaya.

Propagation   By seeds

Parts used   Root and leaves

Uses   The bark of roots is used for leucorrhoea, gonorrhoea, hyperdiuresis, nervous and urinary diseases, fever and stomach complaints. The leaves are good for diarrhoea. Flowers and ripe fruits are useful in relieving burning sensation and pectoral lesions.

Sida cordifolia Linn.

Family   Malvaceae

Indian names   *Badiyalaka, Baladhya* (Sanskrit); *Kharinta, Barial* (Hindi); *Hettutti, Kisangi* (Kannada); *Nilatutti, Cirra muttiver* (Tamil); *Kuruntotti, Velluram, Katturan* (Malayalam); *Antisa, Chiribenda* (Telugu).

Description   A small much branched shrub; minute star-shaped hairs present all over the plant; leaves 2–5 cm, ovate, thick margins toothed, petioles shorter than leaves. Flowers yellow, small, one or a few together; fruits 6–8 mm diameter, divided into 7–10 parts, each strongly reticulated and with two awns on tip.

Flowering and fruiting   June–October

Distribution   Throughout India as a common weed, usually in waste places and open scrub forests; very common in Garo hills, Meghalaya.

Propagation   By seeds

Parts used   Whole plant

Uses   It is used as a general tonic and for improving vigour. Seeds of the plant are particularly credited with this property. Decoction of root with ginger is considered useful in certain fevers. The powder of root bark is given in certain diseases of women, such as leucorrhoea, and in nervous diseases. Root juice is used for promoting healing of wounds. The bark of the root, with sesamum oil and milk, is efficacious in curing certain types of facial paralysis. Seeds of the plant are considered useful in gonorrhoea and colic pains.

Sida rhombifolia Linn.

Family   Malvaceae

Indian names   Queensland hemp, Paddys lucerne (English); *Ahikhanda, Atibala* (Sanskrit); *Bariara, Kharenti* (Hindi); *Bennegaragu* (Kannada); *Anaikurundotti* (Tamil); *Anakkurunthotti* (Malayalam); *Athobalacettu* (Telugu).

Description   Herbs or undershrubs; leaves ovate or obovoid, rhomboid, acute, base narrow, truncate, puberulous above, densely tomentose beneath, usually 3-nerved from base; flowers solitary or 2-few together; carpels puberulous, awned.

Flowering and fruiting   July–February

Distribution   Throughout India, in Meghalaya throughout the state.

Propagation   By seeds

Parts used   Root and leaves

Chemical constituents   Ephedrine, quinazolines, phenethylamines, tryptamine derivatives, malvalic and linoleic acids.

Uses   Roots and leaves are good for rheumatism, flatulence, haemothermia, seminal weakness, arthritis and diarrhoea.

Skimmia laureola (DC.) Sieb. and ex. Zuoc. Ex. Walp.

Family   Rutaceae

Indian name   *Kedarpaiti, Shashra* (Hindi).

Description   An aromatic evergreen shrub, up to 3 m high, branched from the base, all parts glabrous, bark greyish-white smooth, branches with scattered warty lenticels; leaves crowded at the ends of branches, alternate, simple, very variable in size and shape, oblanceolate or oblong, acuminate, softly coriaceous, glabrous, dark green above, pale beneath lateral nerves margined; flowers white or greenish yellow, polygamous, bracts ovate, oblong, acute, ciliolate, petals 5, subimbricate about 0.45 cm long, oblong, much longer than the calyx; stamens 5, hypogynous about as long as the petals, filaments subulate, anthers dorsifixed; ovary ovoid, 2–5 celled, ovule 1 in each cell, style stout, stigma capitate 2–5 lobed; in male flowers, ovary usually of 2–3 rudimentary carpels.

Flowering and fruiting   April–November

Distribution   Khasi and Jaintia hills

Propagation   By seeds

Parts used   Leaves

Uses   Leaf is used in skin diseases.

Smilax ovalifolia Roxb.

Family   Liliacee

Indian names   *Copacini, Maitri* (Sanskrit); *Guti, Jangliushbah* (Hindi); *Malai-tamarai* (Tamil); *Kal-tamara, Karivilanti* (Malayalam); *Konda-tamara, Kummarabaddu* (Telugu); *Gutvel* (Marathi); *Kaihapui* (Mizoram).

Description   A large prickly climber with sparsely prickled or unarmed stems and thick tuberous rhizomes; leaves simple, 3–7 costate, long petioled, petiole narrowly sheathed, not auriculate.

Flowering and fruiting   Throughout the year

Distribution   Entire North-East Region

Propagation   By seeds and rhizomes

Parts used   Leaves

Chemical constituents   Tannin, cinchonin, a steroidal saponin, flavonoid glycosides, diosgenin and rutin.

Uses   Leaf paste is useful in ulcers, allergic skin diseases, cholera, fever, rheumatic diseases, and menstrual complications.

Solanum erianthum D. Don.

Family   Solanaceae

Indian names   *Priyamkari, Vidari* (Sanskrit); *Akra, Ban tobaccoo* (Hindi); *Aarimani, Kallaarthi* (Kannada); *Karimulli, Sundai* (Tamil); *Budama, Rasagadi* (Telugu); *Khimkha Nagong* (Garo); *Dieng sohmon niangkodong* (Khasi).

Description   Small trees up to 8 m high; leaves ovate, lanceolate, acute rounded or truncate, stellately woolly on both surfaces, densely hairy beneath; cymes terminal, panicled, 5–15 cm across; flowers white, 1–1.3 cm across; berries large yellow when ripe.

Flowering and fruiting   Nearly throughout the year

Distribution   Tropics and subtropics and large parts of India; frequent in Meghalaya.

Propagation   By seeds

Parts used   Leaves

Chemical constituents   Leaves contain an essential oil, saponins and alkaloids, solanine, solasodine.

Uses   Fresh leaves are useful for the relief of haemarrhoids and scrofula. A plaster made of concentrated fresh leaf juice cures dermatitis and impetigo.

Solanum indicum L.

Family   Solanaceae

Indian names   *Akranta, Alpaphala* (Sanskrit); *Barhanta, Bhutkataiya* (Hindi); *Badane, Gulla* (Kannada); *Kandal, Karimulli* (Tamil); *Cheruchunda, Chunda* (Malayalam); *Adaviyuchinta, Challamulaga* (Telugu); *Tid-bhagnri* (Assamese).

Description   A shrub up to 3.5 m high; branches herbaceous, bark smooth, pale brown or greenish grey, warty, armed; often with curved prickle; young part of inflorescence thickly stellate, leaves oblong, serrate or obtusely lobed, acute, herbaceous; flowers bluish purple, in extra-axillary racemose cymes; calyx stellate, pubescent outside, usually prickly; corolla tomentose outside, lobes reflexed; fruit globose, smooth yellow.

Flowering and fruiting   June–October

Distribution   Khasi and Jaintia hills

Propagation   By seeds

Parts used   Leaves

Chemical constituents   The plant contains enzymes alkaloids, solanine, solanidine in roots. Leaves and fruits contain 1.8% of alkaloids.

Uses   Fresh leaves are indicated for the relief of haemarrhoids and scrofula. It is better to apply during the night.

Solanum khasianum C. B. Clarke.

Family   Solanaceae

Indian name   *Rulpuk, Athlo* (Mizoram).

Description   The plant is a stout, much branched, under shrub, upto 1.5 m tall, with almost straight prickles; leaves are ovate, lobed and the lobes are lanceolate or triangular, hirsute and prickly on both the surfaces; flowers white, 1–4 flowered racemes; berries are yellowish or greenish globose and 2.5 cm in diameter. Seeds are smooth, brown and compressed.

Flowerinf and fruiting   July–November

Distribution   Entire North-East Region up to 2000 m

Propagation   By seeds

Parts used   Fruits and seeds

Chemical constituents   Berries contain solasodine, solakhasianin and diosgenin.

Uses   The seed is useful for asthma, cold, cough, insect bite, toothache, bronchitis and decay. Fruit is also useful for abortion. Seeds are useful to relieve headache, cold and blocked nose.

Solanum melongena L.

Family   Solanaceae

Indian names   Egg plant, Egg fruit, Brinjal (English); *Angana, Bartaku* (Sanskrit); *Aubergine, Badanjan* (Hindi); *Badane, Badanekaya* (Kannada); *Kattirikkai* (Tamil); *Nilavazhutina, Kattirika, Vazhutina* (Malayalam); *Chiruvanga, Eruvanga* (Telugu).

Description   The brinjal plants are extensively cultivated; fruit grows to a very large size and prickles are reduced to many distinct forms. Fruits are eaten cooked as a vegetable; the seeds are used as a stimulant and the leaves as a narcotic.

Flowering and fruiting   Throughout the year

Distribution   Common in Meghalaya

Propagation   By seeds

Parts used   Roots, leaves and unripe fruits

Chemical constituents   Solasodine from fruits. Arginine and aspartic acid from leaves.

Uses   Roots are useful in inflammations, neuralgia, cardiac debility and ulcers in the nose, leaves are narcotic, and therapeutic and are useful in cholera, bronchitis, asthma, and fever.

Solanum nigrum L.

Family   Solanaceae

Indian names   Black nightshade (English); *Bahuphala, Bahutikta* (Sanskrit); *Gurkkamai, Kabaiya* (Hindi); *Ganike, Kakarundi* (Kannada); *Manattakkali, Milagutakkali* (Tamil); *Karimthakkali, Manathakali* (Malayalam); *Kacci, Kaccipandu* (Telugu).

Description   Annual herbaceous weed; leaves ovate, lanceolate or oblong, entire, sinuate, toothed or lobed, acute or acuminate, thin, glabrous, base narrowed into the petiole; flowers white, small in sub-umbellate cymes extra axillary; calyx sparsely puberulous, 5-toothed, segments rounded; corolla glabrous outside, rotate, 5-lobed. Ovary glabrous; style bearded at base, berry black shining, but sometimes yellow or red when ripe, globose; seeds discoid, minutely pitted.

Flowering and fruiting   June–July

Distribution   Khasi hills up to 1500 m

Propagation   By seeds

Parts used   Whole plant

Chemical constituents   Leaf is a rich source of riboflavin, nicotinic acid and vitamin C, besides these β-carotene and citric acid are present.

Uses   It is useful for rheumatism, swellings, cough, asthma, bronchitis, wounds, ulcers, dyspepsia, otalgia, nasal catarrh, vomiting, leprosy, skin diseases, fever and splenomegaly. The infusion of plant is useful in dysentery, other stomach complaints and fevers; it promotes urination. The juice of the plant is useful on ulcers and other skin diseases. Its fruits are more important. They are tonic, laxative, improve appetite, and are useful in asthma, skin diseases and applied locally on ringworm.

Solanum surattense Burm. f.

Family   Solanaceae

Indian names   Yellow berried nightshade (English); *Kantakari, Nidigdhika* (Sanskrit); *Kateli, Remgani* (Hindi); *Nelagulle, Chikka sonde* (Kannada); *Kantankattiri* (Tamil); *Kantakariccunta, Kantakarivalutana* (Malayalam); *Callamullaga, Nelamulaka* (Telugu).

Description   Perennial undershrubs, woody at the base, with zigzag branches that spread close to the ground covered over with strong broad, sharp, compressed, straight, yellowish white prickles; leaves ovate oblong; flowers blue or bluish purple, on extra axillary cymes; fruits glabrous, globular drooping berry, yellow or white with green veins, surrounded by the calyx; seeds many, small, reniform, smooth and yellowish brown.

Flowering and fruiting   May–September

Distribution   Common in Khasi hills

Propagation   By seeds

Parts used   Whole plant

Chemical constituents   Fruits yield campesterol and gluco-alkaloid, solasodine, solasonine and solanocarpine.

Uses   It is useful in helminthiasis, dental caries, inflammation, arthralgia, constipation, anorexia, leprosy, skin diseases, asthma and cough. The fruit is useful for bronchitis, sore throat, muscular pain, fever, etc.

Solanum torvum Sw.

Family   Solanaceae

Indian names   *Brhati* (Sanskrit); *Tit-baigun, Bhurat* (Hindi); *Borasunde, Kaada kallatti* (Kannada); *Sundakkai, Kottukkattari, Malasundai* (Tamil); *Kattucunta, Puttaricunta* (Malayalam); *Konda vusti* (Telugu).

Description   A shrub up to 4 m high, pubescent, leaves ovate, serrate or lobed, shrortly acuminate, membranous, sparsely stellate, pubescent, above, more closely so beneath; base unequal or rounded; flowers white usually extra-axillary, often branched, unarmed cymes; pedicels slender; fruit pubescent outside, lobes spreading, linear, oblong or lanceolate, berry globose, seated on a persistent calyx and the thickened pedicel.

Flowering and fruiting   September–October

Distribution   Assam, Khasi and Jaintia hills of Meghalaya

Propagation   By seeds

Parts used   Leaves and fruits

Uses   Leaf is useful for wounds and fruit is used as a cure for cough.

Solanum tuberosum L.

Family   Solanaceae

Indian names   Potato (English); *Golakandah* (Sanskrit); *Alu, Aalu* (Hindi); *Alugadde, Batata* (Kannada); *Urulaikkilangu* (Tamil); *Urulakkizhangu* (Malayalam); *Bangaladumpa, Urlagadda* (Telugu).

Description   Annual herbs, stem obscurely angular; leaves pubescent, branches arising from the axils of the lower leaves, a large number of axillary shoots are made to become tuber bearing; flowers about 2.5 cm across, bluish or white in subterminal cymes.

Flowering and fruiting   Cold season

Distribution   Largely cultivated in North-East Region

Propagation   By tuber

Parts used   Leaves and tuber

Chemical constituents   It contains protein, amino acids, amides and nitrogenous bases. The principal protein is globulin.

Uses   The potato tubers are sweet, aphrodiesia, antiscorbutic, aperient, diuretic. It is useful for strangury, constipation, hyperacidity, scabies. The leaf is useful for cough.

Sonerila maculata Roxb.

Family   Melastomataceae

Description   Small herb, stem with spreading hairs; leaves opposite, broadly ovate to lanceolate, serrulate and ciliate on the margins, with spreading hairs on both surfaces or nearly glabrous, membranous, penninerved, base often unequal; petiole 2.5–5 cm long, hairy; flowers 3-merous, about 1.5 cm cross; calyx thinly hairy; petals elliptic, acute; stamens 3, equal, connective not appendaged, ovary 3-celled, inferior; capsule slightly angled, glabrous, about 0.80 cm long, seeds numerous, with a slightly excurrent lateral ridge.

Flowering and fruiting   August–October

Distribution   Khasi Hills up to 5000 m

Propagation   By seeds

Parts used   Root

Uses   Leaf is useful for stomach ailments.

Spondias pinnata (L.f.) kurz.

Family   Anacardiceae

Indian names   Wild mango, Hogplum (English); *Ambaka, Ambarataka* (Sanskrit); *Amra, Jungli-am* (Hindi); *Amatekayi vrksamla, Ambatemarra* (Kannada); *Kattimagirankai, Mambulichi* (Tamil); *Mampuli, Ambazham* (Malayalam); *Adavimamadi, Adavimamena* (Telugu).

Description   Middle sized trees up to 20 m high; crown lax, branches spreading; bark with vertical streaks and horizontal wrinkles; leaves up to 45 cm long, leaflets elliptic, oblong, lanceolate, abruptly caudate acuminate, or acute, obtuse, often oblique at base, glabrous; panicles up to 35 cm long, spreading, conical; flowers creamy white, or yellow; stamens shorter than the spreading petals; drupes oblong, ovoid, greenish yellow when ripe, acidic.

Flowering and fruiting   March–November

Distribution   Common in tropical deciduous forest belt of Meghalaya, often cultivated.

Propagation   By seeds

Parts used   Bark

Uses   Bark refrigerant and used in dysentery and rheumatism, fruit in dyspepsia.

Stachytarpheta jamaicensis (L.) Vabl.

Family   Verbenaceae

Indian names   *Cimainayuruvi* (Tamil); *Maeda balaku* (Telugu); *Jarbo, Jarbas* (Bengali); *Kaadu uttharaani, Kariyuttharaani* (Kannada); *Koraputia buta* (Oriya).

Description   A perennial herb about 1,5 m high; branches mostly dichotomous. Leaves elliptic, ovate, acute or obtuse, serrate or cuneate, almost glabrous, base cuneate; petiole 3 to 15 cm long; flowers bluish, sessile, adpressed on rachis of terminal spikes and nestled in the depression enclosed by the bract; stamens 2, perfect, staminodes 2; ovary 20 celled, style elongate, ovule solitary in each cell. Fruit dry, enclosed in the calyx tube separating into 2 hard, 1-seeded pyrenes.

Flowering and fruiting   July–November

Distribution   Entire North-East India; Common in Meghalaya, particularly in Khasi hills.

Propagation   By seeds

Parts used   Leaves

Uses   Leaf is useful for curing cholera, dysentery, ulcers, intestinal worms, stomach and venereal diseases.

Stephania japonica (Thunb.) Miers.

Family   Menispermaceae

Indian names   *Patakkilannu* (Malayalam); *Molakaranaikkoti* (Tamil); *Duvvathige* (Telugu); *Patha, Rajapatha* (Sanskrit); *Tubukilota* (Assamese); *Akanadi, Chhotopard* (Bengali); *Khaarkha* (Garo); *Thangga-uri-angouba* (Manipuri).

Description   Climber; stem striate; leaves ovate, deltiod, acuminate, base subcordate or truncate, glabrous or nearly so beneath; flowers minute, yellow, drupes globose, red, tubercled.

Flowering and fruiting   March–September

Distribution   Tropical, confined to North-East Region, common in Meghalaya.

Propagation   By seeds and vegetative method

Parts used   Root and leaves

Uses   Paste is used over boil, treatment of fever, diarrhoea, dysentery and urinary diseases.

Sterculia coccinea Roxb.

Family   Sterculiaceae

Description   A medium sized to large tree about 18 m high with reddish brown heartwood, whitish bark and whorled horizontal branches; leaves digitate, crowded at the ends of branches, leaflets 7–9, oblong, lanceolate, acute or acuminate, pubescent when young, glabrous when mature, stipules cadueous; flowers dull orange, fetid smelling in erect racemose panicles, unisexual; calyx tube very short, lobes very narrow, white hairy, free or cohering at the tips; follicle 2–5 scarlet, 3–5×2–3 cm, thinly coriaceous, velvety outside, glabrous and bright red inside; seeds 4–8, ovoid, smooth.

Flowering and fruiting   April–September

Distribution   Entirely in North-East India, common in Garo and Khasi hills, Meghalaya.

Propagation   By seeds

Parts used   Roots, leaves, bark and flowers

Chemical constituents   Thiamine, riboflavin, nicotinic acid, and vitamin C.

Uses   Bark is a veterinary medicine.

Sterculia villosa Roxb. Cor.

Family   Sterculiaceae

Indian names   *Udal, Odal* (Assamese); *Umale* (Garo); *Katira, Godgudala* (Hindi); *Balnaru, Kaithali* (Kannada); *Dieng star* (Khasi); *Vakka, Chavuthi* (Malayalam); *Kottaithanuku, Vakkai* (Tamil); *Kummaripoliki, Gogai* (Telugu).

Description   Lofty deciduous trees; bark brown, smooth; crown lax, spreading; leaves crowded at branch tips, palmate 5–7 lobed, lobes acuminate, remotely serrate, base deeply cordate, tomentose; petioles 30–60 cm long, tomentose; panicles terminal, spreading or drooping; flowers brownish yellow, pedicels slender; male flowers many; calyx tube campanulate, short, lobes 5, ovate, dense; fruits 2–5, free follicles, stellately spreading, brown, tomentose; seeds black, about 1 cm long, oblong.

Flowering and fruiting   April–September

Distribution   Burma and throughout India, found commonly in deciduous forests in Meghalaya.

Propagation   By seeds

Parts used   Wood

Uses   It is useful in convulsion, neuropathy, helminthiasis, erysiphales, skin diseases, fever, ulcers and cephalalgia.

Stereospermum chelonoides DC.

Family   Bignoniaceae

Indian names   *Kastapatala, Patala* (Sanskrit); *Padeli* (Hindi); *Adri, Hadari* (Kannada); *Pathiriver, Ambuvagina* (Tamil); *Karanyavu, Pathiri* (Malayalam); *Gallugudu, Goddalipukusu* (Telugu).

Description   Middle sized to lofty trees up to 40 m high; crown dense, ovoid with spreading branches; bark dark grey, often black patched, rough with concentric plates; leaves 15–30 cm long, leaflets ovate, elliptic, oblong or lanceolate, caudate acuminate, base often oblique, rounded or cuneate; panicle erect, pyramidal; flowers 2–3 cm long, white or shortly purplish, fragrant, capsules 30–60 cm long, spirally twisted, obscurely 4-angled, drooping.

Flowering and fruiting   April–March

Distribution   Indo-Burma, throughout the greater parts of India, very common in Meghalaya in tropical evergreen and deciduous forests.

Propagation   By seeds

Parts used   Bark, roots leaves and flowers

Uses   Used for asthma, stomach ache, cholera and dysentery. A decoction of root is used in intermittent and perpetual fever. The bark is useful for its diuretic and tonic properties.

Strobilanthes divaricatus T. Anders.

Family   Acanthaceae

Description   An erect almost glabrous shrub 1–2 m high, gregarious, leaves lanceolate or elliptic, long, acuminate, serrulate, membranous, obscurely setulose; lateral nerves 4–6 on either half; base cuneate; petiole 1–2.5 cm long; flowers purple in divaricate lax zigzag spikes; bracts very small, caducous, ovate; calyx segments narrow–elongate, caudate; corolla 2.5 – 12 cm, curved, tubular; stamens glabrous; capsule 1.5 cm, glabrate, 4- seeded; seeds 2.5 cm, ovate, pubescent, areolas small.

Flowering and fruiting   October–January

Distribution   Khasi hills, at 1500 m altitude

Propagation   By seeds

Parts used   Seeds

Chemical constituents   Seeds contain glucoside, divaricoside, caudoside and sinoside.

Uses   It is useful to treat acute and chronic heart failure.

Strobilanthes flaccidifolius Nees.

Family   Acanthaceae

Indian names   *Rum, Raspat* (Assamese); *Khum, Khuma* (Manipuri).

Description   Glabrous shrub; leaves elliptic, acute at both ends; flowers in densely lax spikes; bracts petioled, ovate, deciduous; calyx linear, spathulate; spike dense panicle; flowers mostly distant, alternate; corolla 5 cm long, glabrous, capsule 2 cm not included at base.

Flowering and fruiting   December–February

Distribution   Khasi hills, Meghalaya

Propagation   By seeds

Parts used   Bark

Uses   Bark juice is used for parotitis, cut wounds and bruises.

Strychnos wallichiana Steud.

Family   Loganiceae

Indian name   *Naagamushti* (Kannada).

Description   Lianes; bark grey or greyish brown; leaves oblong, oblong lanceolate, shining, corymbs 7–10 cm across; flowers 1–1.5 cm long, tubular, yellowish green often with a purplish tinge; corolla lobes recurved, throat villous; styles emerging; fruit globose, greenish white, beaked.

Flowering and fruiting   April–November

Distribution   North-East India and neighbouring countries; common in Meghalaya, particularly in tropical deciduous forests in Khasi hills.

Propagation   By seeds

Parts used   Bark

Chemical constituents   The stem bark contains alkaloids 5.23% strychnine 2.37–2.43% and brucine 2.8%.

Uses   It is useful for rheumatism, paralytic cramp of extremities, lumbago and diarrhoea.

Swertia chirata Kar.

Family   Gentianaceae

Indian names   *Bhunimba, Chiratika* (Sanskrit); *Chirata, Chiraitu* (Hindi); *Nelabevu* (Kannada); *Cirattakucci, Nilavembu* (Tamil); *Kiriyattu, Nilaveppa* (Malayalam); *Nilavembu, Nilaveru* (Telugu).

Description   A perennial herb, with rooting stem; stem robust branching, cylindrical below, 4 angled upwards containing a large pith; leaves broadly lanceolate, 5 nerved sub-sessile, flowers, greenish yellow in large panicles; capsule egg shaped; ovoid, glabrous, black when ripe; seeds one or two, yellow, circular, not much compressed, 8 mm in diameter, shining with short appressed silky hairs.

Flowering and fruiting   September–January

Distribution   Khasi hills 1600 m altitude

Propagation   By seeds

Parts used   Seeds

Chemical constituents   Ophelic acid, chiratin, swertinin, swertianin, swerchirin, isobellidifolin, friedelin, and β-sitosterol.

Uses   The plant is well known for its bitter, stomachic, astringent, refrigerant, demulcent, emetic, diuretic, digestive and anthelmintic, properties. Seeds are useful for diarrhoea, gonorrhoea, gastropathy, bronchitis, chronic diarrhoea and conjunctivitis. The roots are useful for leprosy.

Swertia dilatata Clark.

Family   Gentianaceae

Description   A glabrous herb, about 1 m high, stem obscurely lineolate; leaves sessile, about 5 cm long, linear, lanceolate, glabrous; flowers in thyrsoid panicles; calyx lobes narrow, lanceolate, acuminate; corolla yellowish; filaments dilated; capsule oblong, ovoid 1.5–3 cm long.

Flowering and fruiting   November–January

Distribution   Khasi hills, Meghalaya

Propagation   By seeds

Parts used   Whole plant

Uses   It is used for treatment of fever, oedema, eruptic boils, haemorrhage and vomiting.

Swertia purpurascens Wall.

Family   Gentianaceae

Description   A small annual herb; stem 4-winged; leaves elliptic, lanceolate, 3-nerved; panicles many flowered; calyx segments oblong, linear, 2 cm long; corolla lobes ovate, acute with one orbicular gland at the base. filaments dilated downwards united into a short tube free from the corolla; style long, stigma sublinear, base narrowed. Capsule 1 cm long, stalkless. Seeds globose, light yellow.

Flowering and fruiting   August–December

Distribution   Khasi hills, Meghalaya

Propagation   By seeds

Parts used   Whole plant

Uses   Decoction of the plant is given for fever.

Symplocos lucida (Thunb.)

Family   Symplocaceae

Description   Middle sized trees, 6–20 m high; crown branches horizontal or nearly so; bark dark brown, vertically fissured and lenticellate; leaves elliptic, lanceolate oblong, acuminate to subacute, base cuneate, repand serrate, glossy green, glabrous; panicles up to 3 cm long; flowers white or yellowish white; fruits ovoid-ellipsoid, brownish red.

Flowering and fruiting   October–July

Distribution   Himalayas, in Burma, common in Meghalaya.

Propagation   By seeds

Parts used   Bark

Uses   Stem bark is used to treat haemorrhage, acne and pimples, leucorrhoea, wounds, skin disorders, hoarseness of voice, fever, menstrual disorders and liver diseases.

Symplocos paniculata Wall.

Family   Symplocaceae

Indian names   *lodhra* (Sanskrit); *Dieng iong*, *Jamiang* (Khasi); *Leiree* (Manipuri).

Description   A deciduous shrub or tree, up to 12 m high; the bark is soft, friable, of light brown colour; leaves obovate, lanceolate or ovate, broadly elliptic, base rounded or cordate, sharply crenate or serrate, pilose, particularly along nerves beneath; panicles 2.5–8 cm long; flowers yellow, fragrant; corolla lobes spreading, nearly free; fruits ovoid, globose, purplish red.

Flowering and fruiting   May–December

Distribution   Eastern–Himalayas to Japan, common in Meghalaya in primary and secondary forests.

Propagation   By seeds

Parts used   Bark

Chemical constituents   The bark contains 2 alkaloids, loturine and colloturine.

Uses   It is cooling and astringent, useful in bowel complaints, diarrhoea, fever and ulcer.

Symplocos racemosa Roxb.

Family   Symplocaceae

Indian names   Lodh bark (English); *Balabhadra, Balipriya* (Sanskrit); *lodh, lodhra* (Hindi); *Pachettu, Baladoddi* (Kannada); *Vellattippattai, Vellilottiram* (Tamil); *Lodduga, Sabaramu* (Telugu).

Description   A small tree, about 6 m high; leaves 8–20 cm long, dark green, leathery usually at tip, margins entire or toothed, leaf stalk small, about 8–20 mm long; flowers small, about 1.2 cm in diameter, white or pale yellow, in small axillary clusters; fruit 1–1.3 cm long, purplish black.

Flowering and fruiting   September–June

Distribution   Common in Meghalaya, nearly throughout the state in evergreen forests

Propagation   By seeds and stem cutting

Parts used   Bark

Chemical constituents   Loturine, loturidine, colloturine, sugars, oxalic acid, phytosterol, 3-monoglucofuranoside, methyl leucopelargonidin.

Uses   Bark is used in diarrhoea, menstrual disorder, indigestion, ulcer, eye disease and tonic. A decoction of the bark is used as a gargle in bleeding gums. It is also used in plasters and applied on wounds for promoting maturation of wounds. It is an astringent and is used in excessive bleeding during menstruation. The astringent properties are utilized also for curing loose motion. It is considered useful in elephantiasis and in controlling fat in urine.

Symplocos theaefolia D. Don.

Family   Symplocaceae

Indian name   *Leiree* (Manipuri).

Description   A middle sized or small evergreen tree, with dense and more or less horizontal branches, branchlets angular; bark dark brown with vertical lines of lenticellate warts, inside very faint, brownish white with very fibrous, distant, broad streaks of white; leaves elliptic, lanceolate, acuminate, sub-entire or shallowly serrulate, coriaceous, dark glossy green, quite glabrous, midrib raised on both surfaces, lateral nerves numerous, sub-parallel, rather inconspicuous, base cuneate; flowers whitish, sessile, in short axillary panicles; bracts and bracteoles broad, oblique; calyx glabrous, segments 5, rather short; corolla lobes oblong; disc densely covered with long white hairs; stamens many; ovary 3-celled; fruit ellipsoid ovoid, embryo straight.

Flowering and fruiting   November–July

Distribution   Khasi hills, 1600 m altitude

Propagation   By seeds

Parts used   Leaves

Uses   Extract of leaves is useful against cancerous cells to some extent.

Syzigium cumini (L.) Skeels.

Family   Myrtaceae

Indian names   Jambolan, Java plum (English); *Jambava* (Sanskrit); *Jaman, Jambhal, Jamun* (Hindi); *Jambuva, Nerale* (Kannada); *Kottainaval, Naval* (Tamil); *Njaval, Njara* (Malayalam); *Neredu, Jambu* (Telugu).

Description   A large evergreen tree; leaves opposite, 8–20 cm long, leathery, smooth, shining; flowers small, dull white, in large bunches; fruit 1.5–4 cm long, ovoid, purplish when young, almost black when ripe, juicy; seed usually one; the fruits are largely eaten raw and the purplish coloured flesh of the fruit leaves the tongue (and lips) dark purple-tinged for several hours.

Flowering and fruiting   March–July

Distribution   Fairly common particularly in lower elevation in Garo hills, Meghalaya.

Propagation   By seeds

Parts used   Bark, leaves and fruits

Chemical constituents   Seeds contain glycoside, jambolin, ellagic acid, tannin and gallic acid.

Uses   Bark, fruits and seeds of the tree are medicinal. The bark is very astringent and is used in sore throats, bronchitis, asthma, ulcers and dysentery. It is also given for purifying blood. The fresh juice of bark with goat's milk is given for diarrhoea. The seeds are useful in diabetes. The fruit juice also has that property, but the effect of preparations from seeds is more marked.

Syzygium jambos (L.) Alston.

Family   Myrtaceae

Indian names   Rose apple, Jambu (English); *Campeyah* (Sanskrit); *Jamun, Gulab-jamun* (Hindi); *Pannerale, Panneer hannu* (Kannada); *Campai, Perunaval* (Tamil); *Champa, Malaykkachampa* (Malayalam); *Jambuneredu* (Telugu).

Description   A large shrub or small tree with spreading branches; leaves simple, opposite, lanceolate, narrowed into short petioles, secondary nerves joined by a prominent looping intramarginal vein; flowers greenish white in short terminal racemose cymes; stamens many, yellowish white; fruits pale yellow to pinkish white, globose; seeds 1–2, grey, in large cavity of the succulent pericarp.

Flowering anf fruiting　April–August

Distribution　Cultivated throughout India up to 1,400 m and in Meghalaya.

Propagation　By seeds and vegetative method.

Parts used　Bark and fruit

Chemical constituents　It contains thiamine, riboflavin, nicotine acid, vitamin C, folic acid, alanine, aspartic acid, cystine, glutamine, threonine and tyrosine.

Uses　The bark is useful for gout, haemorrhage, syphilis, leprosy, dermatopathy, diarrhoea, helminthiasis, wounds and ulcers.

T

Tabernaemontana divaricata (L.) R. Br.

Family   Apocynaceae

Indian names   Crepe jasmine (English); *Nandivrksah, Nandyavartah* (Sanskrit); *Candni* (Hindi); *Kottubale, Nandibatlu* (Kannada); *Nandiyavattam, Adukkunandiyavattai* (Tamil); *Nantyarvattam* (Malayalam); *Gandhitagarapu, Nandivardhanamu* (Telugu).

Description   Shrubs; bark greyish white, minutely fissured; branches dichotomous; leaves elliptic, lanceolate, oblong, ovate, acuminate, base cuneate, often oblique, shining green above; cymes up to 10 cm long; flowers white, sweetly fragrant, in 1–8 flowered cymes at the bifurcations of the branches, lobes of corolla overlapping to right in the bud; fruits follicles, ribbed and curved, orange or bright red within, irregular, enclosed in a red pulpy aril.

Flowering and fruiting   August–February

Distribution   Throughout India, common in entire North-East Region up to a height of 1000 m.

Propagation   By seeds and vegetative method

Parts used   Whole plant

Chemical constituents   It contains coronaridine, voacristine, tabernaemontanine and dregamine and coronarine.

Uses   Leaf paste is useful for headache, fever, boils and sores. Whole plant is powdered and this powder acts as anthelmintic.

Tamarindus indica L.

Family   Caesalpiniaceae

Indian names   Tamarind, Indian date (English); *Cinca, Cincini* (Sanskrit); *Tamrulhindi, Teter* (Hindi); *Amli, Huli* (Kannada); *Puli, Amailam* (Tamil); *Puli, Madhurappuli, Valanpuli* (Malayalam); *Chinta, Chinta-pandu* (Telugu).

Description   A large tree; leaves compound 10–20 pairs, about 1 m long; flowers yellowish with reddish streaks, in small erect clusters among the leaves; fruits 8–20 cm long, 2–3 cm broad, fleshy, pendulous, brown in colour; seeds 3–12, dark brown, shining, embedded in the fleshy, fibrous mass, which is the well-known acid pulp of tamarind.

Flowering and fruiting   April–December

Distribution   Tropics, cultivated throughout India

Propagation   By seeds

Parts used   Roots, fruits and seeds

Chemical constituents   Fruit contains tartaric acid, citric, maleic acid and potassium bitartarate and traces of oxalic acid. Leaves contain flavonoid, glycosides saponaretin, vitexin, orientin and homoorientin.

Uses   Tamarind pulp has laxative properties; its infusion in water is a very refreshing drink; it is useful in fever. As a laxative it is taken singly or in mixture with other purgative drugs. When mixed with other purgative drugs it reduces their laxative property. Fruit is sour, tasty, indigestible, and astringent to the bowels. It cures biliousness, cough and blood troubles. The ripe fruit is an appetizer, laxative, healing tonic to the heart and anthelmintic. It heals wounds and fractures. The seeds are useful in vaginal discharges and ulcers (Ayurveda). The bark has astringent and tonic properties and heals ulcers. The leaves reduce inflammatory swellings. The fruit is sour and sweetish useful in liver complaints, vomiting, thirst, scabies, sore throat, stomatitis causes biliousness and impoverishes the blood. The pulp of the fruit is tonic to the heart, astringent and aperient, aphrodisiac and useful in giddiness. It is applied externally in liver complaints and inflammations (Unani).

Terminalia arjuna Roxb. [Near threatened]

Family   Combretaceae

Indian names   Arjun, Kumbuk, Kahua bark (English); *Arjunah, Chitrayodhi* (Sanskrit); *Arjuna, Jamla* (Hindi); *Arjuna, Bilimatti* (Kannada); *Attumarutu, Kulamarudu* (Tamil); *Attumaruthu, Maruthu, Nirmaruthu* (Malayalam); *Ambotikura, Pulichintaku* (Telugu).

Description   A large evergreen tree 60–80 feet high, with buttressed trunk and spreading crown with drooping branches; bark smooth, grey outside and flesh-coloured inside, flaking off in large flat pieces; leaves simple, usually 8–15 cm, subopposite, oblong or elliptic, coriaceous, crenulated, pale dull green above, pale brown beneath, often unequal, nerves 10–15 pairs, reticulate, suddenly narrowed at the base, often cordate, obtuse or very shortly acute at the apex; petiole rarely more than 0.5 in., often very short, with two glands near its apex; flowers white in panicles of spikes with linear bracteoles; calyx teeth nearly glabrous both within and without; young ovary very short, covered with crisped brown hair. Wings of the fruit usually truncate or suddenly narrowed at the top.

Flowering and fruiting   May–February

Distribution   Deccan, Ceylon and the sub-Himalayan tracts of the North-West Provinces. Indo-Malaya, throughout India, occurs frequently in tropical evergreen forests at lower elevations, particularly along catchments and slopy areas.

Propagation   By seeds

Parts used   Bark

Chemical constituents   Arjunolic acid, tomentosic acid, β-sitosterol, ellagic acid, saponin and leucodelphinidin. Bark contains a crystalline compound arjunine, a lactone, and arjunetin.

Uses   The bark is astringent, sweet, cooling and heating, alexiteric, styptic, acrid, demulcent, cardiotonic, urinary astringent, expectorant, and tonic. Bark reduces "kapha", and is useful in diseases of heart, anaemia, excessive perspiration, tumours; leucoderma and false presentations of the fetus (Ayurveda). Bark is also used to cure asthma and heart diseases (Unani), in preparation of mother tincture and also for blood pressure and heart diseases (Homeopathy). It is useful for fracture, ulcers, diabetes, vitiated conditions of pitta, anaemia, cardiopathy, fatigue, bronchitis, tumours, cirrhosis of liver and hypertension.

Terminalia bellirica (Gaertn.) Roxb.

Family   Combretaceae

Indian names   *Bedaro, Bhirda, Yehela* (Kannada); *Baheda, Bahera, Behasa* (Marathi); *Tuikuk-reraw, Char-vantai* (Mizoram); *Vibhitakamu* (Telugu).

Description   Large deciduous tree up to 40 m high, crown lax, spreading, ovoid; bark grey, pale greyish brown, scaly; young parts rusty tomentose; leaves obovate, orbicular, abruptly acute, sometimes rounded or truncate, base cuneate, glabrous, glaucous beneath; spikes 6–15 cm long, spreading; flowers foetid smelling; disc greenish red; fruits globose, yellowish when dry.

Flowering and fruiting   March–December

Distribution   Indo-Malaya, throughout India; common in Meghalaya in deciduous forests and tropical evergreen forests.

Propagation   By seeds

Parts used   Bark and fruits

Chemical constituents   Fruit contains about 17 per cent tannin and β-sitosterol, gallic acid, ellagic acid, ethyl galate, galloyl glucose and chebulagic acid.

Uses   The bark is useful in anaemia and leucoderma. The fruit is bitter pungent, acrid and digestible and is useful in bronchitis, sore throat, biliousness, headache and diarrhoea. Extract of fruits is also applied to eyes and piles. It is also used to cure asthma and diseases of nose and heart (Ayurveda). The fruit is bitter, astringent, tonic, aperients and antipyretic and is used to cure dyspepsia and headache. Aqueous extracts of juice is useful in diseases of eyes, piles and nose. Fruit is used as a brain tonic (Unani). Fruit in combination with other drugs is prescribed for snake bite (Charaka and Sushruta). The oil is used in case of rheumatism. Oil is considered a good application for hair. Gum is demulcent and used as purgative.

Terminalia chebula Retz.

Family   Combretaceae

Indian names   *Balhar, Hana* (Hindi); *Anilaykayi, Hirade* (Kannada); *Kadukka* (Malayalam); *Haritaki, Haritakipushpam* (Sanskrit); *Kadakai, Kadookai, Kaduk-kai* (Tamil); *Kadukar, Karaka, Karakai* (Telugu).

Description   It is a moderate sized deciduous tree with spreading branches; the bark is thick and dark brown in colour with vertical cracks; leaves ovate, elliptic or obovate, glabrous to villous beneath with pair of large glands at the top of the petiole; flowers 0.4–0.7 cm across, yellow, ellipsoid, nearly smooth; fruit glabrous, shining, ellipsoid or ovoid, drupes, yellow to orange in colour, up to 3.75 cm long; seeds hard, pale yellow.

Flowering and fruiting   May–February

Distribution   Throughout India; occurs frequently in tropical evergreen forests at lower elevations, commonly in Garo hills, Meghalaya.

Propagation   By seeds

Parts used   Ripe and unripe fruits

Chemical constituents   Chebulin from flowers; palmitic, stearic, oleic, linoleic, arachidic and behenic acids from fruit kernels. Fruit contain about 30 per cent of an astringent substance, tannic acid, 20–40 per cent, gallic acid, resin, etc.

Uses   Fruit is a tonic, carminative, expectorant, anthelmintic, antidysenteric and alternative. The fruits are useful in asthma, sore throat, thirst, vomiting, cough, eye diseases, diseases of heart, urinary discharges, bleeding piles, typhoid, itching pain, constipation, anaemia and leucoderma (Ayurveda). The unripe fruit is an astringent useful in dysentery and diarrhoea. The ripe fruit is a purgative, tonic, carminative and enriches the blood. Fruit is used to cure disease of spleen, brain and eyes, paralysis and for the piles (Unani). Aqueous extract of fruits kept for the night is considered a coolant for eyes. The ash mixed with butter forms a good ointment for sores.

Terminalia citrina Gaertn. Flem.

Family   Combretaceae

Indian names   *Silikka, Monalu, Hilikha* (Assamese); *Harra, Haritaki* (Bengali); Aritok, Bolomit (Garo); *Soh handru diengartaki* (Khasi); *Manahi* (Manipuri); *Reraw* (Mizoram).

Description   Large trees up to 30 m high; stems buttressed at base, crown ovoid, spreading; bark grey or greyish brown, vertically fissured, branchlets green with white, elongated specks;

leaves broadly ovate, base truncate, sometimes rounded, subcordate; racemes panicled, flowers yellow, stamens erect, calyx villous within; fruits oblong, lanceolate, obtusely 5-angled.

Flowering and fruiting   April–December

Distribution   Confined to North-East India; common in Meghalaya in the deciduous and tropical evergreen forests.

Propagation   By seeds

Parts used   Bark

Uses   Bark is useful in cardiopathy, wounds, ulcers, haemorrhages, diarrhoea, dysentery, cough, bronchitis, leucorrhoea, gonorrhoea and burning sensation.

Thalictrum javanicum Bl.

Family   Rununculaceae

Indian name   *Makdya ghas, Garbini mamiri* (Hindi).

Description   Stem erect, up to 1 m high, glabrous; leaves several times ternately decompound, leaflets ovate or orbicular, membranous, glabrous, somewhat glaucous beneath, 3–7 toothed, base, rounded or cordate; stipules adnate imbricate. Panicle laxly branched; flowers white, on filiform pedicel; filaments club-shaped; achenes up to 15, strongly ribbed, beaked short.

Flowering and fruiting   July–September

Distribution   Khasi hills up to 1500 to 2000 m

Propagation   By seeds

Parts used   Root and leaves

Uses   Leaves are useful for eye diseases, skin diseases and sores, and has action germicidal. Root is antiperiodic, diuretic, aperient and purgative and used as a bitter tonic during convalescence and dyspepsis.

Thunbergia grandiflora (Rottb.) Roxb.

Family   Acanthaceae

Indian names   Bengal clock vine, Clock vine (English); *Mulluta* (Hindi); Khakkhu (Garo); *Jermi khnong* (Khasi); *Kukualoti* (Assamese); *Vakohrui* (Mizoram).

Description   Large climbers, young parts tomentose; leaves ovate, orbicular in outline, shallowly 5 or more lobed, dentate at base, acute or shortly acuminate, base deeply cordate, 7–11 nerves, scabrous, tomentose; flowers 6–8 cm long, fascicled on usually pendent racemes, white or bluish white, campanulate; capsules 3–5 cm long with 4-quetrous beak.

Flowering and fruiting   April–January

Distribution   Wild in North-East, very common in Meghalaya.

Propagation   By seeds and vegetative method

Parts used   Leaves

Chemical constituents   The flowers contain amino acids, like aspartic acid, serine, glycine, alanine, valine, flavonoids like apigenin 7-glucoronide, luteolin, anthocyanin, malvidin and carbohydrates like saccharose, glucose and fructose.

Uses   The leaves are commonly used against snake bite, the petioles are removed and the juice of 30–50 g of pounded fresh leaves is used to massage the site of the snake bite, from the top downwards.

Tinospora cordifolia (Willd.)

Family   Menispermaceae

Indian names   *Amara, Amrita* (Sanskrit); *Amara, Amrita* (Hindi); *Amrytaballi, Madhuparne* (Kannada); *Amridavalli, Amudam* (Tamil); *Peyamrytam, Sittamrytu* (Malayalam); *Guduchi, Guluchi* (Telugu).

Description   A large succulent climber with corky bark, young shoots ovate, cordate, acute or shortly cuspidate, acuminate, glabrous; petiole 3.5–6 cm long; flowers greenish yellow; glabrous; male fascicled; female solitary on longer pedicels. Bracts boat-shaped, the lower ones often leaflets. Petals cuneate, stamens free; anthers oblong. Drupes, globose, shortly stalked, red and glossy when ripe.

Flowering and fruiting   February–April

Distribution   Throughout India, common in North-East Region.

Propagation   By seeds and vegetative method

Parts used   Stem and leaves

Chemical constituents   The plant contains tinosporin, columbin, chasmanthin, palmarin, berberine, tinosporin, tinosporic acid, tinosporol, giloin, giloisin, substituted pyrrolidine, a diterpenoid, octacosanol, nonacosane and β-sitosterol.

Uses   The stem is bitter, astringent, thermogenic, anodyne, anthelmintic, antiperiodic, antispasmodic, anti-inflammatory and antipyretic, It is useful in vomiting, burning sensation, hyperdipsia, helminthiasis, dyspepsia, chronic fever, cardiac debility, skin diseases, leprosy, anaemia, cough, and asthma. The drug is useful as a tonic and antiperiodic. It is also considered aphrodisiac.

Toddalia asiatica (L.) Lam.

Family   Rutaceae

Indian names   Forest pepper, Lopez root (English); *Dahana, Kancanah* (Sanskrit); *Kanchan, Dahan* (Hindi); *Kara, Kattukarimilaku, Mulakuthali* (Malayalam); *Kaduhakukare, Kadumaraju* (Kannada); *Kattumilagu, Kichilikaranai* (Tamil); *Errakasinda, Kondakasinda* (Telugu).

Description   Large prickly climber, bark lenticellate, dark-brown, leaves 6–12 cm long, leaflets oblanceolate, obovate, elliptic, acuminate, base oblique, narrowed, rounded or truncate, crenate, glabrous; inflorescence up to 8 cm across, yellowish white or yellow; sepals minute, petals lanceolate, spreading, stamens erect, yellow, ovary green, berry beaked.

Flowering and fruiting   September–October

Distribution   Nearly throughout India, common in Meghalaya, particularly in Jaintia hills

Propagation   By seeds

Parts used   Roots, leaves, flowers and fruits

Chemical constituents   Plant contains essential oil and an alkaloid berberine. Roots contain a poisonous resin. Leaves contain toddaline and toddalinine, pimpinellin, isopimpinellin and glycosides.

Uses   The bark is acrid, astringent, tonic, expectorant, anthelmintic, dexurative and antiperiodic. The bark is used for chronic infantile dysentery, cough, bronchitis, intermittent fever and ulcer.

Toona ciliata Roemm.

Family   Meliaceae

Indian names   Australian red cedar (English); *Tun, Toon* (Hindi); *Daevadaari, Gandha garige* (Kannada); *Cantanavempu, Tunamaram* (Tamil); *Vempu, Padukarana* (Malayalam); *Nandicettu* (Telugu).

Description A large tree, 20–30 m high with a large crown; bark lenticellate, dark brown; leaves up to 75 cm long, abruptly pinnate, leaflets obliquely ovate, lanceolate, entire or subentire, glabrous, undulate; panicles large, flowers white, fragrant, in large terminal drooping panicles; petals white, oblong, erect; filaments hairy; disc orange red, stigma free; capsules oblong, ellipsoid, 5-valved; seeds 2.5 cm 1 in each cell and its membranous wings at each end.

Flowering and fruiting November–September

Distribution Tropical and semi-tropical evergreen forests; Indo-Malaya up to Australia, very common in Meghalaya.

Propagation By seeds

Parts used Bark and flowers

Uses The bark is acrid, astringent, tonic, expectorant, anthelmintic, depurative and antiperiodic. The bark is used for chronic infantile dysentery, cough, bronchitis, intermittent fever and ulcer.

Trachelospermum lucidum (D. Don.) K. Schum.

Family Apocynaceae

Indian names *Akhahilata* (Assamese); *Kali dudhi* (Hindi); *Soh kyrmoit kroh* (Khasi).

Description A stout woody climber, bark dark brown, usually warty at base; leaves oblong, elliptic, lanceolate, acuminate, base rounded or cuneate, shining above, pale beneath, nerves closely sub-parallel, looping near margin; flowers white, in lax terminal and pseudoaxillary cymes; calyx small, corolla salver-shaped, lobes 5, spreading, follicles 10–30 cm long, cylindric, incurved.

Flowering and fruiting June–January

Distribution   Temperate regions and sub-Himalayas; common in Meghalaya.

Propagation   By seeds

Parts used   Fruits

Uses   The fruit is useful in dyspepsia, diarrhoea, pharyngitis, rheumatoid arthritis, bronchitis, cough, asthma and strangury.

Trema orientalis Bl.

Family   Ulmaceae

Indian names   Mpesi (English); *Jivani, Jivanti* (Sanskrit); *Jiban* (Hindi); *Bende, Gorklu* (Kannada); *Ambaratti, Kuttippala* (Tamil); *Mallantoddali, Pottama* (Malayalam); *Budumuru, Chakamanu* (Telugu).

Description   A small fast-growing and short-lived tree; branches and branchlets somewhat ascending; twigs adpressed, pubescent; bark thin, greenish-grey or bluish-green, smooth with numerous reddish lenticels; leaves ovate, lanceolate or oblong, caudate, acuminate, rather membranous, scabrid above, more or less grey or white pubescent or tomentose beneath, base oblique, subcordate or cordate; male cymes dense, sometimes lax; female cymes lax; sepals flat, stigma papillose; drupe ovoid, angular.

Flowering and fruiting   February–August

Distribution   In North-East India, frequently in Meghalaya

Propagation   By seeds

Parts used   Bark

Uses   The bark yields a strong fibre. The leaves are chopped for cattle fodder.

Trevesia palmata Roxb.

Family   Araliaceae

Indian names   *Bhotola* (Assamese); *Chena thong* (Garo); *Dieng soh kynthur* (Khasi); *Kawhte-bel* (Mizoram).

Description   Small, unbranched tree, armed with small prickles particularly at top; bark brown. Leaves clustered at the tip, deeply 7–9 lobed, base cordate, lobes oblong or elliptic, acuminate, serrate, petiole 18–40 cm long; panicles up to 50 cm long, brown, tomentose, branches sub-dichotomous; peduncles 5–8 cm long; flowers greenish or yellowish white; petals 8–12, thick, ovate, disc flat, reddish-yellow.

Flowering and fruiting   February–June

Distribution   Burma, Nepal and India, particularly along streamlets in shady areas in Meghalaya.

Propagation   By seeds

Parts used   Root and leaves

Uses   Leaf, root are used in stomach ache, dysentery, diarrhoea, etc. The flowers are eaten in Garo hills and are used for fever.

Trichosanthes dioica Roxb.

Family   Cucurbitaceae

Indian names   *Karkasacchada, Patola* (Sanskrit); *Palval, Parvar* (Hindi); *Kahi-padavala* (Kannada); *Kombuppudalai* (Tamil); *Kattu-potolam, Patolam* (Malayalam); *Kommupotla, Kambupotala* (Telugu).

Description   A climber with perennial rootstock with scabrous and more or less woody stems; leaves upper surface downy, ovate, oblong, cordate, acute, petiole 2 cm long tendrils bifid;

male peduncles paired but not racemed, male flower woolly outside; calyx tube narrow, lobes erect, anthers free; fruit oblong or nearly spherical, often with light green stripes, red when ripe; seeds half ellipsoid, corrugate on the margins.

Flowering and fruiting   March–August

Distribution   Khasi hills

Propagation   By seeds

Parts used   Root and fruits

Chemical constituents   The root has an alkoloid cucurbitacin.

Uses   The root is a strong pugative. It is useful for constipation, fever, skin infections and wounds. The leaf juice is applied to patches on skin.

Trichosanthes tricuspidata Lour.

Family   Cucurbitaceae

Indian names   *Indravaruni, Kakanasa* (Sanskrit); *Indrayan, Mahakal* (Hindi); *Avagude, Haavumekke* (Kannada); *Ankorattai, Korattai* (Tamil); *Kakkattonti, Kattuvellari* (Malayalam); *Avuduta, Aabuda* (Telugu).

Description   Large climbers, woody lenticellate, tendrils 3 or 2-fid; leaves 3–7 lobed, suborbicular, cordate, dark green above, pale beneath with dark-coloured, circular, glands scattered along the lower fringes and gland dotted, 3–7-nerved from base, lobes acuminate or acute, serrate, racemose; flowers 4–5 cm across, unisexual, males in axillary racemes, bracts broadly ovate, many-nerved, fringed and gland dotted, female flowers solitary and axillary; fruit globose, red when ripe with 10 orange streaks; seeds numerous, smooth.

Flowering and fruiting   June–November

Distribution   Throughout India, common in Meghalaya

Propagation   By seeds

Parts used   Leaves and seed

Chemical constituents   Punic acid, cycloeucalenol, hexacosanoic acid, trichonin, santholin, β-sitosterol, cyclotrichosantol, bryonolic acid, cucurbitacine B and D, and trichotetrol.

Uses   Leaf paste is applied on burns. Seed extracts show haemagglutining activity. Root is useful for hemicrania, carbuncles and gonorrhoea. The fruits very valuable in curing otitis and rhinitis and are also for inflammations and weakness of lymphs.

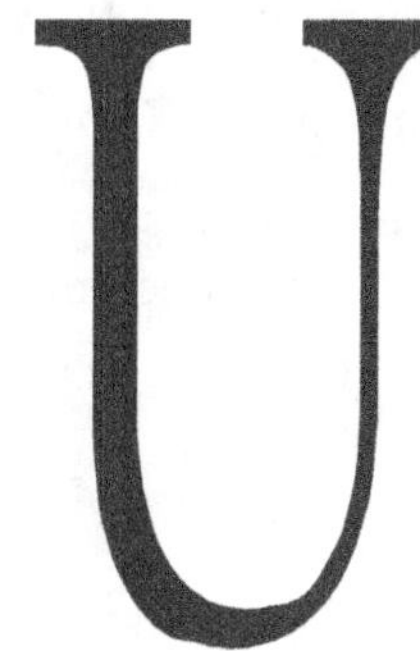

Urena lobata L.

Family   Malvaceae

Indian names   Cousin mahoe, Aramina fibre (English); *Nagabala, Vanabhenda* (Sanskrit); *Bachata, Bachit* (Hindi); *Baralu kaddi mara, Dodda bende* (Kannada); *Ottatti, Ottuttutti* (Tamil); *Udiram, Uram* (Malayalam); *Peddabenda, Nalla benda* (Telugu).

Description   Herb to undershrub, up to 1.5 cm high; leaves ovate, variously lobed, upper ones lanceolate, serrate; inflorescence racemose; flowers pink with a darker centre, bracteoles 5, connate at base, sepals equalling the bracteoles, hairy, globose, lobed; seeds trigonous, angular, vertically grooved.

Flowering and fruiting   August–January

Distribution   All over North-East Region; a weed of forests in Meghalaya, Khasi hills.

Propagation   By seeds

Parts used   Leaf, root and flowers

Uses   Leaf is an expectorant. It is useful in diarrhoea and boils. The root is diuretic. It is used as abortifacient. Flowers are administered as an expectorant in dry cough.

V

Valeriana hardwickii Wall ex. Roxb.

Family   Valerianaceae

Indian names   *Balaka, Tagara* (Sanskrit); *Pind tagar, Asarun* (Hindi); *Balagur* (Bengali); *Asarun, Taggar* (Urdu).

Description   Erect herb or shrub up to 2 m high; radical leaves, ovate, usually undivided; cauline leaves, pinnate or pinnatifid; leaflets or lobes lanceolate, acute, base narrowed, entire, acuminate, the terminal one longest; flowers reddish, inflorescence lax in cymose clusters forming axillary, compound corymbs or panicles; fruits angular and hairy.

Flowering and fruiting   August–December

Distribution   Confined to Himalayas and sub-Himalayas, Meghalaya up to 2000 m

Propagation   By seeds

Parts used   Leaves and root

Uses   The paste of leaf and root is useful for skin diseases, epilepsy, hysteria and neurosis.

Vanda roxburghii R. Br. [Endangered]

Family   Orchidaceae

Indian names   *Gandhanakuli, Nakuli* (Sanskrit); *Rasna nai* (Hindi); *Bandanike* (Kannada); *Maravalai, Akipucam* (Tamil); *Maravazha* (Malayalam); *Badanike, Kanapachettu* (Telugu).

Description   A small epiphytic herb, stem 1–2 ft., climbing; roots white, shiny, silvery grey coloured. The plant developed two kinds of roots, the clinging roots and aerial roots; leaves 6–8 in. narrow, complicate. Peduncle 6–8 in. 6–10 ft.

Flowering and fruiting   July–December

Distribution   This is an epiphytic orchid on mango and other trees. It is formed all over India and in Sri Lanka.

Propagation   By seeds and vegetative method

Parts used   Whole plant

Chemical constituents   The plant contains alkaloids and glycosides. The plant also contains tannin, saponin, sterols, fatty oils, resins and colouring matter.

Uses   The plant is used for rheumatic disorders, local swelling, and indigestion. The root is useful in inflammations, rheumatic disorders, and osteoarthritis. It is also used as a laxative and stimulant to liver. The medicated oil is used for external application in rheumatic pains.

Ventilago madraspatana Gaerth.

Family   Rhamnaceae

Indian names   *Dinesavalli, Kaivartika* (Sanskrit); *Pitti* (Hindi); *Haruge, Kubbila* (Kannada); *Pappili, Surul* (Tamil); *Vempata* (Malayalam); *Errasurugudi, Ettashirattalativva* (Telugu).

Description   A large woody climber with long branches; bark dark grey, furrowed, red in the furrows, branchlets glabrecent; leaves bifarious, elliptic oblong lanceolate, crenate or serrate or nearly entire, thinly coriaceous, glabrous, turning yellow before falling, lateral nerves 5–9 on either side of the midrib, slender, much arched, tertiary nerves transverse to the midrib, very fine, close and parallel, base more or less acute, subsymmetrical; petiole channelled, stipules small, subulate; flowers small, fascicled in group on large leafy pubescent panicles; calyx puberulous outside, glabrous within lobes, keeled within with an incurved pointed apex; petals spathulate, shorter than the calyx lobes, enveloping the stamens; disc glabrous or thinly pubescent; ovary hairy at the base.

Flowering and fruiting   September–February

Distribution   Entire North-East India; common in tropical evergreen forests in Meghalaya.

Propagation   By seeds

Parts used   Bark

Chemical constituents   Trihydroxymethyl anthranol monomethyl ether; emodin monomethyl ether in root bark. Stem bark contains friedelin. Lupeol, β-sitosterol and its glucosides are found in stem fruits and leaves.

Uses   It is useful in dyspepsia, colic, flatulence, erysiphales, leprosy, scabies, purits, and other skin diseases, fever and general debility.

Vernonia cinerea Less.

Family   Asteraceae

Indian names   Vernonia, Purple fleabane (English); *Ardhaprasadana, Gandhavalli* (Sanskrit); *Dandotpala, Sadodi* (Hindi); *Sadodi, Sahadevi* (Kannada); *Naycitti, Puvamkurundal* (Tamil); *Pirina, Puvankurutala* (Malayalam); *Garitikamma, Gharitikamini* (Telugu).

Description   Scandent woody shrub, bark yellowish-brown; leaves elliptic, ovate, lanceolate, acute or acuminate, base cuneate, glabrescent beneath; heads, in leafy panicles, involucral bracts ovate or lanceolate, scarious; flowers white; achenes deeply 10-lobbed, pappus red.

Distribution   Burma, Bangladesh and North-East India; frequent in Meghalaya, Khasi hills.

Propagation   By seeds

Parts used   Whole plant

Chemical constituents   Plant contains α-amyrin acetate; α-amyrin benzoate; lupeol and its acetate; β-sitosterol, stigmasterol, β-spinasterol and KCl.

Uses   The roots are useful in diarrhoea, stomachalgia, cough, inflammations, skin diseases, leprosy, renal and versical calculi, chronic and intermittent fevers. The leaves are useful in eczema, ringworm, guineaworms and elephantiasis. The flowers are useful in conjunctivitis, vitiated conditions of vata and fever. The seeds are useful in ringworm, threadworm, cough, leucoderma, chronic skin diseases, and dysuria.

Viburnum colebrookianum Wall. ex. Cl.

Family   Caprifoliaceae

Indian names   *Bolmichek* (Garo); *Pani phuti*, *Mezonga* (Assamese).

Description   Large shrub; branches horizontal and spreading; young parts stellate, pubescent; leaves drooping, elliptic, oblong, shortly acuminate, crenate, serrate, thinly coriaceous, almost glabrous, lateral nerves 8–9 on either half, arcuate, almost glabrous; flowers white, in erect axillary compound pedunculate umbels; bracts minute, bracteoles wanting; calyx tubes glabrous, teeth minute; corolla short, rotate; drupe orbicular, red; seed albumen uniform.

Flowering and fruiting   February–June

Distribution   Common in Meghalaya ascending up to 1500 m in Khasi hills.

Propagation   By seeds

Parts used   Stem bark

Uses   It is useful for bringing down pains in uterus after menses, cancer of tongue, and menstrual irregularities of sterile females with uterine displacements.

Viburnum foetidum Wall.

Family   Caprifoliaceae

Indian names   *Shirporna-jaya* (Sanskrit); *Guia* (Hindi); *Narvel* (Marathi).

Description   A shrub upto 3.5 m; bark greyish, branchlets, petioles and inflorescence clothed with stellate hairs; leaves rhomboidal, elliptic, lanceolate, mucronate, coriaceous, glabrescent above, pubescent on nerves beneath, lateral nerves 3–4 on either half, lowest pair basal, base cuneate; flowers small in compound pedunculate umbels, bracts numerous; calyx tube glabrous, teeth minute, triangular; corolla short, white; drupes scarlet red; seeds dorsally 2-grooved, ventrally 3-grooved; albumen uniform.

Flowering and fruiting   June–November

Distribution   North-East Region, particularly in Khasi hills up to 2000 m

Propagation   By seeds and vegetative methods

Parts used   Stem and leaves

Chemical constituents   Viburinin, salicin, esculetin, scopoletin, isovaleric acid, salicylic alcohol.

Uses   Plant is astringent and emmentic; leaf juice is used in haemorrhage. Pounded leaves are used to cure old sores.

Viscum monoicum Roxb. ex.

Family   Loranthaceae

Indian names   *Katavi, Mohanika* (Sanskrit); *Kuchle-ka-malang* (Hindi); *Bavanchi badanike, Dodda badanike* (Kannada); *Pulluruvi, Pulluri* (Tamil); *Badanika, Kadisebadanika* (Telugu); *Ottu, Pullurivi* (Marathi).

Description   Drooping bushy shrubs; branches greenish; leaves lanceolate, oblong, often falcate, acute, base narrowed, cuneate; flowers greenish yellow; berries oblong, ellipsoid, 0.5–0.8 cm long, yellowish green.

Flowering and fruiting   August–January

Distribution   Throughout India, common in Meghalaya, particularly in Jaintia hills.

Propagation   By seeds

Parts used   Whole plant

Chemical constituents   Main constituents are fatty acids, nicotine alkaloids, amines and other constituents are flavonoids, chaleone derivatives and flavonone derivatives; terpenoids, beta-amyrin, betalinic acid and ester combinations.

Uses   It is employed as a decoction or tincture in intermittent fevers, splenic and hepatic enlargements, in lumbago and piles.

Vitex negundo L.

Family   Verbenaceae

Indian names   *Indrani, Nilapushpa, Nirgundi* (Sanskrit); *Nengar, Ningori, Nirgandi* (Hindi); *Hulihunice, Teltuppi* (Kannada); *Nallanocci, Nirkkundi, Sinduvaram* (Tamil); *Karunocci, Vallanochi* (Malayalam); *Nallavavili, Sindhuvaruma* (Telugu).

Description   A large strongly scented deciduous shrub or a small tree; bark thin, grey or ashy white; branchlets striate; shoots, inflorescence and undersurface of leaves grey-pubescent; leaves digitately 3–5 foliolate, petiole 2.5–5 cm long; leaflets lanceolate, membranous, glabrescent above, base cuneate or rounded; flowers lavender to blue, 0.5 cm across, in opposite cymes, arranged on an elongated panicle up to 30 cm long; calyx campanulate, 5-toothed, teeth triangular; corolla 1.5 cm long, campanulate outside, 2-lipped; upper 2-lobed, lower longer, 3-lobed; middle lobe longest, crenulated; stamens exerted, ovoid, filaments villous at the base; drupe slightly ribbed, gland-dotted.

Flowering and fruiting   May–August

Distribution   Common throughout North-East India

Propagation   By seeds and vegetative method

Parts used   Whole plant

Chemical constituents   Leaves contain two alkaloids nishindine and hydrocotylene and a pale green-yellow oil.

Uses   The plant is bitter, acrid, thermogenic, anthelmintic, expectorant, carminative, digestive, stomachic, anodyne, anti-inflammatory, antiseptic, cephalic, alternate, antipyretic, diuretic, etc. The juice of leaves is used in foetid discharges. An ointment made from the juice is applied as a hair tonic. The drug is also reported to possess tranquilizing properties. Decoction in used in cough, fever, malaria, chicken pox, pneumonia, leaf vermifuge, and headache, flowers are astringent.

Vitex peduncularis Wall. ex. Schauer.

Family   Verbenaceae

Indian names   *Kakajangha, Paravatapadi* (Sanskrit); *Charaigorwa, Chhagriaruba* (Hindi); Navaladi, *Seeme navilaadi* (Kannada); *Rangri, Shilangri* (Garo); *Ashoi, Khoidoi* (Malayalam); *Navaladi* (Telugu).

Description   A large tree; young shoots pubescent; bark greyish to dark grey, rough, exfoliating in irregular flakes, cream coloured or dull yellow and mottled inside; leaves 3-foliate, leaflets lanceolate, narrow, elliptic, entire long, acuminate, membranous, glabrous, yellowish, gland dotted beneath; flowers pale yellow, in axillary panicled cymes, 15–27 cm long and exceeding the leaves; calyx very minutely toothed, pubescent, with golden or light yellow resinous dots; corolla 0.25 cm long, yellow near base and hairy; stamens upcurved, arching over the throat of the corolla, anther limb purple after dehiscence; drupe obovoid, about 12 cm across.

Flowering and fruiting   April–September

Distribution   Common throughout Meghalaya

Propagation   By seeds and stem cutting

Parts used   Root, leaves and bark

Uses   Leaf decoction is used for black fever, malaria, diarrhoea and dysentry. Bark decoction is used for muscle pain.

Vitex trifolia Linn.

Family   Verbenaceae

Indian names   Three-leaved chaste tree (English); *Indranika, Indrasurasa, Jalanirgundi* (Sanskrit); *Nichinda, Panikisanbhalu* (Hindi); *Karinochi, Lakki* (Kannada); *Sirunochi, Nirnocci* (Tamil); *Karunocchi, Neernochi* (Malayalam); *Ambotikura, Pulichintaku* (Telugu).

Description A shrub or small tree; bark grey, smooth; branchlets obscurely quadrangular, tomentose; leaves with leaflets elliptic or obvate, oblong, entire, obtuse or acute, glabrous above, greenish white, tomentose beneath; base rounded or cuneate; lateral leaflets sessile, terminal leaflets, petioled; panicles 4 cm long; flowers lavender to blue. Calyx very shortly 5-toothed, greyish, tomentose. Corolla tomentose with filaments at the base; drupe ellipsoid, blackish. It closely resembles *V. negundo*.

Flowering and fruiting February

Distribution Very common in Meghalaya; moist wasteland along drains and roads and on river banks.

Propagation By seeds and vegetative method

Parts used Whole plant

Uses The fruit is used for, fever, headache, photopsis, vertigo, ophthalmalgia, glaucoma, rheumatism and neuralgia. Flowers are useful for vomiting and severe thirst. Decoction of root is a febrifuge. Juice of leaf is useful for fever, decoction is used to prevent greying of hair, and extract of leaf is useful for tuberculosis, it also exhibits anti-cancer activity.

Vitis barbata Wall.

Family Vitaceae

Description A large deciduous tendril climber, tendrils leaf-opposed, often bifid; branches, petioles and peduncles covered with numerous long, spreading, glandular, capitate hairs; leaves simple, more or less deeply 3–5 lobed, orbicular, cordate, irregularly toothed, membranous, peduncle flattened, bearing a long forked slender tendril; cymes regularly paniculate as long as the peduncle; fruit pedicellate, seed elliptic.

Distribution   Meghalaya up to an altitude of 1200 m

Propagation   By vegetative method

Parts used   Rhizome

Uses   Decoction of rhizome is used as an expectorant and paste is applied on boils and eczema.

Viola serpens Wall.

Family   Violaceae

Indian names   *Banafsha, Vanaphsa* (Sanskrit); *Giddar tamaku, Banaksha* (Hindi).

Description   Herbs, rarely shrubby below; rootstock slender, stem erect or decumbent; leaves ovate, acute or acuminate, more or less deeply cordate with a narrow or broad sinus, crenate, serrate, glabrous or pubescent; petiole slender, stipules toothed or fimbriate; flowers long, acute, canescent; peduncles bracteate an about the middle or a little above it; stigma on an oblique or crooked trumpet-shaped style; capsule small, globose, few seeded.

Flowering and fruiting   February–April

Distribution   Khasi Hills, Meghalaya

Propagation   By seeds

Parts used   Whole plant

Uses   Plant decoction is useful for cough, malaria, chicken pox and pneumonia. Paste is applied for wounds.

Withania somnifera Dunal.

Family   Solanaceae

Indian names   *Asvagandha, Varahakami* (Sanskrit); *Ashvagandha* (Hindi); *Asan, Ghoda asor, Santhiana-popda* (Gujrati); *Hiremaddinegida* (Kannada); *Amukkiram* (Malayalam); *Amukkira, Amukkirankilangu* (Tamil); *Vajigandha, Pannirugadda, Pulivendramu* (Telugu).

Description   A small or middle-sized undershrub, up to 1.5 m high; stem and branches covered with minute star-shaped hairs, leaves up to 10 cm long, ovate, hairy; flowers pale green, small, about 1 cm long, few flowers borne together in short axillary clusters; fruit 6 mm in diameter, globose, smooth, red, enclosed in the inflated and membranous calyx.

Distribution   The plant occurs in drier regions of India; it is also cultivated.

Propagation   By seeds

Part used   Roots and leaves

Chemical constituents   Roots contain several pyrazole alkaloids like withasomnine and steroidal lactones, withaferin A and withanolides. They also contain starch, reducing sugars, hentriacontane, glycoside, dulcitol, withaniol, an acid and a neutral compound. Withaferin is a bacteriostatic and antitumorous agent.

Uses   The tuberous roots are astringent, bitter, acrid, somniferous, thermogenic, stimulant, aphrodisiac, diuretic and tonic. The leaves are bitter and are recommended in fever and painful swellings. A paste of the roots and bruised leaves are applied to carbuncles, ulcers and painful swellings. Drugs consist of the dried roots of the plant. Ashvagandha is useful in, sexual and general weakness and rheumatism. It is diuretic, and it promotes urination, acts as a narcotic, and removes functional obstructions of body. The root powder is applied locally on ulcers and inflammations. The antibiotic and antibacterial activity of the roots as well as leaves has recently been shown experimentally.

Woodfordia fruticosa Kurz. Syn.

Family   Lythraceae

Indian names   Woodfordia, Fire-flame bush (English); *Madaniyahetu* (Sanskrit); *Dhay, Tavi, Thawi* (Hindi); *Bela, Tamrapushpi* (Kannada); *Dhattari, Jargi, Velakkai* (Tamil); *Tatiri, Tatirippu* (Malayalam); *Dhataki, Dhatupushpika, Jaaji* (Telugu).

Description   A much branched deciduous shrub, attaining a height of about 3–7 m, with many long arching branches and reddish brown bark; leaves and branchlets covered with dots, leaves sessile or subsessile, opposite, sometimes in whorls of three, distichous, lanceolate or ovate, acuminate; puberulous above, grey and pubescent with more numerous glandular dots beneath, lateral nerves 6–12, meeting in an intramarginal nerve, base rounded or cordate; panicles of closely clustered cymes, usually from the axils of fallen or existing leaves; peduncles and pedicels pubescent, usually glandular; lower bracts more or less leafy; calyx scarlet, persistent, tubular, somewhat curved, oblique at the mouth, lobes alternating with 6 rudimentary hairy teeth; petals 6, white, acute, scarcely exceeding the calyx.

Flowering and fruiting   February–May

Distribution   Garo and Khasi Hills, Meghalaya, ascending up to 1000 m

Propagation   By seeds

Parts used   Flowers

Chemical constituents   Flowers contain ellagic acid, β-sitosterol, polystachoside, octacosanol, myricetin-3-galactoside, cyanidin-3,5-diglucoside, pelargonidin-3,5-diglucoside, chrysophanol glucopyranoside.

Uses   Flowers are a stimulant, astringent and tonic, flowers are useful in leprosy, skin diseases, burning sensation, haemorrhages, menorrhagia, leucorrhoea, haemoptysis, erysiphales, diarrhoea, dysentery, foul ulcers, diabetes, bilious fever, hepatopathy and verminosis.

Wrightia arborea (dennst.) Mad.

Family   Apocynaceae

Indian names   *Kutajah* (Sanskrit); *Dudhi, Dharauli* (Hindi); *Kaadu veppaale, Kaadunagalu* (Kannada); *Palai, Karupaalai* (Tamil); *Kutakappala, Nelem-pala, Mylampala* (Malayalam); *Tella pala, Pala, Thellapala* (Telugu).

Description   Small deciduous tree, up to 10 m high; leaves opposite; flowers in terminal or axillary cymes; calyx short, 5-partite with glands or scales within; corolla salver-shaped; tube short, cylindrical with one or two seriate scales in the throat, lobes overlapping to the left; stamens at the top of the corolla tube, filaments short, dilated, anthers sagittate, exerted, conniving in a cone round the stigma and adhering to it, cells spurred at the base; ovules many, stigma ovoid, usually with a toothed basal ring, follicles distinct or connate; seeds linear, compressed with a deciduous coma at the base, albumen scanty or none, cotyledons broad, convolute, radicle short, superior.

Flowering and fruiting   May–December

Distribution   Khasi and Jaintia Hills, Meghalaya.

Propagation   By seeds

Parts used   Bark, leaves and seeds

Chemical constituents   Seeds yield 30–49 per cent fixed oil, β-sitosterol, α-amyrin and its acetate and lupeol benzoate is obtained from the bark.

Uses   Leaves are useful in odontalgia and hypertension. The bark and seeds are useful in dyspepsia, flatulence, colic, diarrhoea, leprosy, psoriasis, haemorrhoids, dipsia, helminthiasis, fever, burning sensation and dropsy.

Xanthium strumarium L.

Family   Asteraceae

Indian names   Bur weed, Cocklebur (English); *Mangalvanamalini, Mangalyakamalini, Arishta* (Sanskrit); *Banokra, Chhotagokhru* (Hindi); *Lokra* (Garo); *Marlumutta, Maruloomathai* (Tamil); *Marlumutta, Maruloomathai* (Telugu).

Description   A coarse annual; stem hispidulous or strigillose; leaves broadly triangular or suborbicular, often lobed, acute, scabrid or hispid; flowers unisexual in single or clustered axillary heads; female flowers, covered with hooked spines; male heads at the top of the inflorescence; anther exerted, base entire; pappus 0.

Flowering and fruiting   December–May

Distribution   It is common in Meghalaya.

Propagation   By seeds

Parts used   Roots, leaves and fruits, and seed oil

Chemical constituents   The sterioisomers, xanthinin and xanthatin. The seeds on solvent extraction yields 30–35 per cent of semi-drying oil, resembling sunflower oil.

Uses   Fruits are slightly narcotic, useful in many diseases, a good diuretic, powerful diaphoretic and sedative. The root is bitter, useful in cancer and scrofula. Its extract is used locally over ulcers, boils and abscesses. The seeds are used for resolving inflammatory swellings.

Xeromphis spinosa (Thunb.) Keay.

Family   Rubiaceae

Indian names   *Gurman* (Bengali); *Thiskeng* (Garo); *Rara, Mendol, Mendula* (Hindi); *Dieng maksingkhlaw* (Khasi); *Gela* (Marathi).

Description   Small trees or shrubs with a dense, compact crown, up to 10 m high, often with axillary spines; bark greyish or dark brown, usually rough; leaves oblong, obovate, spathulate, obtuse or subacute, narrow calyx campanulate, lobes foliaceous; corolla tube much shorter than the lobes; berries fleshy, 1.5–4 cm across, globose.

Flowering and fruiting   March–November

Distribution   Frequent in Meghalaya in secondary forests and forest margins. Khasi Hills.

Propagation   By seeds

Parts used   Fruit

Uses   Fruit juice is useful for boils.

Xylosma longifolium Clos.

Family　Flacourtiaceae

Indian names　*Dandal, Katari, Kandhara* (Hindi); *Dieng kani* (Khasi); *Phulwal* (Garo); *Kutta* (Malayalam); *Kondanerasi, Paddayi* (Telugu); *Kataponial* (Assamese); *Godya* (Oriya).

Description　A tree up to 20 m in height and 1.8 m in girth, thorny when young, often multiplying by root suckers; bark fairly smooth, greenish below the cuticle, inside yellowish brown; leaves oblong, rarely elliptic, lanceolate, acuminate; flowers dioceous, yellow, in short dense axillary racemes, which are fascicled or panicled; male flowers–stamens 15–20; filiform, surrounded by a disk with about 10 pink glandular lobes; female flowers–ovary glabrous, seated on a lobulate glandular disk, style 1, short, ovules few, fruit globose, more less dry, red when ripe with 2–8 seeds.

Flowering and fruiting　November–March

Distribution　Entire North-East India

Propagation　By seeds

Parts used　Bark

Uses　Bark powder is useful to relieve stomach ache.

Z

Zantedeschia aethiopica (L.) Spreng.

Family   Araceae

Indian name   Arum lily (English).

Description   Small perennial herbs with thick rhizomes with shining, ovate or sagittate radical leaves ending in a cusp. Flowers fragrant on a yellow spadix enclosed in white spathe.

Distribution   Entire North-East Region.

Propagation   By seeds and vegetative method

Parts used   Rhizome and leaves

Uses   Juice of rhizome is useful for uterine problems, diarrhoea and dysentery. Leaves are used as poultice on sores.

Zanthoxylum armatum DC.

Family   Rutaceae

Indian names   Indian sorrel (English); *Ahangeri, Amlapatrika* (Sanskrit); *Cukatripati, Amrulsak, Tinpatiya* (Hindi); *Hulihunice, Teltuppi* (Kannada); *Puliyarai* (Tamil); *Puliyaral, Puliyarala* (Malayalam); *Ambotikura, Pulichintaku* (Telugu).

Description   Small trees up to 8 m high; crown lax; stems thorny; bark greyish brown; leaves up to 20 cm long, leaflets oblong, lanceolate, ovate elliptic, acuminate, base cuneate, obscurely serrulate, glabrous, nerves, pubescent, flowers in axillary panicles, polygamous; fruit reddish subglobose, glabrous follicles; seeds solitary, globose, shining, black.

Flowering and fruiting   March–November

Distribution   Himalayas and North-East India, common usually at higher elevations in Meghalaya.

Propagation   By seeds

Parts used   Bark and fruits

Chemical constituents   It contains linalool, linalyl acetate, citral, cinnamatic, limonene, and sabinene.

Uses   The bark and fruits are acrid, bitter, aromatic, deodorant, antiseptic, stimulant, digestive, carminative, stomachic, anthelmintic, liver tonic, diuretic, constipating. They are useful in tumours, odontalgia, cephalalgia, otopathy, dyspepsia, flatulence, helminthiasis, diarrhoea, fever, leucoderma and skin dieases. Fruit is useful for liver complaints and seed is used for stomach ache and indigestion.

Zanthoxylum khasianum Hk.f. [Endemic]

Family   Rutaceae

Description   Climbing shrub, stems greyish brown; young parts tomentose; leaves 10–25, rachis often with prickles, leaflets usually 9–17, ovate or oblong, generally oblique, acuminate, obtuse, cuneate, minutely crenulated; flowers minute, calyx lobes ovate, oblong, petals obovate, ripe carpels, obovoid, rugose when dry.

Flowering and fruiting   May–October

Distribution   Endemic to Khasi hills in high altitudes

Propagation   By seeds

Parts used   Bark, fruits and seeds

Uses   Bark, fruits and seeds are carminative, stomachic and anthelmintic.

Zea mays L.

Family   Poaceae

Indian names   Maize, Indian corn (English); *Kandaja, Mahakaya* (Sanskrit); *Barajuar, Bhutta, Jawdra* (Hindi); *Bottah, Goinjol* (Kannada); *Makkaccolam* (Tamil); *Cholam* (Malayalam); *Makkazonnalu, Mokkajanna* (Telugu).

Description   A tall annual grass having broad leaves arranged in two vertical ranks. The inflorescences are monoecious, i.e., the tassel is staminate and sheds pollen while the ear shoot is pistillate, producing silks or style. The flowers of the tassel are borne in numerous spike-like racemes which together form large spreading panicle, which terminated the stems. A pistillate inflorescence is borne in one or more axils of the leaves. The spikelets are arranged in 8 to as high as 30 rows on a thickened almost woody axis known as the cab being enclosed in foliaceous bracts or husks. The long styles or silks protrude from the tops of the bracts. The spikelets are unisexual. The staminate spikelets occur in pairs in the tassel and are two-flowered. The two glumes of the spiklets are membranous, acute and covered by short hairs. Inside the glumes are present the lemma and palea. The pistillate spikelets are sessile and occur in pairs, consisting of one fertile and one sterile floret. The glumes are broad and rounded or emarginated at the apex. The styles or silks are very long and slender.

Distribution   Cultivated everywhere

Propagation   By seeds

Parts used   Styles with stigmas

Chemical constituents   The styles contain potassium salts. The grains contain starch, glucose, fructose, sucrose, raffinose, fatty compounds, vitamins E, C, K and β-carotene.

Uses   Maize silk (styles with stigmas), is used as a diuretic in the treatment of heart disease, hypertension, cystitis, urethritis, urinary lithiasis, cholecystitis, hepatitis, rheumatism and diabetes mellitus. It is also used in combination with vitamin K as a haemostatic.

Zingiber officinale Roscoe.

Family   Zingiberaceae

Indian names   Ginger (English); *Aadu, Adrak* (Hindi); *Ardraka, Hasishunthi* (Kannada); *Inji* (Tamil); *Inji* (Malayalam); *Allam, Ardrakamu* (Telugu); *Ada* (Assamese); *Sowhthing* (Mizoram).

Description   Perennial rhizomatous herb with erect leafy stem; leaves linear, sessile, glabrous; flowers in spikes, yellowish green with a small dark purple or purplish black tip, oblong, cylindric spikes, ensheathed in a few scarious, glabrous bracts; fruits oblong capsules; the rhizomes are white to yellowish brown in colour, irregularly branched, somewhat annulated and laterally flattened.

Distribution   Cultivated throughout India, also cultivated throughout North-East Region.

Propagation   By rhizomes

Parts used   Rhizome

Chemical constituents α-Curcumene, β-D-curcumene, α-bergamotene, β-phellandrene and camphene, α-bisabolene, β-bourbornene, D-borneol and its acetate, calamene, etc.

Uses   The raw ginger is acrid, thermogenic, carminative, laxative and digestive. It is useful in inflammations. The dry ginger is useful in dropsy, otalgia, cephalalgia, asthma, bronchitis, cough, diarrhoea and stomach troubles. It is a stimulant, rubefacient, expectorant and used in insect bites.

Zizyphus jujuba (Lamk.)

Family   Rhamnaceae

Indian names   French jujube, Jujube, Sedra (English); *Badara, Badari* (Sanskrit); *Beri, Kath ber* (Hindi); *Barihannu, Bogari* (Kannada); *Ilandai, Iratti* (Tamil); *Elantha, Perimtoddali* (Malayalam); *Badaramu, Badari* (Telugu).

Description   Small tree, usually up to 10 m high and 1.5 to 2 m in girth, almost evergreen; bark dark grey, old stems nearly black with long vertical crack, reddish and fibrous inside; young parts rusty tomentose; leaves very variable, obliquely elliptic, ovate or suborbicular, closely serrulate or entire, dark green, glabrous and often shining above, densely rusty; flowers small, greenish yellow; calyx globose or ellipsoidal or obovoid acuminate, with a fleshy, mealy aromatic acid or subacid pulp, red or orange when ripe, stone 2-celled, tubercled.

Flowering and fruiting   September–February

Distribution   Fairly common throughout the plains and up to 650 m in Garo and Khasi hills of Meghalaya.

Propagation   By seeds and vegetative method

Parts uses   Fruit

Chemical constituents   Leucocyanidin from bark, carbohydrate, fat, protein, amino acids, anthocyanins from fruits, seeds and leaves, rutin from leaves, Mauritines A, B, C, D, E and F frangufoline and amphbines B, D and F from the whole plant.

Uses   It is useful for palpitational insomnia; night sweats, hysteria, poor appetite and general fatigue.

Zizyphus mauritiana Lamk.

Family   Rhamnaceae

Indian names   Indian jujube (English); *Badarah, Kolah* (Sanskrit); *Ber, Beri* (Hindi); *Bore, Elanji, Elachi* (Kannada); *Ilantai, Ilantappalam* (Tamil); *Ilanta, Ilantappalam, Peruntutali* (Malayalam); *Gangaregu, Regu* (Telugu).

Description   Small trees up to 10 m high, crown, umbrella-shaped, branches drooping, short spiny, bark smooth or nearly so, with vertical fissures in old trunks; leaves obliquely ovoid, orbicular, rounded at tip, minutely crenate, shining and dark green above, densely brownish, tomentose beneath, 3-nerved from base; inflorescence short fascicles or cymes; flowers 2–3 mm across, yellow; disc 10-lobed; drupe orange yellow.

Flowering and fruiting   March–December

Distribution   Tropical, throughout India, common in Meghalaya, tree cultivated widely in many parts of India

Propagation   By seeds

Parts used   Whole plant

Chemical constituents   Same as in *Z. jujuba*.

Uses   The bark is used to heal ulcers and wounds. The leaves are laxative, used in scabies, throat trouble and burning of the body. The fruit is sweet, sour, expectorant and purifies and enriches the blood, good for chronic bronchitis, fever, enlargement of the liver. The seeds are good for dry cough and for skin eruptions (Unani). The drupes are a good emollient.

Zizyphus oenoplia (L.) Mill.

Family   Rhamnaceae

Indian names   Jackal jujube (English); *Karkandhuh* (Sanskrit); *Makai, Makkay* (Hindi); *Barige, Harasurali* (Kannada); *Surai ilantai, Suraimullu* (Tamil); *Ceriyalantaa, Tudali, Tudari* (Malayalam); *Paragi, Paraki* (Telugu).

Description   Throny shrubs; branches somewhat zig-zig; leaves ovate-lanceolate, acuminate, base slightly oblique, softly pilose beneath, nerves 3–4 from base, spines recurved or straight; cymes axillary, up to 2 cm across; flowers pedicelled; calyx tomentose outside, glabrous within, lobes keeled, broad ovate; petals greenish yellow, tringular, hooded, shorter than the calyx; stamens shorter than the petals and embraced by them; ovary glabrous, 2-celled; drupe, obovoid, globose, black when fully ripe, stone usually one, rarely 2-celled.

Flowering and fruiting   August–February

Distribution   Tropics, nearly throughout India; occurs in lower elevations in secondary forests and wastelands forming thorny thickets.

Propagation   By seeds

Parts used   Roots

Chemical constituents   The root bark contains two new cyclopeptide alkaloids, zizyphine-A and zizpyhine-B, betulinic acid, D-glucose, D-fructose, sucrose and unidentified polysaccharides.

Uses   The roots are useful in hyperacidity, ascaris infection, stomachalgia and healing of wounds.

Conclusion

Today, many medicinal plant species face extinction or severe genetic loss, but information is lacking. For most of the endangered species, no conservation action has been taken. And for most countries, there is not even a complete inventory of medicinal plants. Traditional people, whose very existence is now under threat, hold much of the knowledge on their use. Herbalists now report about having to walk increasingly greater distances for herbs that once grew almost outside their doors. The herbal medicine trade is a booming business worldwide. In India, for example, there are 46,000 licensed pharmacies manufacturing traditional remedies, 80% of which come from plants (Alok, 1991). Another example is Hong Kong, which is claimed to be the largest market in the world, importing over US $ 190 million annually (Kong, 1982). In Durban (South Africa), in 1929, there were only two herbal traders; by 1987, there were over 70 herbal trader shops registered. The species-specific nature of the demand for medicinal plants is responsible for generating long-distance trade across international boundaries. According to Malla (1982), 60–70% of the medicinal herbs collected in Nepal are exported to India, with 85–200 tons exported annually between 1972 and 1980.

The World Health Organisation now is giving serious attention, as is evidenced by the recommendations it gives to medicinal plants (Wondergem *et al*, 1989). It says that proven traditional remedies should be incorporated within national drug policies. The World Health Organisation (WHO) the International Union for Conservation of Nature (IUCN) and the World Wide Fund (WWF) for nature convened an International Convention on the Conservation of Medicinal Plants in March 1988 in Chiang Mai, Thailand. For the first time, the Convention brought together in the same forum, policy makers and scientists from the two areas of health care and nature conservation.

The global sales of herbal medicines in the USA reached US $ 14 billion in 1996. Of this, 26% was generated in Germany, 19% in Asia, 17% in Japan, 13% in France, 12% in the rest of Europe and 1% in North America (Zhang, 1998). The international market of medicinal plant related trade is US $ 60 billion with an annual growth of 7% (Anonymous, 1997). The volume of Germany's export is second only to the China (Lange, 1997). The largest imports are from India, followed by Bulgaria, Poland, Chile, Hungary, Argentina and Albania (Lange, 1996). The current value of trade in Indian systems of medicine and Homoeopathy is around Rs. 4,205 crores, roughly close to US $ 1 billion. There are also strong traditions of herbal medicine in parts of Europe such as Germany, France and Eastern Europe. The herbal sector is growing fast, increasing by 12–15% by value per year in the UK, the USA and Italy (Abrahams, 1992). There are more than 2,000 herbal medicinal companies in Europe and more than 220 in the USA; Germany is the largest market in the world for herbal medicines, with an annual sales of $ 1.2 billion representing nearly 25% of the national pharmaceutical market.

The USA is the largest market with sales of $ 480 million (Thorpe and Warrier, 1992). In India, out of 4,752 communities, as many as 3,226, that is, around 70% of the population are dependent on traditional plant-based medicine (Gadgil and Rao, 1998). These plant-based medicines are used for primary health care needs. Many modern drugs are derived from plants. Demand for medicinal plants is increasing in both developing and developed countries. The practice of traditional medicine is widespread in China, India, Japan, Pakistan, Sri Lanka and Thailand. In China about 40% of the total medicinal consumption is attributed to traditional tribal medicines.

It has been quite known that currently biopiracy of certain high valued bioresources (mostly medicinal plants) has been a major threat to the world. However, this issue has not been able to draw public attention because there is still lack of awarness among the people as to the pattern and consequences of biopiracy carried on in the region. The whole plant or a part is collected for medicine. Many plant species are collected every year from this region although no data are available on the quantity of collection. Generally, plant parts are collected by the forest dwellers, some of which are semi-processed and handed over to the agents of companies situated outside the region. Actually the local people involved get a very small amount of money in return, whereas the companies are illegally earning a lot at the cost of the valuable genetic resources. During collection, the collectors generally cause unnecessary damage to the biodiversity of the concerned localities. In case of roots and rhizomes, collectors uproot the plants and after taking the required parts, abandon the rest just on the forest floor. As collections are generally made during dry winter and early spring, the abandoned parts dry within a short period rendering no regeneration. In the case of bark collection, generally trees are cut down and barks are removed. In some cases, the entire barks are removed from the standing trees, while in the case of seeds, these are collected before their natural dispersal, leaving little chance for regeneration. All these cause irreparable loss of the concerned valuable species and also their habitats.

Many medicinal plants having industrial potential are growing wild in this region, which can be used in pharmaceutical industry. These raw materials have their own chemical entity to be used as a specific drug. They are also being used for suitable chemical change developing derivatives of the original product, refining and enhancing their therapeutic activity. Alternately, the medicinal products can also serve as a model for development of new synthetic pharmaceutical products. The tremendous export potentialities of a large number of medicinal plants from India have been focused by different scientists, which could definitely earn a considerable amount of foreign exchange and uplift the economy of the state (Nath, 2000).

IMPORTANCE OF MEDICINAL PLANTS

Medicinal and aromatic plants have an important role in the ecology and economy of the country; as these cannot be cultivated as sole crops at the cost of other agricultural crops on cultivable lands; these can be grown in degraded lands (by shifting cultivation) or marginal lands in combination with trees. These species can be grown successfully in various agroforestry systems under different agroclimatic conditions of the country.

Preservation of old beliefs of the inhabitants   In the primitive human society, the conservation of plant resources was an ancient tradition, therefore it led to the conservation of medicinal plant wealth. They utilized the resources only as much as required and the rest was preserved for future needs. Many of the medicinal plant species are believed by them to be sacred and ritually important. These religious beliefs played a significant role in proper management and preservation of these resources.

Establishment of medicinal plant sanctuaries   Our ancestors were conscious about the conservation and over-exploitation of the plant resources. It is therefore, upon the present generation to take further steps so that valuable material is not lost for posterity. It is therefore necessary that the areas having useful medicinal plants be protected. The areas having important plant species such as *Dioscorea pentaphyllum, Gloriosa superba, and Rauvolfia serpentina* be preserved as medicinal plant sanctuaries in their natural habitat as these plants are in much demand in the pharmaceutical industry.

Cultivation of medicinal plants in farms   The local inhabitants as well as the Ayurveda, Siddha and Unani systems of medicine have utilized the medicinal plant wealth for various preparations. The recent years have brought a renaissance of the herbal as well as Ayurvedic system, which have received wide attention and importance. A number of pharmaceutical companies are coming forward with herbal preparations. The spurt in demand for Ayurvedic preparation as well as herbal cosmetics has lead to increased demand for medicinal plants. This increased demand cannot be met from the available present plant resources. Therefore if indiscriminate collection is permitted, a time may come when many of the medicinal plants will disappear. The only solution to the problem is cultivation of medicinal plants to meet the present increased demand of the pharmaceutical industry. The per-hectare income generated from growing medicinal plants is much more than any other crop plant. Even wastelands and other areas lying unused around the villages can be utilized for the purpose. This system if properly utilized and exploited can help in boosting the economy as well as open avenues for employment in the Garo Hills.

The plant species which are suitable for the low hills and valleys for cultivation are herbaceous plant species like *Achyranthes aspera, Amaranthus spinosus, Centella asiatica, Costus speciosus, Curcuma aromatica, Curcuma longa, Eclipta alba, Eupatorium* sp., *Gloriosa superba, Leucas cephalotes, Ocimum basillicum, Ocimum sanctum, Oxalis corniculata, Rauvolfia serpentina, Solanum nigrum, Solanum surattense, Sida rhombifolia, Zingiber officinale,* etc.

The shrubby or climing medicinal plants are suitable for cultivation are *Argyreia nervosa, Cissus quadragularis, Ichnocarpus frutescens, Mucuna prurita, Piper longum,* etc. These are in great demand and so their cultivation should be taken up on large scale.

The shrubby medicinal plants which are suitable for cultivation are: *Adutilon indicum, Abroma angustata, Adhatoda zeylanica, Alstonia scholaris, Aphania polytachya, Azadirachta indica, Acacia nilotica, Albizzia lebeck, Anthocephalus chinensis, Bauhinia veriegata, Butea monosperma, Cannabis sativa, Clerodendron indicum, Cassia fistula, Cassia sophora, Cinnamonum tamala, Carataeva nurvala, Croton tiglium, Citrus medica, Dalbergia sissoo, Datura stramonium, Emblica officinalis, Elaeocarpus spharicus, Ficus benghalensis, Ficus religiosa, Gmelina arborea, Hibiscus rosasinensis, Holarrhena antidysenterica, Mallotus philippensis, Melia azadirachta, Mussanda* sp. *Nerium indicum, Oroxylon indicum, Prunus cerasoides, Ricinus communis, Semecarpus anarcardium, Shorea robusta, Solanum indicum, Syzygium cumini, Terminala arjuna, Terminalia bellirica, Terminalia chebula, Vitex negundo, Vitex penducularis.*This will help in fulfilling the present demand of pharmaceutical companies and help the economy of the region as well as to help in conservation of plant species.

References

Abrahams, P. (1992). "Herbal sales set to grow." *Financial Times*, London, UK. 2 October.

Ahmed, M. and Das, N.K. (2003). "Effects of Deforestation on Tree canopy cover and Soil fertility of Goalpara, Assam." *Environmental Biology and Conservation*. Vol. 8. pp. 24–27.

Allen, B. C. (1905). "Gazetteer of Assam state (District Goalpara)." Sree Guru Press, Maligao, Gauhati. 1–70.

Alok, S.K. (1991). "Medicinal plants in India: Approaches to exploitation and conservation." In: *Conservation of Medicinal Plants.* (Eds.). Akerele, O., Heywood, V. and Synge, H. Cambridge University Press. pp. 295–304.

Ambast, R.S. (1990). *Environment and Pollution (An Ecological Approach)* Students, Friends and Co, Lanka, Varanasi, India.

Anonymous. (1997). "Indian medicinal plants: A sector study." Occasional Paper. No. 54. Mumbai: Export Import Bank of India.

Arimitsu. K. (1983). "Impact of shifting cultivation on the soil of the tropical rain forest in the Benakat district. South Sumatra, Indonesia." In: IUFRO Symposium on forest site and continuous productivity. (Eds.). Ballard, R. and Gessel, S.P. USD of Agriculture, Forest Service. 163: 218–222 (6.ed).

Atul, Punam and Sarma, S. (2002). "The medicinal wealth of Western Himalayan agro-ecological region of India: 1. An Inventory of herbs." *Ann. For.* 10: 28–61.

Barley, K.P. (1961). "The abundance of earthworm in agricultural soils and their possible significance in agriculture." *Agronomy*. 13: 249–268.

Barthakur, D.N. (1983). "Shifting cultivation in North East India." *Indian Council of Agricultural Research*. Shillong. pp. 1–2.

Basak, R.K. and Pandit, P.K. (2003). "Role of a Sacred Grove in conservation of Medicinal Plants." *Indian Forester*. 129 (2): 224–232.

Basham, A.L. (1981). *The Origin and Early History of the Khashi-Synteng people.* pp. 1–15.

Bedard, P.W. (1960). "Shifting cultivation benign and malignant aspects." Proceedings of the Fifth World Forestry Congress. Washington, USA; 3: 2016–2021.

Bera, S.K., Basumatary, S.K., Agarwal, A. and Ahmed, M. (2006). "Conservation of forestland in Garo Hills, Meghalaya for construction of road: A threat to the environment and biodiversity." *Current Science.* 91 (3): 281–284.

Bhakta, G.P. (1992). *Geography of North East India*. Akashi Book Depot. Shillong. pp.1–2,

Bisset, J. and Parkinson, D. (1980). "Long-term effects of fire on the composition and activity of the soil microflora of a subalpine, coniferous forest." *Can. J. For. Res.* 58: 1704-1721.

Bora, A.K. (2001). "Physical background." *Geography of Assam*. (Eds.). Bhagabati, Bora, A.K. and Kar, B.K. Rajesh Publication, New Delhi. pp. 18–35.

Chauhan, N.S. (2002). "Integration of medicinal and aromatic plants in agroforestry." Proc. Summer Institute Production technology and Management of agrforestry tools. PAU, Ludhiana. July 10–30. pp. 83–91.

Christensen, N.L. (1977). "Soil micro fungi of dry mesic conifer hardwood forests in North Wisconsin." *Ecology.* 50: 9–27.

Conklin, H. (1954). "An Ethnological approach to Shifting Cultivation." *Trans. New York Academy of Science.* 17: 133–42.

Coulson, A.L. (1942). "Water supply of the Darang district, Assam." Water supply Paper No. 3, Geological Survey of India. V. 76. pp. 1–152.

Cox J.A. (1975). *The Endangered Ones.* Crown Publishers Inc., New York.

Das, S., Dash, S.K. and Pandey, S.N. (2003). "Ethno-medicinal informations from Orissa state, India, A Review." *Journal Human Ecology.* 14: 165–227.

Dhar, U., Manijkhol, S., Joshi, M., Ghatt, A., Bisht, A.K. and Joshi, M. "Current status and future strategy for development of medicinal plants sector in Uttaranchal, India." *Curr. Sci.* (2002). 83: 956–964.

Dikshit, V.K. (1999). "Export of medicinal plants from India: need for resource management." In: *Biodiversity–North-east India Perspectives: People's Participation in Biodiversity Conservation*. (Eds.). Kharbuli, B., Syem. D. and Kayang, H. NEBRC, North-Eastern Hill University, Shillong. pp. 85–88.

Dutta Choudhury, A.N. (1978). "The Brahmaputra and the valley of his blessings." In: *Souvenir on harnessing the river Brahmaputra*. Assam Science Society.

Ehrlich, P.R. and Ehrlich, A.H. (1981). *Extinction: The causes and consequences of the disappearance of species.* Victor Gollancz Ltd., London.

Englang, C. M. and Rice, E.L. (1957). "A comparison of the soil fungi of a tall grass prairie and an abandoned field in central Oklahoma." *Botanical Gazetteer.* 118: 186–190.

Evans, P. (1936). Trans. of Mineralogical and Geological Institute of India. V. 27. pp. 161–253.

Gadgil, M and Rao, P.R.S. (1998). *Nurturing Biodiversity: An Indian Agenda.* Centre for Environment Education, Ahmedabad.

Gill, A.S., Bisaria, A.K., and Shukla, S.K. (1998). "Potential of agroforestry as source of medicinal plants." *Current concept of multidisciplinary Approach of Medicinal Plants.* (Ed.). Govil, J.N. Today and Tomorrow Printers and Publishers, New Delhi. pp. 1–28.

Gill, A.S., Handa, A.K. and Neelam Khare. (2004). "Medicinal value of important tree species in agroforestry." *Ind. For.* 130(6): 615–629.

Gogoi, K. (1984). *Geology and Mineral Resources of Garo Hills, Meghalaya. Garo Hills Land and people.* (Ed.). Gassah, L.S. Omsons Publication, Guwahati, pp. 40–51.

Goswami, D.C. (1982). "Brahmaputra river, Assam (India). Suspended sediment transport, Valley aggregation and basin denundation." Unpublished Ph.D. thesis.

Goswami, D.C. (1985). "Brahmaputra river, Assam (India). Physiography basin denudating and channel aggregation." *Water Resources Research.* American Geographical Union. V. 21 (7). pp. 959–978.

Goswami, D.C. (1997). "Brahmaputra river basin forestry (background paper)." In: *Forestry and key Asian water sheds.* (Eds.). Myint, A.K. and Hofer, T. ICIMOD. Kathmandu. pp. 32–37.

Gupta, R. and Chadha, K.L. (1995). "Medicinal and aromatic plants in India." In: *Advances in Horticulture, Medicinal and Aromatic Plants.* (Eds.). Chanda, K.K. and Gupta, R. Malhotra Publishing House, New Delhi. pp. 1–44.

Hayden, H.H. (1904). "Geology of spite." *Mem. Geological Survey of India.* 34 (1).

Herman, R.P. and Kuchera, C.L. (1984). *The Am. Midland Nat.* 101: 13–20.

Jackson, N.L. (1967). *Soil Chemical Analysis.* Prentice Hall of India Pvt. Ltd., New Delhi. The Eastern Economy edn.

Jalaluddin, M. (1969). "Micro-organic colonization of forest soil after burning." *Plant and Soil.* 30: 150–152.

Jha, M.M., Badola, K.C. and Pandey, P. (1980). "Soil changes due to Podu cultivation to the state Orissa." Forest Research Institute. Dehradun.

Jorgansen, J.R. and Hodges, C.S. Jr. (1971). "Effects of prescribed burning on the microbial characteristics of soil." In: Prescribed Burning Symposium Proceedings. April 14–16. Charleston, South Carolina.

Kamboj, V.P. (2000). "Herbal medicine." *Current Science.* 78: 35–39.

Kaushik, Purushottam and Dhiman, A.K. (2000). *Medical Plants and Raw Drugs of India.* Bishen Singh Mahendra Pal Singh, Dehra Dun.

Kong, Y.C. (1982). "The control of Chinese medicine–A scientific overview." In: *Yearbook Pharm. Soc.* Hongkong. pp. 47–51. cited by Farsworth, N.R. "Screening Plants for New Medicine." pp. 83–97. In: *Biodiversity.* (Ed.). Wilson, E.O. National Academy Press, Washington, DC, USA.

Krishnan, M.S. (1956). *Geology of India and Burma.* Higgins Brothers Pvt. Ltd., Madras.

Kumar, Y., Haridasan, K. and Rao, R.R. (1980). "Ethnobotanical notes on certain medicinal plants among some Garo people around Balphakram sanctuary in Meghalaya." *Bull, Bot. Surv. India.* 22: 161–165.

Lange, D. (1996). "Medicinal Plant Market Study in Germany–Germany, state of project." *Medicinal Plants Conservation Newsletter of the IUCN species Survival Commission.* Medicinal Plants Specialist Group. 2: 9–10.

Lange, D. (1997). "Trade in plant material for medicinal and other purpose–A German case study." *TRAFFIC Bull.* 17: 20–32.

Lanly, J.P. (1983). "Assessment of the forest resources of the tropics." *Commonwealth forestry Review* 6 (6): 137–62.

Lillesand, T.M. and Kiefer, R.W. (2000). *Remote Sensing and Image Interpretation,* 4th edn. John Wiley and Sons, Inc., New York.

Lucarotti, C.J., Kelsey, C.T., and Auclair, A.N.D. (1978). "Microfungal variations relative to post-fire changes in soil environment." *Oecologia.* 37: 1–12.

Maheshwari, J.K. (ed). (2000). *Ethnobotany and Medicinal Plants of Indian Subcontinent.* Scientific Publisher, Jodhpur.

Malla, S.B. (1982). *Medicinal plants of Nepal.* FAO Regional Office for Asia and the Pacific, Report No. 64. Bangkok, FAO.

Mukherjee, P.K. (1966). *A text book of geography.* The World Press Pvt. Ltd.

Mukherjee. (1974). "Problems of shifting cultivation." *North Eastern Affairs* 3(1): 49–54.

Nath, S.C. (2000). "Diversity of the higher plants of medicinal value growing in Northeast India in relation to conservation and sustainable use." *J. Assam Sci. Soc.* 41(4): 267–288.

Nye and Greenland (1960). "The soil under shifting cultivation." Technology Commonwealth Agricultural Bureau, Farnham Royal, Bucks.

Oldham, R.D. (1999). Report on The great earthquake of 1897." *Mem. Geological Survey of India.* 29, 1899.

Pareek, S.K. (1996). "Medicinal plants in India. Present status and future prospects." In: *Prospects of Medicinal Plants.* (Eds.). Gautam, P.L. *et al.* Indian Society for Plant Genetic Resources, New Delhi. pp. 5–14.

Phillipson, J.D. (2001). "Phytochemistry and medicinal plants." *Phytochemistry.* 56: 237–243.

Pilgrim, G.E. (1910). "Tertiary fresh water deposits of India." *Record Geological Survey of India*. V. 3.

Podder, M.C. (1952). "Preliminary report of Assam earthquake 15 August 1950." *Bulletin Geological Survey of India*. V. 2. pp. 11–12.

Postel, S. and Heise, Z. (1990). "Reforesting the Earth." *Span Magazine*. 6(60): 11–18.

Prajapati, N.D., Purahit, S.S., Sharma, A.K. and Kumar, T. (2003). *A Hand Book of Medicinal Plants*. Agrobios (India).

Prakas, A and Sing, K.K. (2001). "Use of medicinal plants by certain tribal people in North India." *J. Trop. Med. Plants*. 2(2): 225–229.

Prater, S.H. (1965). *The Book of Indian Animals*. Bombay Natural History Society. Bombay.

Rao, R.R. (1981). "Ethnobotany of Meghalaya: Medicinal Plants used by Khasi and Garo tribes." *Economic Botany*. 3: 4–9.

Rao, U.R. (1994). "Space for Sustainable Development with special emphasis on Himalayan Region." G.B. Pant Memorial Lecture: IV. G.B. Pant Institute of Himalayan Environment and Development, Kosi-Katarmal, Almora.

Richter, F.C. (1969). *Elementary Seismology*. Eurasia Publishing (Pvt.) Ltd., New Delhi. pp. 19–54.

Sanchez, P.A. (1976). *Properties and management of soil in the Tropics*. John Wiley and Sons, New York.

Singh, H. B. (2000). "Alternative source for some conventional drugs plants of India." In: *Ethnobotany and Medicinal Plants of Indian Subcontinent*. (Ed.). Maheshwari, J.K. Scientific Publisher, Jodhpur, India. pp. 63–78.

Smith, A.G. (1981). "The Neolithic." In: *The environment of British prehistory*. (Eds.).Simmons I.G. and Tools, Duckworth, M.J. London. p. 334.

Smith, F.A. and Smith, S.E. (1981). *Air Pollution of Forest*, 2nd edn. Springer. New York.

Spurr, H.S. and Barnes, V.B. (1980). *Forest Ecology*, 3rd edn. John Wiley and Sons, New York.

Taher, M. and Ahmed, P. (2001). "Population." *Geography of North East India*, 2nd edn. Manik Prakash, Panbazar, Guwahati. pp. 233–34.

Thorpe and Wrier. (1992). "Competitive positioning: Who's doing what in the herbal medical industry." Consultancy Report. Private publication, U.K.

Tulaphitak, T. (1983). "Soil fertility and tilth in shifting cultivation, an experiment of Namphram, North East Thailand and its implication for upland Farming in the monsoon tropics." (Eds.). Kyuma, K. and Pairintra, C. Ministry of Science, Technology and Energy, Bangkok. pp.63– 83.

Uniyal, M.R. and Uniyal, R.C. (2002). "Utilization of medicinal plants by pharmaceutical industries in India." Paper presentation in the Workshop on Vanaspati Van, 24-25 June, WII, Dehra Dun.

Wadia, D.N. (1953). *Geology of India*. Macmillan and Co.

Warcup, J.H. (1957). "Studies on the occurrence and activity of fungi in a wheat field soil." *Trans. Bureau Mycological Society.* 40: 237–263.

Wondergem, P., Senah, K.A. and Glover, E.K. (1989). "Herbal drugs in primary health care: Ghana: An assessment of the relevance of herbal drugs in PHC and some suggestions for strengthening PHC." *Zimbabwe Science News*. Royal Tropical Institute, Amsterdam.

Wright, E. and Bollen, W.S. (1961). "Microflora of douglas-fir forest soil." *Ecology*. 42: 825–828.

Zhang, X. (1998). "Significance of traditional medicine in human health care." WIPO Asian Regional Seminar on Intellectual Property Issue in the Field of traditional Medicines, New Delhi. pp. 6.

Index